CW01080848

A PIONEER OF CONNECTION

A PIONEER OF C

Recovering *the* Life *and* W

SCIENCE AND CULTURE IN THE NINETEENTH CENTURY
Bernard Lightman, Editor

ONNECTION

ork *of* OLIVER LODGE

Edited by James Mussell *and* Graeme Gooday

UNIVERSITY OF PITTSBURGH PRESS

Published by the University of Pittsburgh Press, Pittsburgh, Pa., 15260

Manufactured in the United States of America
Printed on acid-free paper
10 9 8 7 6 5 4 3 2 1

Cataloging-in-Publication data is available from the Library of Congress

ISBN 13: 978-0-8229-4595-6
ISBN 10: 0-8229-4595-9

COVER ART: *Sir Oliver Joseph Lodge ('Men of the Day. No. 907. "Birmingham University"')* by Sir Leslie Ward, chromolithograph, published in *Vanity Fair,* 4 February 1904.

COVER DESIGN: Joel W. Coggins

CONTENTS

PART THREE: SCIENCE, SPIRITUALISM, AND THE SPACES IN BETWEEN

ACKNOWLEDGMENTS

THIS BOOK was developed from an Arts and Humanities Research Council (AHRC)– funded research network called "Making Waves: Oliver Lodge and the Cultures of Science, 1875–1941" (AH/K006223/1 and AH/K006223/2). The project was led by James Mussell (principal investigator) and Graeme Gooday (co-investigator) and ran from 2013 until 2015, first at the University of Birmingham and then at the University of Leeds. The project held four workshops: "Civic Science: Oliver Lodge, Physics, and the Modern University," at the University of Birmingham in November 2013; "Wireless: Oliver Lodge, Science, and Spiritualism," at the Royal Society in April 2014; "Science, Pure and Applied: Oliver Lodge, Physics and Engineering," at the University of Liverpool in October 2014; and "Scientific Lives: Oliver Lodge and the History of Science in the Digital Age," at Leeds Art Gallery in March 2015. The project also funded two public lectures, both delivered in March 2015: "Why Did Scientists Come to Write Autobiographies?" by Graeme Gooday, in Leeds; and "Civic Life: Oliver Lodge and Birmingham," by James Mussell, in Birmingham (the Cadbury Research Library Annual Lecture). We would like to acknowledge the contributions made by all the participants at the workshops, whether speakers or those in the audience. On the organizational side, Peter Rowlands and Robert Bud were excellent hosts: Peter organized the workshop in Liverpool and Robert arranged an extra meeting at the Science Museum in July 2015. We are particularly grateful to Helen Williams and Georgina Binnie, who worked as the project administrators at Birmingham and Leeds, respectively, and those in the research offices at both institutions.

The Lodge family has been supportive of both the project and subsequent book; we thank them for how they have welcomed our research. We were particularly excited to be invited to take part in the event to commemorate the hundred-year anniversary of the passing of Raymond Lodge, which took place in Saint George's

Church, Edgbaston, in November 2015. We would also like to acknowledge the contribution of Jane Darnton, both to the project and subsequently, and celebrate her achievement in getting a plaque dedicated to Oliver Lodge installed on the site of the Lodges' house, Mariemont, in 2018. In February 2017 contributors to the volume took part in an event at the Royal Institution called "Spirits in the Ether: Oliver Lodge and the Physics of the Spirit World." We would like to thank the chair of the event, Samira Ahmed, and the organizers at the Royal Institution.

We also want to acknowledge the assistance and support of all at the University of Pittsburgh Press. We thank the two anonymous reviewers for their valuable suggestions, our copyeditor, Amy Sherman, and our editor, Abby Collier, for the many ways in which she has improved the book as it went through production. Finally, we thank all of our contributors, whose hard work and patience have made this book what it is.

A PIONEER OF CONNECTION

INTRODUCTION

OLIVER LODGE

CONTINUITY AND COMMUNICATION

JAMES MUSSELL AND GRAEME GOODAY

> It may be felt in any single page of Forbes's writing, or De Saussure's [or Tyndall's either], that they love crag and glacier for their own sake's sake; that they question their secrets in reverent and solemn thirst: not at all that they may communicate them at breakfast to the readers of the *Daily News*—and that, although there were no news, no institutions, no leading articles, no medals, no money, and no mob in the world, these men would still labour, and be glad, though all their knowledge was to rest with them at last in the silence of the snows, or only to be taught to peasant children sitting in the shade of the pines.
>
> **John Ruskin, quoted by Oliver Lodge, *Past Years***

OLIVER LODGE quotes the English art critic and intellectual John Ruskin in the penultimate chapter of his memoir *Past Years* (1931). Titled "Scientific Retrospect," this chapter is a version of a talk Lodge delivered at the fiftieth anniversary of the Physical and Chemical Society at University College, London (UCL) in 1926. After reviewing the progress of science since the society's foundation in 1876, Lodge promoted a model of natural science as a disinterested enterprise carried out for its own sake rather than for personal gain. This view of science as a "pure" intellectual study drew upon a tradition deriving from Ruskin's position in a major geological controversy nearly a century before the publication of *Past Years* and involved Lodge's early inspiration for physics, John Tyndall.

The text by Ruskin that Lodge quotes originated in an extract from *Fors Clavigera* (1871–1884) later appended to a translation of Rendu's *Theory of the Glaciers of Savoy* edited by George Forbes and published in 1874.[1] Forbes had published the translation in response to renewed allegations of plagiarism made against his father, James David Forbes, by Tyndall in his *Forms of Water* (1872), and the younger Forbes included the extract from Ruskin as part of his defense of his father's position as a disinterested participant. As the original dispute over the movement of glaciers had erupted in 1857, it was already sixteen years old when Ruskin addressed it in *Fors Clavigera,* and seventy-four years old by the time of the publication of *Past Years.*[2] Looking back at the end of his life, Lodge returned to the Victorian context of his youth, setting out a version of unworldly scientific practice that contrasted markedly with the account of his career he describes elsewhere in the book.

Delivered in a speech to students on the anniversary of a scientific society, such sentiments were not unusual; however, coming at the end of a volume that documents Lodge's long career, they are not so straightforward to explain. Lodge might have espoused scientific research for its own ends in *Past Years,* but he was deeply interested in applying science to real-world problems, starting companies, and holding patents, as well as carrying out a range of consultancy and other advisory work. And he was no solitary researcher: while Lodge seemed to approve of Ruskin's celebration of the individual scientist, he nevertheless recognized that his own career had been shaped by those around him. In *Past Years* he acknowledges many of those with whom he collaborated as well as the contributions of some of his assistants; the book also makes clear how science was entangled with family life and celebrates the many friendships that sprang from his scientific work.

In 1931 Lodge was one of the most famous scientists of his day, yet he was best known for work carried out many years before. As he himself admits in the preface to *Past Years,* his scientific career had reached "a sort of climax" in the 1890s: in 1898 Lodge's work was acknowledged by the Royal Society, who awarded him the Rumford Medal, and in 1902 he was knighted.[3] Nevertheless, thirty years later Lodge remained a respected scientific authority, able to find substantial audiences in print and person. His career had begun in London in the 1870s and he had grown up with figures like Forbes, Tyndall, and Ruskin. For many of those who read his words or heard him speak, this link to what seemed a prior age made Lodge a reassuring figure of continuity with Britain's Victorian heyday. He had not only witnessed the rise of Darwinism, the advances in electromagnetism, and the wonders of radioactivity but had rubbed shoulders with key figures, written about and discussed their innovations, and, in many cases, played a direct part himself. While Lodge might have appeared to condone Ruskin's attack on those who communicated science "to the readers of the *Daily News,*" Lodge had written for the public from the very beginning, producing books and pamphlets as well as articles for both commercial and academic presses. Indeed, he understood such work to be part of a scientific career and, as his career progressed, so did the nature of the invitations. Lodge not only explained the latest scientific advances to his audiences but increasingly speculated as to their significance, too.

Lodge was considered a spokesperson for science, but as the contents of *Past Years* make clear, Lodge worked in a number of fields that were to him interconnected. Just as there are three chapters about his days at Liverpool, those heady days between 1881 and 1900 when he had most of his scientific success, so there are three chapters that detail his psychical research. Lodge had become interested in spiritualism in Liverpool in 1883, joined the Society for Psychical Research (SPR) the following year, and served as its president from 1901 to 1903. Lodge's interests were initially in thought transference, but by 1889 he had become privately convinced that the mind could survive the death of the body.[4] While always cautious about the way he presented his spiritualist views, he was not afraid to defend them, even declaring his belief in survival in his presidential address to the British Association for the Advancement of Science in 1913.[5] Lodge's scientific work focused on the ether, the imponderable, perfect medium necessary to account for electromagnetic radiation, and this elusive substance provided a ready explanation for what would otherwise be occult phenomena. Poised between matter and metaphor, the ether became central to Lodge's broader philosophy: not just the medium of electromagnetic radiation or spirit, it became the point at which the two met. In person and in print Lodge was always reasonable yet never afraid to address areas of controversy. Not only did his work explain the latest scientific advances as well as the claims of spiritualism, it offered to reconcile both within a recognizable Christian framework.

The way Lodge presents his career in *Past Years* underscores continuity, whether over time, bridging the gap between the Victorian period and the generations that followed, or between those rival and contested bodies of knowledge that would account for the workings of the universe. Lodge was someone with whom his readers had grown up, and his very appearance corresponded with how they thought an eminent scientist should look. While there were scientists, especially fellow physicists, who found Lodge's engagement with spiritualism objectionable, and spiritualists who thought he remained too skeptical, for many others Lodge's position offered a comfortable middle ground, open to new ideas while engaging seriously with the evidence. Yet while Lodge's ideas had appeal, his authority rested on the way those ideas were communicated. Throughout his career, Lodge addressed an audience beyond that of his immediate peers. As he became more eminent—as a scientist, as a spiritualist, as the head of an important civic university—he was able to address larger subjects and, at the same time, could command more cultural authority.

It is between these two poles, continuity and communication, that we present Oliver Lodge in this volume. Through "continuity," the book explores Lodge's long career as well as how it challenges the way we divide and make sense of the period. Under this theme, our contributors examine Lodge's participation in the various fields, scientific and otherwise, in which he was active, as well as how this activity was interconnected. Focusing on communication opens up Lodge's practice as a researcher, teacher, and popularizer, as well as how cultural authority was linked to celebrity. Our contributors look at Lodge's work in communications, whether his pioneering research in wireless telegraphy or his experiments with thought transference, as well as how Lodge presented himself in person and print. Together, both themes provide the means to understand the neglect of Lodge in scholarship from the mid-twentieth century onward. In the decades after his death, Lodge became seen as an anachronism: rather than a reassuring figure from the past, he was a Victorian who lived too long; his opposition to developments such as relativity and quantum mechanics and his staunch defense of the ether associated him with a bygone age; his enthusiastic embrace of spiritualism marred his scientific legacy; and finally, because Lodge's reputation was due as much to the way he popularized science as to his original research, significant aspects of his career appeared subsidiary in a history of science that acknowledged only innovation and discovery. This book argues that the reverse is true: that Lodge's long career offers the opportunity to reflect on both historical periodization and the construction of legacy, that the way he worked across fields exposes the historicity of disciplinary knowledge, and that the communication of scientific ideas is part of their history.

APPROACHING THE LIFE OF OLIVER LODGE

Much scholarly research on Lodge has focused on just a few aspects of his broad-ranging career. Typically these explore connections between two parts of life and work; for example, his ether physics and his spiritualism,[6] his applied physics and his commercial patenting,[7] or his spiritualism and cultural authority.[8] The binary focus of such research undoubtedly brings simplifying clarity and explanatory power, yet the necessary restriction makes it difficult to view the many (more or less) interconnected strands that quickly become apparent when we take a broader view of his life and career. More than that, such accounts cannot tell us why Lodge was such an extraordinarily well-known and respected authority in British culture and beyond. Instead, they tend to leave the impression of Lodge as a somewhat eccentric, cranky, or inconsistent individual with an unmanageably broad array of interests, largely a bit-part player in other people's stories of science and its popularization.

Even Lodge himself in *Past Years* could only manage the telling of his own life by compartmentalizing his accounts into discrete chronological or thematic chapters.

In so doing he said very little about some major areas of his life; notably his popular writing, journalism, commercial consultancy, patenting activity, radio broadcasting, and municipal socialism. These features were not, after all, part of the preferred public identity of the professional physicist, who, after the First World War, tended to present him- or herself on the basis of expertise and "pure" knowledge rather than Victorian commitments to public service and useful science.[9] Lodge was more gifted than other physicists in his capacity to cultivate relationships across different social strata but, while he acknowledged some of those who influenced him, others are missing entirely or have their roles underplayed. So, while Lodge's debts to the spiritualist Frederic Myers are given more than a cursory treatment in the spiritualist sections of *Past Years* and occasional references made to the inspiring popular lectures of Irish physicist John Tyndall, much less is made of the enormous influence upon him of Alexander Muirhead, his collaborator in the telegraph industry, and, during his Birmingham days, the politician Joseph Chamberlain.[10] Throughout, Lodge presents his life as a series of loosely integrated accounts of his physics, spiritualism, and family life and himself as a self-effacing individual, proud of what he has achieved but aware that he might have done more. Yet there are passages where this performative modesty is dropped in an attempt to reclaim credit, most notably against the better-known claims of Guglielmo Marconi for the invention of wireless.[11]

How then can we capture Lodge's episodic yet by no means fragmented life in a single chronological structure? Simplifying the narrative of *Past Years,* this section follows the four-stage career pattern outlined in W. P. Jolly's 1974 biography *Sir Oliver Lodge*: education and early life (1851–1880); the professorship of physics and mathematics at University College Liverpool (1881–1899); the principalship of Birmingham University until its end just after the First World War (1900–1919); and then his active retirement between two world wars (1919–1940).[12]

To start with the early years, Lodge grew up in the Potteries district of Staffordshire. After a classical grammar school education memorable only for its vicious corporal punishments, followed by two years as a private pupil, Lodge left school at fourteen to join his father's clay business. His passion, however, was for experimental science and before long he sought to escape the drudgery of pottery sales. Inspired by lectures in London, particularly by Tyndall at the Royal Institution, he began to study at local institutions—the Stoke Athenaeum, the Wedgwood Institute at Burslem, and the Potteries Mechanics' Institute—and carrying out experiments at home. After a more rigorous training in chemistry and physics at the Science Schools in South Kensington 1872–1873, Lodge returned to London in 1874 and began to work with George Carey Foster at University College, London. Having shown a strong aptitude for writing and lecturing in popular and academic forms of physics, he secured the first chair in that subject at Liverpool's recently opened University College in 1881.

There, at northwest England's Atlantic port city, Lodge developed an interpretive approach to Maxwellian ether physics that earned him the respect of mathematicians as well as the broader public; however, alongside his growing fame, Lodge also deployed his Maxwellian wisdom in commercial projects for lightning conductors and telegraph cables. In *Past Years*, Lodge traces the origins of his interest in lightning conductors to the Mann Lectures delivered at the London-based Society of Arts in 1888 and frames this work in Maxwellian terms. Rather than understanding lightning as a direct current—the received view (and one supported by William Preece, the chief engineer at the post office, among others)—Lodge maintained that it was better understood as a rapidly oscillating electrical discharge, and so self-induction (electromagnetic inertia) was more important than resistance in the design of conductors.[13] However, Lodge does not mention in *Past Years* the patenting he enacted to profit from this aspect of his Maxwellian expertise. Central to this was his long-distance collaboration with the London-based telegraph manufacturer Alexander Muirhead, which is entirely omitted from *Past Years.* Their partnership began in April 1887, with the two men first working on lighting protection apparatus, then on submarine telegraph cables, and then the Lodge-Muirhead wireless telegraphy syndicate, until it was wound up in 1911. In a very extensive correspondence of hundreds of letters, we can see that it was to Muirhead that Lodge turned for commercial and patent advice in electrotechnical enterprise—expertise that Lodge, busy working within the university while pursuing psychical research, had little time to develop.[14]

Prior to his awkward encounters with Marconi and his wireless patents, Lodge's intermittent recourse to patenting in the late 1880s and early 1890s related to power supply and submarine telegraphy. However, while recognizing the role patents played in the commercial development of technology, Lodge was more interested in sharing his knowledge about electricity through lectures and textbooks. It was in this context that he first presented his research into Hertzian waves and the means of generating and detecting them. Heinrich Hertz died on January 1, 1894, and, as the leading public speaker on the relationship between Maxwellian topics and Hertz's experimental work, Lodge was the natural choice to give the memorial lecture at London's Royal Institution in June 1894. Lodge published the lecture as *The Work of Hertz and Some of His Successors* later that year; however, stunned by Marconi's patent for a means of wireless telegraphy (and one that used an improved version of his apparatus) in 1897, he revised the title of the second edition to *Signalling across Space without Wires* (1898). With this more Marconian title, Lodge recategorized his early work on the coherer as if it were the invention of wireless rather than an exploration of Hertz's research.[15]

Notwithstanding his subsequent feud with Marconi over priority in the invention of wireless—or perhaps because of the moral high ground that he won from it—by 1899 Lodge had established a reputation as a public authority on electromagnetic theories and technologies, and much else besides. For example, after the Royal Society awarded Lodge the Rumford Medal, the Lord Mayor of Liverpool marked his achievement with a banquet in his honor. This drew the attention of Chamberlain, who saw in Lodge the ideal candidate for principal of his new Birmingham University. Undeniably in tension with his ever-growing interest in spiritualism, this position enhanced Lodge's visibility as a public figure, providing opportunities for him to address the issues of the day both locally and nationally. The demands of running the university meant less time for research; his tenure also encompassed the difficult years of the First World War and the death of his son, Raymond, killed at Ypres in 1916 and with whom Lodge sought spiritual contact long thereafter. Upon Lodge parting company with Birmingham the year after the war ended, his retirement saw him become a renowned expert on the new media technologies of "wireless" radio broadcasting in the 1920s and then of television in the 1930s. Although he controversially retained older commitments to the ether and spiritualism in an era that cherished Einstein's secular theory of relativity, Lodge's final years nevertheless saw him as a much respected authority, an intellectual on par with George Bernard Shaw and Winston Churchill.[16]

To refine such an episodic account, however, it is important to emphasize the nature and range of continuities in Lodge's long life. His deep family connections are arguably most fundamental, especially with two women: his aunt, Charlotte Anne Heath, and the woman he married, Mary Marshall. Heath had served as a Woman of the Bedchamber to Queen Adelaide (consort to William IV) and, after inheriting a significant sum on Adelaide's death, settled in London and dedicated her life to educational and religious causes. She was Lodge's most attentive aunt, nurturing the youngster's curiosity and intellect, not least by introducing him to Royal Institution lectures as a child, and then hosting him in London for his early years as a student. It was in London that Lodge courted Marshall, whom he had known since childhood, as their families were long acquainted in the Potteries. A talented artist, she started at the Slade School in London just as Lodge approached his last year studying for his doctor of science degree (DSc) at UCL. They were forbidden to marry until Lodge was earning £400 a year; within a few years of finishing his degree his teaching work at UCL and Bedford College was bringing in twice as much, and they married in 1877.

It was a long and productive marriage. Between them, they raised twelve children (two more died as infants), six boys and six girls, born between 1878, when they were in London, and 1896, when they had moved to Liverpool. The sheer costs of managing and educating such a large household in itself readily explains why Oliver took on so many remunerative duties of science directed to utilitarian ends in his working life. These included writing numerous books, public lecturing, telecommunications consultancy, and patenting far beyond anything required by his professorship at Liverpool or his principalship in Birmingham. Mary influenced Oliver's work, too: although her inspiration and guidance was often tacit, it was all too evidently lost after her death in 1929.[17] Their children in turn built on Oliver's research, transforming some particularly valuable innovations into long-lasting commercial businesses. Francis Brodie (born 1880) and Alec (born 1881) patented their own improved system of high-tension ignition in 1903, developing the results of Leyden jar discharges that their father had explored in his laboratory research at University College Liverpool, and set up the Lodge Plug Company to produce spark plugs for cars and airplanes in 1904. Their younger brothers Lionel (born 1883) and Noel (born 1885) developed Oliver's research on dust extraction, also carried out in Liverpool, into an electrostatic technique for removing smoke, setting up the Lodge Fume Deposit Company Limited with their sister Norah (born 1894) in 1913.[18]

Continuity across the generations was not only manifest in the financial and technological capital that Oliver bequeathed to his children. He himself had inherited a religious framework from his senior family members. As Lodge tells us in *Past Years*, both of his grandfathers had fulfilled the double role of vicar and headmaster. His maternal grandfather was Reverend Joseph Heath of Herefordshire; his paternal grandfather, the Reverend Oliver Lodge, was born in Ireland, and, with three successive wives, produced twenty-five children, the youngest of whom were raised at a vicarage in Barking in south Essex (Oliver Lodge's father was the twenty-third). Young Lodge, then, entered a relatively conventional family tradition of Anglicanism, sustained by a very large number of uncles, aunts, and cousins.

While Lodge never lost his faith, he certainly took it in directions not previously explored in the Lodge dynasty. His first encounter with psychical research came as a student in London, far from the Potteries or grandpaternal vicarages of his childhood. At University College, Lodge met Edmund Gurney, who was then preparing material for his *Phantasms of the Living*, published posthumously in 1886. It was through Gurney that Lodge met Myers, who would eventually become his close friend and mentor. Lodge was initially unimpressed by Gurney's work, dismissing it as a "meaningless collection of ghost stories," and, when Gurney, Myers, and the Cambridge philosopher Henry Sidgwick founded the SPR in 1882 Lodge did not join.[19] Drawing its membership from the middle and upper classes and rich in intellectual and social capital, the SPR was dedicated to seeking experimental evidence for psychical phenomena of various kinds. Not until Lodge carried out his own experiments to test thought transference in Liverpool in 1883 did he begin to accept the possibility of such communications and the legitimacy of scientific methods to establish evidence for them. Lodge joined the SPR that year and, to the end of his days, it remained a major organizational commitment, shaping his approach to integrated physics and spiritualism.

Lodge was by no means the only physicist among its membership—William Fletcher Barrett, who had stayed with the Lodge family in the Potteries, was a member, as were Lord Rayleigh and J. J. Thomson—but he remained circumspect about how his research might be received by fellow physicists. With the publication of *Raymond* (1916), however, in which Lodge recorded details of apparent contact with his son, killed in the trenches the year before, Lodge's psychical research became much more visible and prompted some to question his judgment.[20] Yet this strand of Lodge's work was integral to his broader understanding of the universe, and after his retirement he embraced the opportunity to focus upon it untrammeled by the professional norms of academic science. While there were plenty of critics, Lodge's late work allowed him to set out his views regarding the philosophy of the ether to a much wider audience, granting him status as a public authority that he would not otherwise have achieved.

Although controversial, Lodge's work with the SPR nevertheless reflected a sustained practice in experimentalism he first developed as a teenager in 1869. Even as he contemplated the complexities of abstract ether theory, laboratory investigation—usually working with assistants or collaborators (or delegating to them)—was a major feature of his career at both Liverpool and Birmingham. Yet Lodge was not content for his work to remain in the laboratory. While at Liverpool, for instance, he gained a reputation for developing innovative demonstration models of James Clerk Maxwell's theory of electromagnetism, which he employed both in the lecture hall and in his publications; meanwhile, Lodge gave talks and wrote articles about the very innovations in communication media that he was working to patent with Muirhead. Over his career, his writing appeared in specialist scientific and technical journals, in newspapers, and in general periodicals; he maintained extensive correspondence with friends and collaborators in all the domains in which he was interested; he wrote popular books, technical treatises, and textbooks; gave lectures in the university at prize givings, for various societies, and at a wide range of institutions, large and small; filed dozens of electrotechnological patents; and appeared on both radio and television. Lodge, then, was adept at communicating in a range of media while at the heart of much of what he said and wrote was his specialist research in modern *means* of communication.

Perhaps Lodge's role as a public communicator is best seen in the lectures he gave on these technical topics to broad and often nonspecialist audiences. Both his Mann Lectures on lightning conductors in 1888 and those on the characteristic discharge of the traditional Leyden jar the following year established him as the leading interpreter of Maxwell's work and led directly to the lectures and demonstrations he gave in 1894 that, he later claimed, showed the potential for electromagnetic waves to be used for wireless signaling. These latter talks were held in a range of venues to different audiences, specialist and nonspecialist: the first was the memorial lecture for Hertz at the Royal Institution on June 1, 1894; Lodge then demonstrated his apparatus again at the annual Ladies Conversazione at the Royal Society on June 13; and finally he gave a paper at the British Association meeting in Oxford on August 14.[21]

While the move to Birmingham meant that Lodge had less time to spend in his laboratory, it brought more opportunities to speak. He continued to be active in those organizations with which he was close, like the British Association and the SPR, but he also took advantage of other opportunities. In 1903, for instance, Lodge gave a lecture on radium at Birmingham Town Hall (with proceeds going to the new university) and then repeated it again for working men, writing it up as "Radium and Its Lessons" for the highbrow monthly *Nineteenth Century and After* and including it in the third edition of *Modern Views of Electricity*.[22] His position also allowed him to broaden the subjects he addressed. The next year, for instance, he gave a talk to the Ancient Order of Foresters in Birmingham titled "Public Wealth and Corporate Expenditure," which was published by Beatrice and Sidney Webb as Fabian Tract 121, *Public Service versus Private Expenditure*, in 1905.[23] The string of books Lodge published in his first decade at Birmingham also demonstrates his growing confidence to tackle subjects beyond physics. *Life and Matter* (1905), for instance, is an attack on the materialism of Ernst Haeckel; *The Substance of Faith* (1907) offers a nondenominational catechism aimed at introducing children to the essence of religion; *Man and the Universe* (1908) attempts to reconcile science and faith; and *The Survival of Man* (1909), dedicated to the SPR, sets out Lodge's belief in the survival of personality after death.[24]

Nevertheless, his new role placed considerable constraints upon his work. Lodge had to oversee the development of the new campus at Edgbaston, which grew out of Mason College, then based in the center of Birmingham, as well as establish the university's new, broader, curriculum. He also had to negotiate his civic role, representing the university at various events in the city and beyond. Lodge's position as principal was made more difficult by the influence of Chamberlain, the university's founder and chancellor. While there was a great deal of correspondence between their views on education, they differed as to how best to establish the university. To make matters more delicate, Lodge frequently had to pander to Chamberlain's resilient celebrity in Birmingham. Shortly after coming to Birmingham, for instance, Lodge was forced to tolerate the imperialistic tenor of the citywide celebrations to mark Chamberlain's departure for South Africa when he found himself on the organizing committee in 1902. Chamberlain's stroke in 1906 allowed Lodge more independence, but at the same time it deprived the university of one of its most effective fundraisers and lobbyists.

There were, as already mentioned, significant continuities across Lodge's long life. Although he had less time to practice science, he maintained his links with scientific institutions and, after he became one of the editors of the *Philosophical Magazine* in 1911, scientific communications. He continued to speak and write, and on an increasingly wide range of topics. And his commitment to psychical research was undiminished. Yet there were also some discontinuities. It is evident that while Lodge was in Birmingham he never secured the sympathy and respect of the local civic and academic community that he obviously had in Liverpool. And while Tyndall's lectures at the Royal Institution inspired Lodge into taking up physics, he turned against both Tyndall's materialism and methods of teaching.[25] His adherence to ether theory and his willingness to situate the ether in spiritual terms, although consistent in itself, put Lodge at odds with many of his contemporaries, and his skepticism toward the more radical interpretations of Einstein's work, particularly that based in advanced mathematics, made him seem increasingly out of touch. Yet while Lodge might have exasperated some of the younger physicists working after the First World War, his authority as a man of science remained undiminished: he remained a respected and influential figure within scientific circles, and the wider public often looked to Lodge for guidance.

There were also some tensions within Lodge's principles and practices that he managed with a certain degree of pragmatism, sometimes so that he could fulfill what he saw as his civic responsibilities, sometimes for financial benefit. For instance, Lodge was adept at managing relationships with those from across the political spectrum: at the same time he was publishing Fabian tracts with Beatrice and Sidney Webb, he was also working with the Liberal Unionist Chamberlain and spending his Easters at Clouds, the locus for the aristocratic group known as the Souls that included his friend Arthur Balfour, then the Conservative prime minister.[26] He was a supporter of women's suffrage, addressing suffragist groups in Birmingham, but he also defended the suffragette Christabel Pankhurst's right to speak, hosting a debate at the university in Birmingham in 1907 after students disrupted a meeting in the Town Hall.[27] Lodge opposed the jingoism of the period, chairing a meeting of the Quaker delegation to Germany in 1909.[28] When war broke out, however, he had no difficulty dedicating himself to British government war committees and accepting royalties from the sales of many wireless sets during the conflict. In this way, Lodge offers a stark contrast to the actions of his close friend, ally, and regular correspondent Silvanus Phillips Thompson, the Quaker head of Finsbury Technical College in east London. In 1911 Thompson's righteous courtroom defense of Lodge's intellectual rights over wireless tuning against Marconi's impertinent claims was crucial in resolving that priority dispute in Lodge's favor. Yet equally robust and earnest was Thompson's refusal to participate in any aspect of the Great War, especially against the many Germans whom Thompson continued to hold in high esteem. Whereas Thompson died in near obscurity as a conscientious objector in 1916, Lodge's personality and connections allowed him to not only achieve success among scientific and technological organizations but also mix with all manner of people of all social classes. Lodge's capacity to communicate with such people is a central issue in what follows.

CONTINUITY AND COMMUNICATION

In the quotation from Ruskin given as our epigraph, Lodge inserts "[or Tyndall's either]" in the short list of people whose writing reveals them to be disinterested lovers of science. In the dispute to which Ruskin refers, Tyndall was the antagonist, claiming that Forbes had not given Rendu due credit for his ideas about the movement of glaciers. Ruskin, while defending Forbes, was unwilling to take sides, arguing that disputes about priority were beside the point. In *Past Years*, Lodge quotes Ruskin to this effect:

> I do not in the slightest degree care whether he [Forbes] was the first to see this, or the first to say that, or how many common persons had seen or said as much before. What I rejoice in knowing of him is that he had clear eyes and open heart for all things and deeds appertaining to his life; that whatever he discerned, was discerned impartially; whatever he said, was said securely; and that, in all functions of thought, experiment, or communication, he was sure to be eventually right and serviceable to mankind, whether out of the treasury of eternal knowledge he brought things new and old.[29]

By arguing that it was the honest and disinterested pursuit of science that was important, Ruskin avoided having to arbitrate in the dispute. Lodge, both in his address to the Physical and Chemical Society at UCL and in the subsequent chapter of *Past Years*, endorses Ruskin's sentiments. However, alert to Ruskin's sleight of hand, Lodge ensures that Tyndall, too, is part of this pantheon of Victorian sages, erasing the dispute in the name of scientific progress.

In *Past Years*, Lodge attempted to do something similar in the way he represented his own career. In his account of the discovery of wireless, he credits Marconi with developing the commercial technology of wireless telegraphy based on Lodge's prior experimental researches. Whereas, according to Lodge, he, George Francis FitzGerald, and Lord Kelvin were "satisfied with the knowledge that it could be done," Marconi worked "enthusiastically and persistently . . . until he made it a practical success." While he draws a distinction between the men of science concerned only with knowledge and the more practical Marconi, Lodge also concedes that he simply did not recognize "that such a method of telegraphy would be, or might be, of international importance."[30] All the same, Lodge was sure to register his contribution to the development of wireless in *Past Years,* stating clearly that he demonstrated the potential for his equipment to be used for signaling at the 1894 British Association Meeting (and so two years before Marconi appeared on the scene). Because wireless "is of so much interest and importance" he also appended a list of significant dates that traced its origins back to Maxwell's "Dynamical Theory of the Electromagnetic Field," presented at the Royal Society in 1864.[31] Lodge's rendering of the history of wireless placed it firmly in the Maxwellian tradition and, as such, preceded both Lodge and Marconi's individual contributions. As Marconi did not have Lodge's authority as a scientist, however, let alone as an expert on Maxwellian interpretations of electromagnetism, he was further marginalized, presented as a parvenu in a continuous line of invention.

A lecture that became a chapter of his autobiography, Lodge's "Scientific Retrospect" exemplifies the way he was able to move between media, as well as his canniness when it came to making the most of his work. In another way, however, it exemplifies the difficulties in understanding Lodge's life. If we take either Lodge's model of scientific progress or the way he constructs the history of wireless, then Lodge himself becomes a marginal figure in a much larger story. But if we take Lodge as our subject, a much richer notion of the scientific life emerges. There is nothing in *Past Years,* for instance, about the circumstances in which Lodge sold his patent rights to Marconi in 1911, dissolving the Lodge-Muirhead Syndicate and accepting a well-paid (and largely honorary) position as a consultant in his company. For Lodge, with his large family, this provided much-needed financial security while ensuring that his reputation as a man of science remained intact; however, in selling his patent—by then the most significant wireless patent still in effect—he ceded authority to Marconi, allowing his rival to control the way that wireless was developed and so understood.[32] Lodge's account in *Past Years* was an attempt to reinsert himself into the history of wireless and so, in his view, reclaim the moral and scientific credit for its invention.

Our book attempts to restore the complexity of Lodge's life that is not by any means fully apparent in *Past Years.* In tracing lines of continuity, we argue, it becomes possible to understand the relationships between Lodge's different activities as well as how he adapted, or not, to the times in which he lived. Focusing on communication, we discover the source of Lodge's authority as well as the reasons for his eclipse in subsequent scholarship. As these two themes run continuously through the chapters that follow we have organized them into three thematic parts. The first part, "Lodge's Lives," opens with an examination of Lodge's life as represented in *Past Years*, before providing detailed accounts of the two key phases of Lodge's career, in Liverpool and Birmingham. David Amigoni's chapter, "Communication, (Dis)Continuities, and Cultural Contestation in Sir Oliver Lodge's *Past Years*," looks at the way Lodge made sense of, and represented, his life in the broader context of early twentieth-century life-writing. Situating the book between Lodge's claims for physics and an understanding of inheritance in evolutionary terms, Amigoni traces the various paths to selfhood that Lodge himself identified and communicated. Peter Rowlands's chapter, "Becoming Sir Oliver Lodge: The Liverpool Years, 1881–1900," considers how Lodge built a scientific reputation among his peers while at the same time establishing himself in the city and beyond. Di Drummond's chapter, "Lodge in Birmingham: Pure and Applied Science in the New University, 1900–1914," turns to the second of Lodge's adopted cities, examining how he shaped both the University of Birmingham and the idea of the modern university itself.

The second part, "Science and Communication," focuses on Lodge's work as a researcher and popularizer. Bruce Hunt's chapter, "The Alternative Path: Oliver Lodge's Lightning Lectures and the Discovery of Electromagnetic Waves," revisits this important moment in Lodge's career, locating his work in electromagnetism in the commission from the Royal Society of Arts to lecture on lightning conductors. Lodge was much more comfortable with physical models than with mathematical ones, and Matthew Stanley's chapter, "Lodge and Mathematics: Counting Beans, the Meaning of Symbols, and Einstein's Blindfold," examines the extent to which Lodge credited mathematics as a faithful representation of nature. The final two chapters in this part consider Lodge's relationship to what became the new physics. Bernard Lightman's chapter, "The Retiring Popularizer: Lodge, Cosmic Evolution, and the New Physics," considers how Lodge consolidated his position as a leading scientific commentator through a string of popular books in which he explained the latest developments in physics and his own position with regards to them. Rather than opposing them outright, Lodge believed that these developments allowed him to articulate a more convincing version of his ether philosophy, one that could reconcile his faith with the results of scientific and psychical research. Imogen Clarke, in her chapter, "The Forgotten Celebrity of Modern Physics," looks more closely at Lodge's reputation in this period. While Lodge is often considered to have been on the wrong side of the debates regarding first relativity and then quantum theory, Clarke shows how his view was courted both in specialist forums and more widely.

The final part, "Science, Spiritualism, and the Spaces in Between," focuses on the continuities between Lodge's scientific and psychical research. Richard Noakes's chapter, "Glorifying Mechanism: Oliver Lodge and the Problems of Ether, Mind, and Matter," explores the philosophical basis for Lodge's ether theory. Drawing on his research into electromagnetism, Noakes argues that Lodge's ether was mechanistic, but in distinct, nonmaterialistic ways. Christine Ferguson's chapter, "The Case of Fletcher: Shell Shock, Spiritualism, and Oliver Lodge's *Raymond*," situates Lodge's book in the genre of spirit soldier biographies published during and after the First World War. Comparing the disrupted speech characteristic of shell shock to that of the séance room, Ferguson argues that Lodge conceived of the dead as traumatized survivors united by impaired communication. Georgina Byrne's chapter, "Beyond *Raymond:* The Theology of Spiritualism and the Changing Landscape of the Afterlife in the Church of England," uses Lodge's book to examine the Anglican church's response to spiritualism in the wake of the war. While many in the church resisted spiritualists' pronouncements about the afterlife, others began to describe it in similar terms. The final two chapters address Lodge's occult and uncanny ideas about communication. David Hendy, in his chapter, "Oliver Lodge's Ether and the Birth of British Broadcasting," traces the remarkable similarities between the way Lodge described the ether and the context within which the BBC was founded. James Mussell, in his chapter, "'Body Separates: Spirit Unites': Oliver Lodge and the Mediating Body," uses Lodge's ideas to interrogate materiality more broadly, describing the way Lodge conceived of the body in his time as well as asking what Lodge has to offer our own wireless age.

Lodge was born in the year of the Great Exhibition and died just before the London Blitz; his career thus encompassed the rise of the first mass communications media. As one of the inventors of wireless, Lodge's story is inseparable from what would become one of the defining conditions of twentieth-century modernity. As one of the thinkers of wireless, elaborating on the significance of connection, Lodge was also one of its prophets. He was at the center of much of what made the period distinct, yet his life and career also teach us about how such distinctness is constructed. In trying to understand Lodge, we have to reconsider notions of periodicity, why people and ideas appear to belong to one moment rather than another; we have to recognize the operation of disciplinarity, the way certain domains of knowledge or expertise appear separate or discrete; and when we trace Lodge's legacy, we learn much about who is remembered, and why. Oliver Lodge shaped his period and how it came to understand itself; studying him today can help us understand how we understand it too.

PART ONE

LODGE'S LIVES

THERE ARE many ways to tell Oliver Lodge's life through the themes of continuity and communication, although none of these ways can effectively be relayed simply as a linear narrative. The many intersecting strands to Lodge's work and relationships require instead a pluralistic treatment, telling different partial stories that add up to a broader if not necessarily unitary account. In this first part of this volume we see geographically specific studies of where he spent the major part of his career: Liverpool (1880–1899) and Birmingham (1900–1919), prefaced by a fresh analysis of Lodge's own retrospective narrative of his life, *Past Years: An Autobiography*, published in 1931.

In some respects, we can see *Past Years* as a conventional autobiography of the interwar period: Lodge conscientiously grounds his pedigree in a lineage of respectable ancestors, then demonstrates how he attained success by dint of hard work, shrewdly following advice, and capitalizing upon fortuitous opportunity. Consistently aiming thereby to represent himself as an authentically Victorian practitioner of science who maintained the same values and beliefs over an entire career, Lodge nevertheless maintained a commitment to public communication of his research that was by no means as fashionable at the close of his career as it had been at the start. Further tensions appear as Lodge treats some of his dealings with spiritualism apologetically while proclaiming the unity of that practice with his physics. And for all his manifest virtues as an experimentalist, he was evidently preoccupied with his inability to secure lasting credit for his inventions in wireless telegraph (radio). Had his life been more successfully conducted in these respects, we might wonder whether Lodge would have felt he needed to write an autobiography at all.

In surveying Lodge's life in University College Liverpool, we see the maturation of his independent skills both as a laboratory researcher and as a public communicator in the physics of the ether. A narrative focused on both the academic institution and the wider city reveals how congenial this setting was for Lodge, his creativity a result of both Liverpool's largesse and the demands of his increasing family. Lodge's life was productively entangled with Liverpool more deeply than any other location in which he lived, and he was remembered there continuously for many years even after his death.

A narrative focused on his time at Birmingham as the university's first principal tells of Lodge's significance on another civic stage, albeit without the easy pursuit of his own research or the fertile social support networks that had made his time in Liverpool so successful. Although there were strong continuities, Lodge was foremost an administrator and so his communications tended to be oriented to colleagues in the university or on behalf of the university when addressing groups in the city and region in which it was situated. It is thus easy to see why his retirement from Birmingham in 1919 gave him refreshed opportunities both to work again on wireless technology, the ether, and their popularization, and to find time to write *Past Years*. The somewhat fragmentary nature of the latter, composed after the passing of his life partner, Mary, in 1929, reveals how much she had been a vital ally in his continuous energetic activities at Liverpool and Birmingham, and in supporting his diverse but cogent work as a communicator throughout his long career.

ONE

COMMUNICATION, (DIS)CONTINUITIES AND CULTURAL CONTESTATION IN SIR OLIVER LODGE'S *PAST YEARS*

David Amigoni

If my children or if any disciples of the late Francis Galton care to know anything about my ancestry, I will jot down the facts as follows.

Oliver Lodge, *Past Years*

THE OPENING words of Sir Oliver Lodge's autobiography, *Past Years* (1931), announce an ostensibly plain communicative intention: to "jot down" his "somewhat uneventful" life history in case anyone should "care to know" it, whether members of his family or students of heredity. Despite the casual air, the words resonate with the complexities surrounding the uses of autobiographical communication to narrate the continuities constituting a personality and career in science, at a post-Victorian juncture of cultural, intellectual, and technological transition.

Lodge begins with the supposition that his large family of twelve children will be the main beneficiaries of the story of his life and ancestry: to this extent, Lodge was following in the footsteps of eminent Victorian scientists such as Charles Darwin who wrote their lives primarily for the sake of their children.[1] Lodge and Darwin might thus be said to have "affirmed everyday life" through their identification with fatherhood, or the productivity and making constitutive of family life. This orientation recalls one of Charles Taylor's key "sources of the self" characterizing the subject matter of self-narration in pursuit of social and moral goods that, from the seventeenth century, shaped the formation of modern identity. In *Sources of the Self*

(1989), Taylor conceives an affirming, everyday selfhood as the overarching outgrowth of another source of the modern self: the practice of "productive efficacy" embedded in empirical scientific inquiry.[2] So, it is perhaps fitting that Lodge the father *and* career scientist identifies the disciples of the eugenicist Sir Francis Galton as a *specialist* group with a possible calling on his ancestral story: a group who, in the earlier twentieth century and following Galton's lead, were systematically investigating the genealogies of those who had achieved scientific eminence in pursuit of social goods.[3]

Thus, Lodge's opening words begin with a group of family readers but very quickly shift to identify a group of scientific specialists. What was Lodge's attitude to the goods associated with this specialism, balanced against the claims of familiarity? To answer this, my approach in this chapter will acknowledge and explore the way in which scientific autobiography constitutes a distinctive approach to the complex justification of knowledge acquisition and management in the course of a scientific career. Lodge's *Past Years* represented a life whereby knowledge had been acquired via distinctively nineteenth-century contexts of the everyday structures of work, gendered family life, and industrial productivity; followed by a scientific career that embraced innovative experimentation, the adaptation of science into technological advances, and its management through the patent system, which could (and did) generate disputes. Lodge's career as a researcher and educator was simultaneously committed to popular scientific exposition through lecturing and textbook production, itself a key approach to knowledge management that fed back into the discourse of his autobiography as a story about scientific advance and social progress.

Despite its casual guise as quotidian "jotting" of facts, Lodge's *Past Years* thus subtly contributed to a conflicted field of scientific popularization and public understanding in which debates about the sources of the self in the form of distinctive personality and its survival and continuity across time were divided between Lodge's articulation of the claims of a physics in which a superadded but imperceptible ether was ever present as a force and the alternative claims of biological heredity and eugenics. Thus, Lodge's autobiography contributed to a debate about science and social goods, articulated through Lodge's affirmation of family, heterodox psychical research, and spiritualism.

Lodge frames these competing claims in part through a literary reflexivity that takes its cue from the title of his work and the poetic source from which it originates. Of course, "past years" underlines the autobiographical mission to narrate continuities between the past and present, but it has Romantic-aesthetic resonances as well. The phrase "past years," as the title page indicates, derives from William Wordsworth's "Immortality Ode"; thus, reflections on the experiences of past years offer the prospect of "perpetual benediction" in the present; while the extended quotation from the ode reflects on "obstinate questionings of sense and outward things, / . . . Blank misgivings of a Creature / Moving about in worlds not realised." Thus, the autobiography's epigraph gestures toward other, transcendent sources of selfhood in the "soul, "immortality," and "worlds not realised."[4] Lodge's literary allusiveness draws upon what Charles Taylor has described as the "subtler language" of selfhood's sources—nature, inwardness, expressivism, aestheticism—that challenged what Taylor referred to as a disengaged "scientism" born of detached, enlightened, ethical rationality. Such an erasure of subjectivity was also linked to intense professional specialization, and despite the magisterial coverage of *Sources of the Self*, the effects of scientific specialization on the self is one that Taylor does not have the space to tell.[5] The relationship between these "subtler languages," including this transcendental strain, at work in Lodge's autobiography, and the material drives of field and reputation management that these languages elide, shows the complex competing claims and cultural forces to which Lodge's scientific career was subject.

Lodge's "past years" were the decades of the Victorian period after 1850, so the appeal to Wordsworth in the epigraph to the autobiography aligns him with the mid-Victorian veneration of Wordsworth's aesthetic among the scientific naturalists of the period. Such veneration is evident in, for example, the quotation from the poet that adorned the masthead of the journal *Nature*; and John Tyndall's appeal to Wordsworth, through lines from "Tintern Abbey," at the end of the "Belfast Address" of 1874.[6] Lodge was aware of himself as a Victorian survival: in a work that frequently returns to gender as a shaping force in education, Lodge recalls that his first appointment as a teacher in higher education was to the all-women Bedford College in 1876: the teaching space was so designed that the male lecturers were segregated "from the atmosphere of femininity" and "precautionary chaperones," occupied with knitting or needlework, attended the lectures. Lodge observed that this constituted management "on Victorian lines"—with which, he added, "I have no sort of complaint."[7]

Lodge's sense of his being a continuous survival from a passing Victorian age produced discontinuities between the generic narrative models that he used to communicate with his readers in 1931. Lodge used his preface to differentiate between the 1890s, when his scientific work on electricity and radio waves had been publicly prominent (and where his legacy was still under active discussion), and his "earlier struggles" to become established, which, he sensed had come to be of limited usefulness to younger readers because "the conditions now are very different from what they were then" (5). The emphasis on "struggles" is telling, for it situates Lodge's autobiography among generic models from an earlier cultural phase: though published in the 1930s, Lodge's book owes much to the tradition of Victorian working-class, autodidact writing that has been identified by David Vincent and that shaped the autobiography of the evolutionist and spiritualist Alfred Russel Wallace. At the same time, it inhabits the space of the middle-class autobiography of self-helping enterprise traced by Donna Loftus. It also works along the lines of classic Victorian conversion narratives of faith and methodologically self-aware scientific autobiography analyzed by Linda Peterson.[8]

At times the more recognizable generic models that Lodge followed can be seen to overdetermine the highly distinctive and nuanced story of social and professional selfhood that Lodge's narrative relates. Because Lodge's scientific autobiography can be seen as a particular contribution to popularization that at times elides Lodge's material investments in science as a corporate practice, it can be theorized in terms of Regenia Gagnier's account of autobiography as an ideologically complex rhetorical construction of class-based, gendered, and professional subjectivities. Gagnier's approach helps to identify the discontinuous time frames and contexts that Lodge negotiated as he moved from his family genealogy to a mid-Victorian story of his own youthful "apprenticeship"; the multiple sources of selfhood that he explored and evaluated ranging from the familial to the scientific-professional and the political, aesthetic, and spiritual.[9] This complex mix of purposes enabled Lodge to invite his readers into subtle cultural contests around the social goods propounded by particular branches of science.

Reproduction and the making of families are Lodge's genealogical preoccupation. He "jotted" a lineage of fecund middle-class clergymen and schoolmasters. However, the story develops less as a narrative of Galtonian lineal descent and reproductive abundance and more often as an account—often digressive—punctuated by mortality, poverty, geographical mobility, and transitions driven by economic pressures; and moved on by social and professional networks. Lodge's maternal grandfather, the Reverend Joseph Heath, had been headmaster at Lucton School, Herefordshire, where he "reared a large family of both boys and girls."[10] His paternal grandfather, the Reverend Oliver Lodge, lived a long life initially in Ireland, latterly in England, during the course of which he fathered twenty-five children with three wives.

For the grandson Oliver Lodge at the beginning of his autobiography, one narrated event and a narrated tension stand out among the anecdotes about grandparents, uncles, aunts, cousins, and brothers. The first is the untimely, early death of his maternal grandfather, Heath, which left his widow and children "in a state of comparative poverty" (11). Despite the poverty inflicted on Mary Heath née Marshall (Lodge's maternal grandmother and wife had the same maiden name), Lodge's autobiography advances an important narrative of women's intellectual resourcefulness and educational influence. The narrated tension introduces a conflict between Lodge's own story of diverse areas of "mental growth" and the theme of "strenuous [working] life" endured by his father, another Oliver Lodge, who, as the offspring of the elderly but supremely fertile Reverend Oliver Lodge, was situated disadvantageously "close to the tail of an immense family" and consequently "not given some of the advantages" enjoyed by his brothers who went on to become, in their turn, clergymen and schoolmasters (18). The second Oliver Lodge launched, lived, and improved his "strenuous" working life through the railway and pottery industries in the north Midlands, first as an assistant cashier for the North Staffordshire Railway, then as a clay agent. Oliver Lodge the physicist may have become one of the most eminent public scientific figures in the professorial class that staffed and led the new British civic universities in the late nineteenth and early twentieth centuries, but his scientific career was a social product of the Victorian Potteries, its professional and industrial networks, as well as the autodidact tradition of "self-help" reading and (gendered) family-led education.[11]

Contextualizing this generic dimension of Lodge's *Past Years* places the autobiography in a substratum of scientific life writing that operated at a distance from the elite social milieu that shaped, by contrast, Darwin's autobiography. In some ways Lodge's work was more like the autobiography of Alfred Russel Wallace, *My Life*, which had appeared first in 1905. Although Wallace also started with the presumption that he would "write some account of my early life for the information of my son and daughter," he was persuaded into print by the promise of a wider audience lured by the prospect of "the diversity of my interests."[12] This diversity included both radical politics and alternative sciences. Like Lodge, Wallace was an early proponent of psychical research from within the Victorian naturalist scientific tradition. If Lodge made the case from the research community of the physical sciences, Wallace, who had independently theorized evolution by natural selection, did so from a position secured through the Darwinian evolutionary life sciences.

Wallace's diversity of interests and commitments was also captured in a life story about survival on the social margins, driven by family fragility and hardship: beginning with a father stranded between faded gentility and school teaching, and fatally undone by misappropriated investments, Wallace's childhood was marked by care among older siblings rather than parents, and a career in science that emerged out of apprentice employment in surveying and, following a largely ineffectual education, a sustained program of autodidact reading. This reading was shaped by principles of self-improvement, grounded in the mutuality of Owenite socialism, spread by the London "Halls of Science" to which Wallace was exposed as an apprentice builder (43–57). Wallace's reading was formative but also a hard-won cost in circumstances where resources were scarce: having paid a small subscription to Leicester town library, Wallace recalls very precisely reading Thomas Malthus's *Principles of Population* to harbor it as a "permanent possession" (123). Wallace's experience contrasts with Darwin's reading Malthus for "amusement," as Darwin famously recalls it in his *Autobiography* as a decisive contribution to the theory of natural selection.[13] Wallace's experiences of Leicester also opened him to the "diversity" that characterized his life, for in his narrative he switches focus immediately from Malthus to his discovery in Leicester of mesmerism and psychical research, which "played an important part in my mental growth."[14]

Lodge's capacity for "mental growth" is represented as being checked by his father's disciplined commitment to the mid-Victorian rigors of self-help and apprentice work. Clerical employment as a cashier in the railways moved Oliver Lodge the elder from an apprenticeship in medicine and employment in London to North Staffordshire.[15] Lodge the younger recalled his home as a place where his father's work constantly intruded: his father would bring the company books to work on during the evening at home. The realization that there were "wider openings" in the ceramics industry led to his father's appointment as a clay agent for the sale of blue clay from Dorset. The business turned out to be lucrative, resulting in a move to a larger house and ultimately a comfortable life. Yet the work remained relentless: "From early morning till late at night without intermission; and one would think there was time for nothing else" (25). Lodge was boarded at a grammar school, unhappily, at Newport in Shropshire, where classics and cruelty were the educational staples. At the age of fourteen, his formal education appeared to have been concluded: following an accident in which his father broke an arm, Lodge was effectively apprenticed into the business, touring the Potteries in a phaeton securing deals on and supplying clay. His father "was proud of his business of potters' merchant . . . and wanted me to inherit it. He had an idea, common in those days, that a boy was no good in business unless he entered it at fourteen, began low down, and went through the whole mill" (51).

As it had proved for Wallace, apprenticeship was the route into work for Lodge. It was also the model by which his father had been inducted into work by a sibling, though initially and, significantly for his son, into the field of medicine. And yet the apprentice narrative delivers more than relentless work: it "discovers" another kind of story based on professional social networks and opportunities. Oliver Lodge the elder was originally apprenticed "in the old-fashioned manner" to his elder brother, Charles, a medical practitioner in London, rather in the manner in which T. H. Huxley recalled commencing medicine "under a medical brother-in-law."[16] Though Oliver Lodge the elder moved away from medicine and into the world of the clerk and clay agent following his migration to Stoke-on-Trent, the medical training provided him with access to an important and influential social network in the Potteries.[17] In presenting this, Oliver Lodge the son and autobiographical subject focused on a leading figure representing an earlier generation of medical practitioner: the surgeon and founder of the North Staffordshire Naturalist Field Club, Robert Garner, who was also attached to radical Owenite circles. Garner's presence richly supplements the generic limits of the autobiography of middle-class enterprise: Garner is the "venerable" obstetrician who "brought [Lodge] into the world" (19). In this instance, the birth was intellectually symbolic as well as a practical fact. Lodge recalls that Garner was one of the figures who nourished his frustrated interests in scientific experimentation. These experiments were designed to test controversies associated with the work of the unorthodox science of Dr. Henry Charlton Bastian, and his theories of spontaneous generation—which Wallace had also championed.[18] The overdetermined autodidact and enterprise narratives cannot entirely obscure the networks of radical science behind the book-loaning vicars and classes at the Burslem Wedgwood Institute (68–69).

Lodge presents his scientific education as being sustained and developed by diverse sources. Thus, he emphasizes the influence of his maternal grandmother, the feminine family line she represented, and broader Anglican social and educational networks in which that feminine influence exerted authority. Mary Heath had been widowed and "retired to the house of refuge called Bromley College," an almshouse of the Church of England for the widows of clergy established in the diocese of Rochester and near to London. Lodge attributes to his grandmother "the first educational influence that I remember," which included the reading aloud of Thomas Day's improving work of children's literature, *Sandford and Merton*.[19] The key figure in this female lineage is Charlotte Anne Heath, Mary's daughter and "ever memorable" aunt to Oliver Lodge. She had been educated at the Clergy Orphan's School in Saint John's Wood, where she was noticed by the dowager queen Adelaide, and appointed to court as a Woman of the Bedchamber. Following the death of the dowager queen in 1849, Anne (as she was called) retired from the court with some "relics" and a sum of money that she invested to support an independent life in Fitzroy Square, London. Her social circle included the Reverend James Moorhouse, Vicar of Saint John's, Charlotte Street. Lodge's aunt Anne provided for him a base in London, and a means for fashioning a life in science away from his father's business in Stoke. Moorhouse, who was to become bishop of Manchester, had been educated for the clergy following an evangelical conversion, having originally been expected to take over his father's trade as master cutler in Sheffield. Lodge recalls listening "rapt to his conversation," and attending classes that he organized for working men of the parish.[20] Lodge's autobiography thus narrates two conversions: one to the cause of science, the other (eventually) to spiritualism.

It was the educational example of his aunt Anne that presented, paradoxically, a highly charged domestic path to advanced learning in the physical sciences; under Anne's tutelage, Lodge acquired study skills that caused him to devise his own system of shorthand. This moment was remembered as an act of *making* derived from the mid-Victorian autodidact tradition of print culture: so that Lodge "took up the subject from hints about early shorthand in an article in the *Penny Cyclopedia.* I copied out large portions of Smiles's *Self-Help*" (66). Lodge's story of Anne's influence was a means for fashioning an integrated story of both escape and a professional contribution to science. Lodge recalls her efforts to get him to understand and explain the phenomenon of a harvest moon with a model comprising "an orange, a knitting needle, and a piece of card-board with a hole in it," concluding that "her ingratiating method was not so much to tell us facts as to set us problems to work out for ourselves" (43). What could be described as a method of problem-based learning instilled in Lodge a commitment to improvisation in the pedagogic art of experimentation and demonstration, which came into its own when Lodge had to teach physics in an anatomy theater in Liverpool while his lab was being built, with only a blackboard at his disposal (158).

Lodge's developing interest in astronomy was also supported by his aunt's gift of Ormsby McKnight Mitchel's *Orbs of Heaven* (1860), a work of transatlantic scientific popularization. Lodge recalls that this was "a subject which I amplified, years afterwards, in my *Pioneers of Science*" (43–44). *Pioneers* (1893) is a work of popularization based on a series of public lectures, delivered when Lodge was Lyon Jones Professor of Physics at Liverpool in 1887–1888. The lectures consisted largely of biographical material and there was no attempt at originality; instead, the emphasis was on presentational "vividness" in illustrating for a lay audience the place of key astronomers in "the progress of thought."[21] Lodge reflects on the cultural phenomenon of the "popular lecture" and its relationship to the press in *Past Years.*[22] As Stathis Arapostathis and Graeme Gooday have argued, commercial popularization, or lecturing and publishing for middle-class audiences who sought access to the unfolding history of modern science in books such as *Modern Views of Electricity* (1889), constituted Lodge's preferred strategy of knowledge management.[23] This delivered royalty earnings that contributed to the financing of his lab: Lodge seemed to express mild annoyance that *Modern Views* had been "allowed" by the publisher Macmillan to "go out of print."[24]

Gendered discourses of popularization were recycled and absorbed back into the story Lodge would tell about his own identity as a "pioneer" of science in *Past Years.* For instance, Lodge's quest for "vividness" meant that gendered and domestic discourses were persistently referenced in *Pioneers.* This was manifest in Lodge's popular analogies, derived from the fiction of the mid-Victorian period: thus Isaac Newton's mother was held to be "a sensible, homely, industrious, middle-class, Mill-on-the-Floss sort of woman."[25] This also meant that women were presented as sources of intellectual authority: Lodge's account of William Herschel was drawn from the work of Agnes Mary Clerke, the leading historian of astronomy in the period.[26] Sometimes, however, the intellectual authority went unrealized. Lodge's own "domestic" path to science shaped a keen recognition of the limitations that family obligations could impose: looking at the history of Caroline Herschel's career as a woman in science, Lodge comments sympathetically in *Pioneers* that it was "pitiful ... to find that her domestic obligations still unfairly repressed and blighted her life."[27] Such discourses fed back into Lodge's act of autobiographical reflection in *Past Years*. The obligations to his father's business are represented as blights upon his intellectual potential. Winter residence at his aunt's produced "an odd life," making him "a rebel from home."[28]

Lodge's rebellion against home is presented positively as a conversion to science through the Royal Institution, which he visited from his aunt's house in Fitzroy Square in the late 1860s, when he was around the age of sixteen. This aspect of Lodge's story was crafted through discourses of scientific popularization that emphasized science's connection to both the everyday and the sacred as sources of renewal and conversion. The autobiography devotes a whole chapter to the experience of attending the course of lectures by Tyndall, whose own entry in the *Dictionary of National Biography*, written by his wife, Louisa Tyndall, in 1898, noted his own humble social origins, and familial links (through William Tyndale) to reformation and martyrdom.[29] The Royal Institution became, Lodge recollects, "a sort of sacred place, where pure science was enthroned to be worshipped for its own sake. Tyndall was in a manner the officiating priest, and [Michael] Faraday a sort of deity behind the scenes."[30] Lodge attended Tyndall's lectures on heat and read his *Heat Considered as a Mode of Motion* (1863), but the main currency in mediating the sacred value of science to Lodge was Tyndall's own contribution to popularization through his life of Faraday, *Faraday as a Discoverer* (1868). Although Tyndall and Faraday held different chairs at the Royal Institution, Tyndall's written life of Faraday played a role in cementing a kind of ethical succession.

The relations that defined life writing as an act also bestowed a dual authority on the lives of those who discovered the secrets of the universe. In the first instance, there was the deistical aura around Faraday, whose life held a particular resonance for Lodge: apprenticed as a bookbinder, he had been enabled by Sir Humphry Davy "to quit trade, which he detested, and pursue science, which he loved."[31] Faraday's story was realized in Tyndall's words that integrated scientific experimentation and self-formation in that "a favourite experiment of his own"—the expulsion of all impurities from water in the process of crystallization—"was representative of himself," in so far as "beauty and nobleness coalesced, to the exclusion of everything vulgar and low."[32] Secondly, such words, and the cultural categories that they mapped, were validated by "Tyndall's own life," or the ethically attuned experimenter and discoverer who could embody "the high ideal for the attitude of a scientific man towards human life."[33] The inspiration of Tyndall's "great discourse" bestowed on Lodge a sense of the material universe "such as ordinary people must be unaware of"; "After a discourse by Tyndall . . . I have walked back through the streets of London . . . with a sense of the unreality of everything around, an opening up of deep things in the universe, which put all ordinary objects of sense into the shade, so that [railings, houses, carts, people] seemed like shadowy unrealities, phantasmal appearances, partly screening, but partly permeated by, the mental and spiritual reality behind" (78). This passage echoes Wordsworth's "Immortality Ode," the epigraph that frames *Past Years*, especially its sense of "those obstinate questionings / Of sense and outward things" experienced by "a Creature / Moving about in worlds not realised." Thus Wordsworth becomes another means of affirming everyday life while also glimpsing its veiled spiritual possibilities.

In some sense, Lodge posits the educational experience of the Royal Institution as a conversion experience comparable to Thomas Carlyle's "Everlasting Yea" in *Sartor Resartus* (1837), and another perspective on Tyndall's conviction, as Frank Turner has observed, that Carlyle's challenging prose was a "moral source" for the new Victorian men of science.[34] If Tyndall's inspiration enabled Lodge to present himself as possessing privileged insight into a transcendental vision of the material world that gave him access to the higher physics (Lodge was convinced that his practical, observational skills were weak), it also resulted in a fevered program of study and examination through the London University extension system, resulting in matriculation and a DSc degree in 1877. In fact, when Lodge attended his first meeting of the British Association for the Advancement of Science in Bradford in 1873, he was balancing commitments to visionary scientific study with work for his father, traveling as far afield as Leeds to sell clays and leads to potters, which enabled attendance at the association's meeting. Thus, the rich experience of complex and overlapping social networks and practices was again generically overdetermined to be seen as a more decisive break, or "conversion." Attendance at the British Association in Bradford was "the means of my finally breaking away from business."[35]

Like the Royal Institution, the British Association was career-defining subject matter deserving of its own chapter. Lodge records that he attended the British Association meeting in Belfast the year following (1874) and so was present for Tyndall's infamous Belfast Address; looking back in *Past Years*, Lodge describes the social and theological atmosphere as "sulphurous" (139). Lodge's autobiographical retrospective is thus divided between an account of goal management, achievement, and field contests on the one hand; and a transcendent vision of a life—indeed "life" and "matter" itself—with an eye on "higher" purposes, including psychic research. The discourse of "transcendent vision" could be used to smooth over real professional difficulties regarding questions of priority in competitive scientific, technological, and commercial contexts.

Lodge's recall of his entry into the competitive cultural and institutional contests of Victorian science in the last quarter of the century was thus overdetermined by a discourse of Romantic literary transcendence that elevated him beyond both his early struggles to escape the pottery industry and the competition that characterized academic and commercial advancement. Increasingly publicly prominent for his science, Lodge was unsuccessful in his application for a chair at Owen's College, Manchester (he lost out to the physicist Arthur Schuster), but he was successful at the very new University College Liverpool, where he was appointed as its first professor of physics in 1881 (151–53).

Lodge's account of his working life in the newly established University College is, in many ways, the centerpiece of the autobiography, tracing as it does the professional physicist's important scientific achievements in the field of electromagnetic wave theory; and, following Faraday and Tyndall, the forms of public engagement and popularization, including lectures at the Royal Institution, that communicated and displayed those findings, often dramatically. Indeed, Lodge's lecture at the Royal Institution, "On the Discharge of a Leyden Jar," delivered in 1889 and published in *Nature* in 1890, provided sensational visual evidence of Heinrich Hertz's theory of unmediated spreading and looping radio waves: the electrical discharges that Lodge used in his experimental demonstrations unexpectedly caused the wallpaper of the lecture theater, decorated with flakes of metal representing foliage, to spark "in sympathy with the waves," as he described it in *Past Years*.[36] The use of the term "sympathy" arguably connects Lodge to a nineteenth-century aesthetic and ethical tradition associated with George Eliot and John Ruskin. Waves and sparks working in sympathy were qualitatively different from the way in which Lodge described the same phenomenon in his paper "Experiments on the Discharge of Leyden Jars," read at the Royal Society in June 1891, when wave experimentation was at its height: here, the "heavily gilt wall paper sparkled brightly, *by reason of the incident radiation*."[37] If rationality explained the phenomenon in 1891, Lodge's autobiography drew on a discourse of mutuality that cautioned against a quest for prioritization in scientific discovery. Lodge notably appealed to the authority of Ruskin in support of this eschewal of priority, and assertion of sympathetic "clear eyes and [an] open heart."[38]

As Lodge acknowledges in the preface to *Past Years*, much of the autobiographical focus on his career in the 1890s is on the incremental discoveries relating to radio waves that dominated the physical sciences during that period and which depended on interactions between a wide variety of British and continental researchers, including Hertz and, of course, Guglielmo Marconi. Thus, competitive questions about priority are negotiated in this section of Lodge's life story. The view that Lodge had been "trumped" by Marconi on the extended transmission of radio signals as a communicative medium, given Lodge's claim that he had been first to succeed over a more limited distance, is part of the story that was apt to be retold in reviews of Lodge's life.[39] Lodge's own account tends to emphasize the positive and cooperative intellectual communications that were maintained between scientists in the quest for discovery, including Lodge's memorial lecture to Hertz at the Royal Institution, following the latter's early death in 1894 (published in *Nature*). While Lodge provides a prose commentary on this competitive pursuit of wireless telegraphy, in which his own contribution is foregrounded, he concludes by eschewing narrative and instead casts the events in a chronicle format, beginning with James Clerk Maxwell's equations leading to wave theory and concluding in the foundation of the BBC in 1922, a fact that starkly declares the scale of the prize that eluded Lodge's priority status.[40] As Arapostathis and Gooday have shown, Lodge's autobiography is a rather problematic reconstruction of his claim for priority in developing what Lodge himself called "a very infantile kind of radio telegraphy" (232). They see this claim as a late episode of rewriting, extending earlier recastings of his published understanding of the applications of Hertz for the purposes of wireless transmission; and a manifestation of the tendency for autobiographical narration to elide the role of *corporate* activity in discovery and technological adaptation.[41]

Lodge's autobiography strikes a tone of equanimity in these matters, conveying a sense that science at its best was a noncompetitive pursuit of progress and perfection. If Lodge presents himself as being content to play his role in an incrementally cumulative process, and as particularly proud of the patent, obtained in 1897, for a selective tuning device that contributed "to the perfection of the transmitters now in use by the B.B.C.," then this is somewhat at odds with Lodge's actual behavior in pursuit of patents in the wake of Marconi's act of patenting in 1896.[42] Arapostathis and Gooday show how Lodge, who had never been a heavy user of the patent system that took shape after 1883, applied for five patents in 1896: Lodge and his competitors realized that "knowledge management for wireless . . . was no longer open ground but ripe for enclosure."[43] For Lodge, the autobiography became, by contrast, an unenclosed space in which to rewrite the history of wireless communication endeavors. For instance, he expresses gladness that in Britain a great deal of attention was paid to the German Hertz's achievement and legacy in 1894, following his early death. Lodge himself had been invited to take a lead in honoring these achievements, he and Hertz having "conducted a very friendly correspondence."[44] Writing for a reader in the early 1930s, this was an important reminder of the chauvinisms that afflicted all aspects of culture in the lead-up to 1914 and beyond.

Thus, the autobiography was articulated in a spirit of mutuality that colored Lodge's approach to politics, economics, and technological adaptation. This clearly went deep in Lodge and was another articulation of his Wordsworthian Romantic and Victorian Ruskinian sympathies. It is manifest in his reading of Ruskin's critique of the political economy of acquisitiveness in *Time and Tide* and *Unto This Last*, and Ruskin's alternative argument for economic pursuits based on human health, well-being, and happiness. Lodge perceived himself to be running against the orthodox liberal grain, observing how Ruskin was disbelieved "contumely" at the time. As we have seen, Ruskin helped Lodge to negotiate the place of prioritization in scientific discovery and his own career and self-image beyond that, while Lodge held that Ruskin's vision of mutuality and sympathy ought to be the "the aim of all our industry" (218). Such aims were ethically and environmentally realized when Lodge worked closely with the emerging energy and automotive industries, as he did extensively throughout his career, in collaboration with his family and many others.

Despite these successes, it is notable that Lodge's autobiographical voice sought to offer the alibi of a capacity for transcendent vision as an explanation for his failures to secure priority and press an advantage in other contexts. Thus, Lodge was aware that he "neglected" his early advances in wave experimentation: he confesses that he "did not follow them up as I ought to have done. I left them to Hertz, mainly" (230). Lodge analyzes this trait of deferral in chapter 9, "A Personal Retrospect," which is, in a sense, a confession of the strengths and shortcomings shaping the professional, scientific career that he would go on to narrate. Thus, Lodge, who at this point had narrated his transcendental experiences arising from his visits to Tyndall at the Royal Institution, claimed that an almost visionary excitement could overtake him when he became aware of new questions and openings in the field of the physical sciences: "Whenever I read modern physics, I acquire a sort of vision, and perceive in a short time a sort of vista of unexplored territory, which causes me to wish to clear the ground of all other engagements so that I can follow it up," but added that he had "not always returned to it" (113). In this, Lodge contrasts himself with Faraday, who "suffered from no such disability" (113). Lodge accounts for this trait in part because of the volume of "engagement" with which he was burdened, including teaching in a new institution of higher education and the subsequent duties of founding university principal at Birmingham. However, such capacity for engagement also owed much to Lodge's openness to intellectual and spiritual movements that were marginal to the Victorian scientific and intellectual mainstream: "I have taken an interest in many subjects and spread myself over a considerable range—a procedure which, I suppose, has been good for my education, though not so prolific of results. As far as I can understand it, it was due to a feeling that if I pursued physics too exclusively, other things would be forgotten. I wanted to clear away everything else, and then concentrate on my own subject: the result was that the time for uninterrupted concentration seldom or never arrived" (112).

Spiritualism was one of the most important and controversial of these diverse, attention-seizing discoveries and interruptions. Toward the end of *Past Years*, three chapters are devoted to the work of the Society for Psychical Research (SPR), prior to Lodge's account of his move to take up the role of principal at Joseph Chamberlain's new University of Birmingham in 1900. Wallace's autobiography records the hostility to which his investigations into spiritualism were subject among X Club members Huxley and Tyndall.[45] Lodge knew that his psychic research, too, "would be unpopular" and so made it a condition of employment that he be permitted to continue.[46] The unpopularity and skepticism that attached to this aspect of Lodge's career became an important part of his legacy both in Britain and the United States. It figured, for example, in Walter R. Brooks's review of *Past Years* in the New York–based magazine *Outlook*. In the review, Brooks tellingly notes the success of Lodge's autobiographical strategy of the representation of gradual step taking toward a spiritualist conversion: one designed precisely to address that unpopularity and skepticism.[47]

Psychical research prompted Lodge to draw upon a theological and spiritual model of autobiography to justify its place in his life. The concluding chapter of the autobiography is revealingly titled "Apologia Pro Vita Mea," a clear allusion to J. H. Newman's mid-Victorian autobiographical defense of his "steps" to spiritual conversion, *Apologia Pro Vita Sua* (1865). Lodge may be said to have countered his own context of unpopularity and skepticism through strategies that were similar to Newman's account of his gradual, circumstantial conversion to the Catholic faith. Newman's answer to Charles Kingsley's attack on the integrity of his position aimed to show, with great narrative subtlety, that his conversion was in no sense a statement of prescribed Catholic doctrine; rather, as Linda Peterson argues, his narrative demonstrates a gradual *historical* transformation, a growing out of established Anglican theology, a blending of Evangelical Protestant and Catholic-Augustinian conversion traditions.[48]

Thus, for Lodge, encounters with the psychic are presented as emerging from the normal course of conducting his scientific work: for instance, he met Edmund Gurney during the mid-1870s when lecturing to a class on mechanics at the (deeply secular) University College, London. When Lodge visited the "striking" young classicist at his home, he was introduced to the materials that would become the two-volume *Phantasms of the Living* (1886), but which Lodge was inclined to see as "a meaningless collection of ghost stories" amounting to little more than "baseless superstition."[49] Sensational performances by the American mentalist Irving Bishop that took place in Liverpool in 1883 were judged to be based on no more than "muscle reading." Shortly afterward, however, when Lodge was approached in his role as professor of physics to verify experimental work undertaken on employees who showed signs of telepathic capacity at the most prosaic location of the drapery store George Henry Lees in Liverpool, he agreed to be the test representing "scientific authority" and went with a fellow scientist from the university, a biologist. Following extensive verification on the experimental regimes to which the young women employees had been subjected by their employer, the "result was gradually to convince me that the faculty of thought-transference . . . was really a faculty possessed by certain people" (274). Lodge reports that while this experimental work was published in the new house journal of the SPR, *Proceedings of the Society for Psychical Research,* he ensured that aspects of his experimental work were also sent to *Nature.*[50] Lodge consequently links his conversion to spiritualism with what had become an arbiter of scientific authority. This strategy is consistent with Lodge's presentation of his gradually acquired associates in the spiritualist enterprise—Gurney, F. W. H. Myers, Henry Sidgwick—as rich in recognized cultural capital (literature, the classics) and, in the case of Sidgwick, Cambridge-validated philosophic rigor. Lodge concludes this section of the autobiography with a section titled "Caution," arguing for open-minded, intellectual plurality: "The subject still bristles with difficulty. All I plead for is study."[51]

If autobiography is the generic medium for narrating both the sources of selfhood and the distinctiveness of human personality, then an innovative convergence of objectives emerges between these generic functions and Lodge's canons and caveats for studying communications that evidence human survival in spiritual form. In a sense, written autobiography becomes a proxy of spiritual survival and communication. Thus, in his statement of "Caution," Lodge holds that "only when we receive unmistakeable evidence of personality are we entitled to treat the messages as conclusive" (289). Lodge recalls his experience of the return of an unmistakable personality through a medium when his dead aunt Anne, who inspired him to return to education, "spoke a few sentences in her own well-remembered voice." For Lodge this was highly "unusual . . . but very characteristic of her energy and determination" (277).

Recalling the first words of his autobiography, Lodge conceived it as a way of "jotting" information about his life for his children. However, the network of communication that his text actually held in play was significantly more complex: it established relations between his personality and the personalities of the living and the dead, including Lodge's wife, who had died two years before the autobiography's publication and to whom it was dedicated, and it also drew consumers of popularized scientific discourse into its network.

"Personality" was a key word for Lodge and his spiritualist colleagues as they contemplated continuity or what survived between the generations.[52] As I noted at the beginning of this chapter, Galton and the question of ancestral inheritance is the first scientific frame of reference to be encountered by the reader. In 1931 Galtonian eugenics, premised as it was on records of inherited health and ability within families, was the other aspirant science of selective survival, attempting to prescribe what should be profitably passed on and what should be selected out. This was a legacy that Galton's disciple Karl Pearson, chair of the Galton Laboratory, recalls in his monumental official biography of Galton that, in volume 2 (1924), also looks back to the 1880s and 1890s and the formation of the new Anthropometric Laboratory at University College, London. The fourth and final volume of the biography was published in 1930, just a year before *Past Years*.[53] The written memorial life could thus become a forum for the subtle contestation of the meanings and goods associated with processes of selection.

How did this work? Pearson, following from Galton, was an early theorist of the hereditary consequences of assortative mating. Lodge, by contrast, wrote about the romance in which his family lines had been perpetuated through a different set of selective affinities. His account of his courtship of Mary Tomkinson of Brampton House, Newcastle-under-Lyme, stresses the importance of personal pronoun affinities from his maternal line: because she had been adopted through her mother's remarriage following the early death of her father, Lodge is keen to remind the reader that her name was "really Mary Marshall," the same name as his all-important maternal grandmother. In addition, the story of the death of his wife's father led to Lodge's first encounters with the stories of psychic visitations from spirits.[54]

When Lodge came to write specifically about his family of twelve children and his wife in the chapter titled "Family Life," he adopted a way of writing about the family that was at variance with Galton's concern with the ancestral and the testing grounds for hereditary principles that it was held to provide. Lodge was especially conscious of modes of speech in his choice of words about his children; he remarks, for instance, on the way in which his wife and himself "half humorously expressed gratitude that such delightful people thought it worth while to come and live with us" (247). Lodge especially celebrates his twin girls, the youngest of his children. Twins were, of course, the perfect objects of comparative measurement and scrutiny for Galton's research into nature and nurture in work that had begun in the mid-1870s.[55] By contrast, in "Family Life," in looking at his twins Lodge chooses not to look at comparative life chances and marvels instead, and through metaphor, at the fact of the late (to the parents) and equal (to the children) opportunity presented by their birth itself: "When the door was closing they just managed both of them to squeeze through."[56] Lodge was well aware that autobiographical narration was grounded in communicative choices and a rhetorical relationship with his readers: in this chapter he asks the reader if "this [is] a reasonable mode of expression," and then answers, "I don't know" (247).

Lodge could recycle popular theories of evolutionary "recapitulation" to explain his own acts of childhood cruelty. At the beginning of *Past Years,* for instance, he suggests that "some of the faults of children must be merely retracing the steps of their ancestry" (29). However, the book is notable for the way in which Lodge uses the occasion of his autobiography to refuse to endorse a "sure understand[ing of] the problem of heredity." Indeed, he staunchly resists the principle "that every individual organism is merely the product of its parents":

> When two cells unite, the result may be better than either: at any rate, it is different. It seems as if something was superadded, something carrying with it powers and personality which were not in the material, but something of which incarnation was rendered possible by the material provided. . . . The resulting personality in human offspring, how does it come about? It seems in some way independent, as if it had sufficient likeness to be attracted, and yet had other aptitudes, not really transmitted, some of them latent, ready to be developed by circumstance. It is as if some choice or selection, a kind of unconscious selection, was exerted by the nascent individual. (247)

This is a striking passage because of the way that it places stress upon "superadded." Lodge's pioneering experimental work in the fields of electricity, magnetism, and wave and light theory arguably took the shape that they did because of his conviction about the existence of the imperceptible yet universal superadded feature of ether.[57] In seeking to express a view about the "superadded" vital features of distinctive human personality that were both materially transmitted and inexplicably "attracted" into existence, Lodge expressed himself through a wide range of conceptual vocabularies: from the theologically resonant idea of "incarnation"; to "circumstance" and "selection" from the fields of evolutionary theory; and "latency" as an echo from the new field of genetics and its identification of the mechanisms for inheriting dominant and recessive traits. Lodge articulates a view of personality as "nascent individual" that could exert powers of "unconscious selection" in becoming absolutely different—an absolute difference that would persist into the realms of spiritual survival and unique communicative signs; indeed, the whole autobiography might be seen as charting diverse paths to the discovery of the sources of selfhood and the means by which they were communicated.

For Oliver Lodge, birth was about newness and possibility rather than inheritance: he felt that he had "inherited a sound constitution from my parents and nothing else."[58] He validated this confidence with an allusion to the poetical religious language of Psalm 23:6: in the final chapter of *Past Years* Lodge claims to have been guided by "goodness and mercy." Identity and personality might be as open, forward-looking, and unknown as the future of knowledge acquisition hailed in the rousing closing words of the autobiography, despite the fact that some knowledge (such as spiritualism) might be seen as "strange and unprofitable" (352). In 1931 this was a brave social and individual moral good to affirm: one of the striking contexts for Lodge's noble survival as a Romantic-Victorian spiritualist-radical into the 1930s is recorded in Brooks's review of *Past Years*: also featured among Brooks's survey of new books for March 1932 is the much less noble, satiric tone of Aldous Huxley's *Brave New World* as it confronted the reproductive politics of the 1920s and 1930s.[59]

TWO

BECOMING SIR OLIVER LODGE

THE LIVERPOOL YEARS, 1881–1900

PETER ROWLANDS

IN 1898 Oliver Lodge was awarded the Rumford Medal by the Royal Society "in recognition of his researches on radiation and on the relations between matter and ether."[1] In March of the following year, the Lord Mayor of Liverpool hosted a banquet in his honor to celebrate the award. Among the guests were the scientists George Francis FitzGerald and William Crookes; the spiritualist and philosopher Frederic Myers; and the politicians Arthur Balfour, then First Lord of the Treasury and Leader of the House of Commons; and George Wyndham, the under-secretary of state for war (both of whom were also prominent members of the aristocratic set the Souls). Lodge was, by this time, very well connected: his scientific achievements had brought him both fame and scientific recognition, but he had also developed a set of personal relationships that linked him with some of the leading figures of the day. Liverpool provided Lodge with the opportunities to establish himself as both scientist and public figure; the banquet, in turn, acknowledged what Lodge had given back to the city.

Lodge lived and worked in Liverpool from 1881 to 1900. He was there between the ages of thirty and forty-nine, the prime years of his scientific life. Though the period was marked by a concentration on teaching and research, Lodge used the time to hone the communication skills, both as a writer and a lecturer, on which he would depend for the rest of his life. Alongside scientific books and articles, he contributed popular pieces on a wide range of subjects and gave extracurricular talks and lectures, sometimes drawing huge crowds eager to hear from one of the leading scientists of the day. And Lodge extended his range beyond his immediate academic environment by involving himself with many organizations and societies. He remained an active participant at the meetings of the British Association, which he had been attending since 1873, long before he began to study science more formally in pursuit of a scientific career, but, in 1884, he also joined both the Fabian Society and the Society for Psychical Research. In these years he built up his personal wealth and became upwardly mobile, moving house twice. The Lodge family had grown considerably. Oliver married Mary Marshall in 1877 and they had three sons before leaving for Liverpool; by the time they left the city, the family had grown to twelve children. The Lodges made their home in Liverpool; in turn, it was through Liverpool that Oliver Lodge became known.

COMING TO LIVERPOOL

Lodge's rise to prominence was quite remarkable for a man without a public school or Oxbridge education, who only began his own higher education at the age of twenty-two. The fact that he was at Liverpool was of immense importance to his progress, because University College Liverpool had been expressly founded at the instigation of the city's most prominent citizens, mainly professional people and gentlemen who had made their fortunes in shipping or sugar, who took a very strong interest in what happened in the city. Lodge was appointed as the Lyon Jones Professor of Experimental Physics, a particularly significant position, as it had been founded with the express requirement of providing the physics training that, since a change in regulations in 1877, had been necessary for the University of London medical degree. As Liverpool had a long-established medical school with a large number of students, Lodge's new position meant that he was an important figure within the college. As one of the few salaried positions for physicists in the country, it also gave Lodge an acknowledged position in scientific culture more broadly.

In the nineteenth century Liverpool rose to become the primary port of the United Kingdom. In many people's view it was Britain's second city after London, and the second city of the rapidly expanding empire. Its wealth on occasions exceeded even that of London itself, and its custom house certainly contributed more to the British Exchequer than any other institution.[2] Described in 1851 as "the New York of Europe," Liverpool was the only provincial city to have its own office in Whitehall (from 1812), and, beginning in 1790, it had had a US consul for longer than anywhere else in the world.[3]

The appointment in 1881 was not Lodge's first experience of Liverpool. In 1935 the *Liverpool Daily Post* recalled that, fifteen to twenty years before coming to Liverpool as a professor, the teenaged Oliver Lodge had ended up stranded in the city without any money. "He landed here from the Isle of Man on a Sunday, out of pocket-money, and had to wait a long time before he could get any. . . . Hungry and lonely, he wandered around the streets, and as he once confessed, "just hated the smug people who were having their dinners, though there was no reason why I should.""[4] His next experience of Liverpool was by proxy. He avidly followed the British Association proceedings in Liverpool in 1870, the third after the meetings of 1837 and 1854, by reading newspaper accounts from his home in Hanley. Lodge, then nineteen, lived with his family at Chatterley House, Old Hall Street, and was working with his father supplying china clay and other pottery materials. He read about T. H. Huxley's "great pronouncement" on biogenesis, and James Clerk Maxwell's presidential address to Section A.[5] Three years later, in 1873, he met some of these luminaries when he attended a British Association meeting for the first time, at Bradford, and decided on a scientific career.

Lodge's time at school (from eight to fourteen) had not been a happy experience. In his memoir, *Past Years*, he recalls that his "school days were undoubtedly the dullest and most miserable part of my life," a fact that contributed significantly to his later interest in "progressive" forms of education.[6] It would be more than eight years after leaving school before he would be in a position to take up a more conducive form of formal education. Once he did, his progress was rapid. In January 1874 he enrolled at University College, London. He obtained the BSc in 1875 and became demonstrator to Professor George Carey Foster. He obtained the DSc in 1877 and met FitzGerald, his great scientific friend and mentor, at the Dublin meeting of the British Association the year after. Then in June 1881 he was appointed professor at Liverpool ahead of fifteen other candidates. On October 18, University College Liverpool received its charter.

EARLY DAYS

Before taking up his post Lodge spent the summer touring Germany, Austria, and France in search of equipment, and, in the process, met Heinrich Hertz, who would be so significant to his future scientific career. During the early years of his residence in Liverpool he lived at 21 Waverley Road, near Sefton Park, and about a half mile walk to his laboratory; he always preferred to walk, but there were no trams at that time in any case. On October 3, 1881, he delivered the opening address at the Royal Infirmary School of Medicine, stressing the medical and other technological spinoffs that frequently came from fundamental physical discoveries, including, for instance, the apparatus used in physiological laboratories.[7] Then, in December, he lectured at Saint George's Hall on the possibility of using electrical technology to reduce smoke pollution, a problem he would continue to work on throughout his career. Teaching began at University College in January 1882, with the work for the medical faculty. The first real course began in October.

Lodge was a great communicator from the beginning. He was very tall, six feet four inches, and made a strong impression on any audience. His skills as a communicator were often mentioned in the obituaries published after his death. J. Ambrose Fleming, for instance, remembered that he "always commanded serious attention in any meeting at which he spoke."[8] For William H. Bragg, "he was a magnificent lecturer; his matter and style were of the first order, and his expositions derived further force from his impressive personality."[9] Fleming wrote that "he had an excellent speech delivery and impressive manner and always held attention when he spoke at scientific meetings," while R. A. Gregory and Allan Ferguson reported that he was noted for his "clear speaking and characteristic and musical voice," which was "pleasing and sonorous."[10] "In conversational style, whether in a lecture or discussion," they said, "he presented the essential points of his subject naturally and with such enlightened insight that he always made close and friendly contact with the minds of those interested by him."[11] This was something also noted by Fleming, who remarked that Lodge had an "attractive and unassuming manner and charming interest in the work of others."[12]

Lodge's assistant, Ben Davies, who came from Cardiganshire in 1881 with relatively little knowledge of English, attended extracurricular evening (and possibly regular day) classes given by Lodge. Davies noted when writing to his mother that "Dr. Lodge is very gentle with the students in his class" and that he was "a young man who is very likeable," conscientious and enthusiastic as a lecturer, and frequently arrived early to prepare demonstration experiments.[13] Lodge's great friend, the college principal, G. H. Rendall, looking back over sixty years, considered him "the finest personality with whom it has been my good fortune to be intimate."[14]

As professor, Lodge's main job was as a teacher, at first mostly to the medical students studying for the University of London external degree. His salary was £400 per annum along with two-thirds of the students' fees. To raise more income he invented new courses, in general science, electrotechnics (now called electrical engineering), and astronomy, though he did offload the mathematics teaching he was also supposed to do after the first year. His weekly lunchtime lecture-demonstrations were very popular with members of the general public who were willing to pay six pence each at the door for admission, while teachers appreciated his Saturday morning classes at one pound ten shillings per session.

A series of extracurricular evening classes, which were based on his regular daytime physics lectures to the college students, were open to members of the public, whether men or women, who could afford the price of admission tickets at six shillings per term. Lodge, who was a great advocate of women's education, was here able to draw on significant previous experience in teaching science to women at Bedford College, and all his courses at University College would be open to women on the same footing as men. Wednesdays during the autumn term were devoted to heat and Thursdays to electricity. The lectures began at eight p.m., but Lodge could often be found answering questions until as late as nine thirty. The term would end with a set of questions carrying a five-pound prize for the best answer.[15]

The lecturing and administrative load on Lodge at the beginning was enormous. As well as giving four or five lectures per day in physics and mathematics, he was singlehandedly setting up and supervising the new laboratories. Looking back in *Past Years* he complains: "I had to prepare my own lectures, as, indeed, I had already done at Bedford College, performing my own experiments and clearing up again. This was all very well during a First Year; but when a Second Year course was necessary as well, and when I had to take First and Second year students in mathematics also, and likewise in mechanics, it became too much of a good thing. I got some laboratory boys in to help."[16]

This method of payment, however, was lucrative, and soon Lodge's salary attracted notice. At £1,200, his total income greatly exceeded the £1,000 earned by the principal, and so his contract was renegotiated to £600 and one-third of the students' fees. It was not long, however, before he overtook the principal again. He needed to: his family continued to grow during the 1880s, with four more children joining the three Oliver and Mary brought to Liverpool; five more would be born in the following decade.

The early days were difficult, and at one point Lodge considered a career change. In 1884 he became involved in a consultation concerning appointments at the City and Guilds Institute in South Kensington. He seriously thought about taking up one of the positions himself and so receive a considerably higher salary than his current one at Liverpool. He claimed that he was elected without an interview "and without my actual consent," but was persuaded by his friends in Liverpool to withdraw his application, although William Preece told him that he had made an extremely foolish decision.[17]

Whatever reasons compelled Lodge to reject a seemingly attractive return to London, the workload at Liverpool continued to increase. By 1886 there were also courses on molecular physics and elementary mechanics. In *Past Years,* Lodge acknowledges that such pressures affected his relationships with his children:

> I find now that at Liverpool the children were somewhat afraid of me. I have asked them to explain what was the reason, if they can. Truly I fear I did not get to know them properly when they were small. My work was a fearful grind, and, as a result, perhaps I was sometimes irritable—not often, I hope. But I deeply remember one

> incident, and perhaps I ought to confess it as a warning. I hope that the elder children concerned have forgotten it. I perceive now that it is rather like "The Ancient Mariner." I had come back home to 21 Waverley Road after a hard day at college, and tried to settle down to a thick batch of examination papers, after the children (there were only three or four then) had gone to bed with instructions to be quiet. Presently, from their bedroom above, came a stream of water, dripping on to my window-sill. I rushed upstairs. They had just got back into bed, and said they had been watering a plant outside on their window-sill. I learned too late that it was one they had been trying to cultivate and were fond of. God forgive me, I flung it out of the window. I heard it smash on the ground below; and, after that, I heard a quiet sobbing from the beds as I shut the door of the room. It was my son Lionel's, I think. Doubtless he has long forgiven me. But when overworked, as I was in those days, I found myself liable occasionally to these ebullitions of temper.[18]

In general, Lodge had very good relations with his children, and these became even better when the family moved to a larger house in Grove Park in 1892.[19] The fact that he remembered this single incident so vividly suggests that his occasional irritability seldom led to such extreme bouts of temper.

RESEARCH AT LIVERPOOL

Typically, Lodge, during the mid-1890s, was required to do five lectures a week, along with five laboratory supervisions and a number of evening classes. He also had significant duties as head of the department and as a member of several college committees, leaving relatively little time or energy for research. Yet, by sheer determination, no doubt supported by the high level of physical fitness he maintained throughout his life, he managed his responsibilities with great efficiency and enthusiasm. At the same time, he occupied whatever evenings he had free as an active member of such organizations as the Society for Psychical Research, the Fabian Society, the Liverpool Literary and Philosophical Society, the Medical and Surgical Society, and many other bodies that valued his significant input, and not only as an ordinary member (he served as president for at least ten societies over his lifetime). Work meant a great deal to Lodge: when asked by the editor of *Modern Wireless* for an article on "At Work and Play" for his seventy-ninth birthday, he replied that he had "always found it easier to work than play, though, in his younger days, he relaxed by playing tennis, dancing or cards."[20]

Research was clearly not the priority for Lodge, but he still accomplished a great deal, partly because he had help from a number of capable assistants (whom he always acknowledged), including J. W. Clark, Arthur P. Chattock, James Howard, Ben Davies, and Edward Robinson. Of these, Clark committed suicide in Heidelberg as a young man; Chattock had a successful career as a professor at Bristol, becoming a Fellow of the Royal Society (FRS) in 1920 though he had no degree; Howard, lecturer in physics under Lodge, died of typhoid fever while still at Liverpool; while Davies and Robinson, Lodge's most valued personal assistants, went with him to Birmingham. Davies became the first nongraduate Fellow of the Physical Society. Lodge's early work was based on his enthusiasm for Maxwell's electromagnetic theory and was fortified by his friendship with FitzGerald. All of his training was in the fields of magnetism and electricity, and this was where he felt his expertise lay. Trying to make sense of Maxwell's theory and at the same time keeping up his London connections, he gave a lecture titled "The Ether and Its Functions" to the London Institution on December 28, 1882.[21] The ether was, at this time, very much in the forefront of orthodox science as the medium that enabled the transmission of light through otherwise empty space, and Lodge was here following in the footsteps of such prominent physicists as Maxwell and Sir William Thomson (and, ultimately, even Isaac Newton) in emphasizing its role as a unifying medium for physical forces. His own contribution was to emphasize its relative nonmateriality or abstractness, an idea he continued to develop for the rest of his career.

Without a great deal of time to plan his research, it would often result as a response to something that had happened previously or to some request from outside. For instance, electricity was a major source of technology and people came to seek his advice on practical matters. He had consultancies with various firms, in particular, the Electric Power Storage Company and Fawcett & Preston, both on Merseyside, and Sellon & Volckman, of London. His earliest scientific work in Liverpool involved improvements in electrical storage batteries, and he made significant improvements to the design and the materials used in this fast-growing industry. He began a long series of publications on these matters on May 19, 1882. Under the title "Electrical Accumulators or Secondary Batteries," they were published over the next year in nine issues of the *Engineer*; as was common in the period, these were then reprinted in ten issues of *English Mechanic* and seven of the *Electrician*.[22]

With his interest in smoke pollution, Lodge reacted to a reprint of an article by Lord Rayleigh in *Nature* in June 1883, reporting experiments in which dust was reduced in the presence of hot bodies, and, with his interest in electricity, he was able to show (with Clark's assistance) that smoke and dust could be reduced by the action of electrostatic machines. Lodge tried, without success, to raise £10,000 from the Port of Liverpool for a pilot scheme to remove the fog on the River Mersey by installing giant electrodes on the promenade at Wallasey and at Bootle. However, the process was rescued when Dr. Alfred Walker of the Deeside Lead Works received news of it. Walker suggested collaboration, with the aim of removing the fumes from his lead-smelting plant at Bagillt without requiring the usual two miles of flues. The collaboration led to a patent filed on August 9, 1884, and small-scale experiments, based on an eighteen-inch Voss machine as precipitator. The industrial-scale trial that followed used two five-foot Wimshurst machines driven by a one-horsepower steam engine, but it failed because the insulation proved to be inadequate, and the electrostatic generators produced too little power. However, the idea was revived successfully by Lodge and his sons Lionel and Noel twenty years later, when further experimental work led to the Lodge Fume Deposit Company, established in Birmingham in 1913. The company was amalgamated with an American enterprise in 1921, by which time its precipitators were used in more than half of tin ore smelters across the world. After various subsequent transformations, it became incorporated into Lodge Cottrell, in which the Lodge name is preserved to the present day.

The work on precipitation led to Lodge's first success on an international stage when he gave a brilliant and well-regarded lecture "Dust" at the British Association meeting in Montreal, August 29, 1884.[23] This made a particularly big impression on an audience that included a considerable number of delegates from the United States (notably, J. Willard Gibbs). He also gave another successful presentation on the chemical and physical processes involved in storage batteries, bringing him into conflict with Sir William Thomson (later Lord Kelvin). In this controversy, Lodge maintained that the potential of a metal surface must be defined on the metal itself, rather than in the air just outside it, as Kelvin believed. Lodge was admired for holding his ground, but he also managed the dispute in such a way that it did not affect his friendship with Thomson, who enthusiastically responded to Lodge's lecture on dust that followed.

It was also in Liverpool that Lodge became interested in spiritualism. In 1883 a stage magician, Irving Bishop, caused a popular sensation with several feats of "thought-transference" and mind reading. The people who tried to emulate the tricks included a group of shopgirls from George Henry Lees drapery store, who practiced willing each other to guess selected dates from the calendar; two of the girls apparently showed remarkable abilities. One of the partners in the business, Malcolm Guthrie (a relative of Lodge's former London teacher, Frederick Guthrie), reported the results to both the newly founded Society for Psychical Research and the Liverpool Literary and Philosophical Society. After some preliminary experiments, Guthrie approached Lodge and a biologist at University College, William Herdman, to act as scientific advisors, observers, and controls. Lodge made it clear in his report that these were not his experiments and that he was not responsible for the evidence produced. However, he decided to conduct his own tests using specially prepared cards, and wrote up his results for both *Nature* and the Society of Psychical Research.[24] He joined the society shortly afterward, membership in which brought him into contact with people such as Balfour, with whom he enjoyed playing occasional rounds of golf, and the Wyndhams (he spent Easters at their house, Clouds); and also led to friendships with Crookes and Myers, who became his mentor in matters both philosophical and psychical.

Lodge's clear understanding of the complexities involved in electrolytic processes, shown at the Montreal meeting, led to his being invited to chair a committee reporting on the subject at the Aberdeen meeting of the British Association, held between September 9 and 16, 1885, and involving such eminent physicists as FitzGerald, Crookes, Rayleigh, and Thomson. He became co-secretary to this Committee on Electrolysis, writing their report for the Birmingham meeting in 1886, and describing a method for making the migration of ions visible by decolorizing an alkali indicator and for measuring the velocities of ions when a current was passed through an electrolyte. Soon afterward, on June 9, 1887, he was elected FRS, largely on account of his outstanding work on the precipitation of dust and his important role on the Committee on Electrolysis.

Then there was another external stimulus. In 1887, Trueman Wood, the secretary of the Society of Arts, asked him to give two lectures in honor of the late Dr. Robert Mann, who had been an enthusiastic supporter of lightning rods in South Africa. In order to produce something like a demonstration of lightning within the lecture, he investigated the discharge of the Leyden jars he had brought back with him from Germany. The experiments, performed at the last minute with the help of Chattock and Robinson, led to the discovery of electromagnetic waves along wires, which was reported in the Mann Lectures given on March 10 and 17, 1888. We can be sure that Lodge also used the occasion to make an impression on a new audience. The experiments were spectacular, and Lodge himself claimed that the lectures "aroused a good deal of attention."[25]

Further experiments after the lectures led to many incidental discoveries, another lecture-demonstration to the Physical Society at South Kensington in May, and publication of the early results in June.[26] These included the first actual publication of any verification of Maxwell's theory and the first published statement that electromagnetic waves had been discovered, with the correct wavelengths if their velocity was assumed to be that of light, as Maxwell's theory required. Lodge's priority in this has been consistently overlooked, partly because he himself gave the impression that he had been forestalled by Hertz in his later *Philosophical Magazine* article of August 1888, and partly because the Society of Arts publication of June was omitted from his official bibliography.[27] It was followed by the often-reported drama of the Bath meeting of the British Association in September, which included the famous confrontation with Preece over whether lightning conductors were required to protect buildings against inductive as well as conductive current surges, and FitzGerald's announcement of Hertz's discovery of electromagnetic waves in free space.[28]

Lodge was stimulated now to work on Hertzian waves in 1889, in the process discovering syntonic Leyden jars and the coherer effect. With his growing reputation as a public lecturer, he had begun to be sought after by the organizers of regular lecture programs, and he now had the opportunity to lecture before a "great crowd" at the Royal Institution on March 8, with spectacular demonstrations. In *Past Years* he recalls that when he demonstrated the resonant Leyden jar, the walls of the lecture theater, "which were metallically coated, flashed and sparkled, in sympathy with the waves which were being emitted by the oscillations on the lecture-table . . . the sparking clearly showed the propagation of waves freely in all directions through space."[29] Lodge thought the large crowd followed the reputation he had won as a result of his success at Montreal in 1884. Looking back in his book *Advancing Science* (1931), he (perhaps immodestly) quotes Arthur Rücker, who remarked, "It isn't often a young man comes up from the provinces and sets all London agog."[30]

Similar results occurred on April 25, 1889, at the Institution of Electrical Engineers' building in Great George Street, with Thomson in the audience. In the middle of the lecture, the caretaker of the building came into the lecture theater white-faced, saying that there was sparking between the gas and water pipes in the basement. This was an effect that Lodge had discovered at Liverpool a month before, and so he had the opportunity of looking completely unperturbed amid the commotion. It was an ideal opportunity for Lodge to display the showmanship that had become an intrinsic part of his public lecturing style, and he invited Thomson and other members of the audience to go down and see it.

Lodge's position as professor of physics at Liverpool was unusual in that the industrialists, shipowners, and other wealthy citizens had been largely responsible for creating it, and, as he became more of a celebrity, so the interest in physics in this community developed—so much so that, on November 6, 1889, a group of a hundred people met in the Physics Theatre of University College and resolved to set up a Liverpool Physical Society, with Lodge as president. It was an interesting idea, but seemingly doomed to fail: 150 people came to hear Lodge's opening presidential address on December 16, but, a year later, the average attendance had dropped to 40 and then to 28.[31] There were some notable successes, however, where Lodge's lectures on new scientific sensations drew huge attendances and massive public interest. In 1895 Lodge gave a lecture on argon, the "recently discovered Constituent of the Atmosphere," to 700 people in the Arts Theatre of the College. Lodge was ill at the time; he had to get out of bed to drive half a mile to the college, and he returned straight to bed as soon as the lecture was over.[32] The next sensation was X-rays, and Lodge lectured on this to the Liverpool Physical Society on January 27, 1896. This drew such a crowd that many were unable to gain admission, so a repeat lecture was organized for the Arts Theatre the following week. Nearly 1,000 attended but many others were turned away. Lodge recollected later, "When I arrived at the building I found a queue outside in the street; I had to fight my way up the staircase through the crowd, and then lecture in the midst of a lot of people all standing crowded behind the table, as well as completely filling the theatre."[33]

A lecture at the Liverpool Medical Institution followed another approach to Lodge, this time by the surgeon Robert Jones and the consultant Charles Thurstan Holland, to do one of the first radiographs taken for medical diagnosis, on February 7, of a boy with a bullet embedded in his wrist. Holland subsequently set up the first dedicated medical X-ray unit, while Lodge and his assistants investigated improved X-ray tubes. Despite the massive audiences at the lectures on X-rays, however, the meeting of the Liverpool Physical Society a year later, on March 22, 1897, turned out to be the last, with only twelve attendees. Soon afterward the society merged into a newly formed student Physical Society, which still exists.

Lodge's success with public lectures furthered his career, but it tended also to obscure the novel developments that sometimes became incidental to the "big picture" that he usually set out to present. From early in 1889, discussions with his friend FitzGerald propelled him in another direction. FitzGerald famously said that Hertz's waves had "enslaved the all-pervading ether" and he also discussed the Michelson-Morley experiment in Lodge's house, coming up with the hypothesis of the FitzGerald contraction to explain the null result.[34] Albert A. Michelson and Edward W. Morley had set out to detect the earth's motion with respect to the ether using an interferometer in which a light beam had been split into two parts, which were sent on different paths before being combined and producing an interference pattern. A shift in the pattern could be interpreted as a change in the speed of light in one of the beams as the apparatus moved with the earth through the ether. According to FitzGerald, however, movement of an apparatus through the ether produced electrical forces that contracted its dimensions by an amount that exactly nullified the effect that the "drift" through the ether was expected to have on the velocity of light. It was still possible that a moving object could "drag" the ether close to it, and so change the speed of a light signal in its vicinity. During 1890 Lodge began his new experiment to test for a drag effect on the ether, the funds coming from his shipowner friend George Holt and the technical help from Davies. It was an experiment on a monumental scale, in which an interferometer similar to that of Michelson and Morley's would be used to measure a shift in the interference pattern as the light was sent between two rapidly spinning metal discs, driven by an electric motor, which had to be secured into the foundations of the building. This difficult, dangerous, and expensive experiment occupied Lodge for several years before eventually, like Michelson and Morley's, producing a null result.

Lodge gave preliminary results from his ether drag experiment in his presidential address to Section A of the British Association at the Cardiff meeting on August 19, 1891. He also advocated the founding of a National Physical Laboratory and introduced psychical research into his public discourse for the first time.[35] The address is also notable for containing a discussion of the fourth dimension and an early version of the world-line concept. This latter concept has since proved particularly important: Lodge is often thought to have been at odds with relativity, but the world-line concept provided a way of tracing the path of an object in four-dimensional space-time, showing all possible positions at all possible times.[36] The ether drag results themselves, which also contributed, indirectly, to relativity, were first fully presented in a paper titled "Aberration Problems" communicated to the Royal Society on March 31, 1892.[37] Here, Lodge made several significant theoretical contributions, including an anticipation of the very subtle Sagnac effect. First formally observed by George Sagnac in 1913, this describes how the interference patterns of beams of light, traveling around a closed loop in opposite directions, register a time of travel distance when rotated. This has now become an important aspect of the positioning of the global positioning system, with the earth functioning as the rotating apparatus. The Royal Society paper was followed on April 1, by a Friday evening discourse at the Royal Institution. Lodge sent reprints of his paper to practically every physicist of importance in Europe and the United States, including Hermann von Helmholtz, Hertz, Michelson, Hendrik Lorentz, and Joseph Larmor. Larmor responded, leading to a whole new series of experiments and conjectures relating to Larmor's developing electron theory.

During this period, Lodge's growing family made it necessary for him to move from 21 Waverley Road to 2 Grove Park, again near Sefton Park. There were major changes also at University College. The arts students' move to the new Victoria building, which opened in autumn 1892, allowed Lodge's physics department to expand to fill the northern half of the old building. Lodge was able to begin courses and set up a laboratory in electrotechnics, which was closely related to his own research areas of lightning conductors, wireless, and electric storage batteries. Because of its technological significance, electrotechnics, or electrical engineering, was a particularly lucrative area for industrial investment in education, and in 1895 W. P. Hartley, a jam manufacturer, gave £5,000 to add a classroom to accommodate seventy-five students. Lodge again saw his salary of £1,045 overtake the principal's.

A description of Lodge from about 1892 is given by the medical student John Campbell Hay, who later became a medical professor. In an address delivered at Saint George's Hall in 1934, he remembered Lodge as the "most lucid teacher under whom it was ever my privilege to sit[;] I can see him now—his hands in the deep pockets of his velvet coat—pacing slowly up and down the length of the lecture room as he explained the most difficult problems in the simplest of language."[38] It is precisely the ability to communicate deep and difficult ideas (some of which were original to him) in the clearest and most transparent manner that remains the most remarkable characteristic of Lodge's scientific writings.

The death of Hertz on January 1, 1894, was another external event that led Lodge to a new research area, for he was asked to give a memorial lecture and his attempts at repeating Hertz's experiments led to work on coherers, decoherence, radio, and radioastronomy. There were three occasions on which he presented the results with spectacular demonstrations: the Royal Institution lecture "The Work of Hertz" on June 1; an annual Ladies' Conversazione at the Royal Society on June 13; and the Oxford meeting of the British Association on August 14. A form of the lecture showing the detection of electromagnetic waves by coherers, followed by automatic decoherence and the detection of further signals, was published in several journals, including *Nature,* and became very influential.[39] The apparatus it described was immediately taken up in Italy by Augusto Righi, in Russia by Alexander Popov, and in India by J. C. Bose, who all acknowledged Lodge's influence. In addition, Righi's work was the immediate inspiration for that of Guglielmo Marconi, and Bose's for that of the Royal Navy's Henry Jackson, who pioneered the use of signaling by radio waves at sea.[40]

Again, a member of the audience was galvanized into making a new proposal for investigation. Alexander Muirhead, a telegraph engineer, persuaded Lodge to take up the idea of radio telegraphy. An abandoned patent of April 23, 1895, suggests that they had considered commercial development a considerable time before Marconi. Several subsequent patents, including one for tuning from 1897, which later became a serious problem for the Marconi Company, did lead to such developments, in particular the foundation of the Lodge-Muirhead Syndicate in 1901. This company had some commercial success, receiving contracts mainly from various parts of the empire, until its patents were bought out by its rival, Marconi, in 1911. The Lodge apparatus was also featured at the British Association meeting held in Liverpool between September 16 and 23, 1896. At the same meeting Preece announced the system that Marconi developed from the prior work of Righi, who, in turn, had taken his inspiration from Lodge's paper of 1894.[41] The Jubilee celebrations of July 1897 included a public display of radio signaling by Lodge from the Victoria tower to the tower of Lewis's store half a mile away. Lodge, however, was equally interested in an inductive method of signaling, which he explored simultaneously with radio. This led to a patent for a moving coil loudspeaker, dated April 27, 1898, and several public lectures. Lodge's loudspeaker had no immediate impact, but it became the basis for communications systems incorporated into telephones, radio, and television when electronic amplification became advanced enough to give good-quality reproduction of music and the human voice. Even today the devices most often used to generate sound from electronic systems are moving coil loudspeakers.

By his final years in Liverpool, Lodge was well regarded by both the scientific community and the public as a man of science. November 1898 saw his achievements recognized by the civic banquet held in his honor at the town hall. The mix of local dignitaries and national politicians, influential citizens of Liverpool as well as his scientific friends, nicely captures Lodge's position in both the city and the nation. He continued to develop the various lines of research he had developed in the previous decade. The two methods of wireless telegraphy, for instance, were exhibited together at the Royal Society's annual Ladies' Conversazione on May 11, 1898, while continuing research on the nature of the ether led to another major paper in October.[42] Larmor's parallel work led to an electron theory, which included contributions from Lodge, such as an estimate of the electron's size, which is close to the value now known as the classical radius. In his last few months in the city, Lodge saw the establishment, on January 1, 1900, of the National Physical Laboratory for which he had long campaigned. The first director, Richard Glazebrook, had spent the previous year as a professor at Liverpool alongside Lodge. Lodge also gave a lecture to the Literary and Philosophical Society at Liverpool's Royal Institution on March 5, 1900, and followed this on March 28 with a lecture to the Liverpool Engineering Society on the two methods of wireless telegraphy. At Easter, however, he was offered the position of principal at what was to become the new University of Birmingham by way of a telegram from Joseph Chamberlain, which arrived when he was on the golf course.

AFTER LIVERPOOL

Lodge stayed at Birmingham from 1900 to 1919, between the ages of forty-nine and sixty-eight. On the whole, this was a less happy period than the nineteen years at Liverpool. Though he did manage to continue some of his research, much of his time was inevitably taken up with administration. The loss of two very close friends, FitzGerald and Myers, in January and February 1901, marred the early years of transition, while the First World War and the death of his youngest son, Raymond, in action, clouded the later years before his retirement.

He made many later visits to Liverpool, and maintained links with the city for many years. On November 16, 1900, he spoke at the Liverpool Physical Society on ions and electrons; on October 12, 1901, he gave the inaugural address at the reopening of University College, where he stressed the importance of links with the local community. On November 27, 1901, the Oliver Lodge Prize to reward the year's best student was inaugurated, and on December 3 the Oliver Lodge Fellowship, "for the promotion of Research in Physics," was launched. The first was funded by the for-

FIGURE 2.1. Oliver Lodge on the Victoria Monument, Liverpool, 1906. Photograph by James Mussell, 2017.

mer professor's "friends and colleagues," and the second by the widow of his great friend and benefactor George Holt.[43] On December 6, Lodge himself returned to give a talk on secondary education at the Liverpool Institute. Shortly after he was knighted, in June 1902, Charles Allen, a sculpture instructor at University College from 1894 to 1905, began work on the Victoria Monument in Derby Square, Liverpool, using his former colleague as the figure of Education (fig. 2.1). On November 14 Lodge was at the Liverpool Philomathic Society, speaking on the opportunities at Birmingham, which was now a university in own right (soon to be followed by Liverpool). On February 10, 1903, he spoke to the Liverpool Physical Society on electrons, and on September 11 he supported Rutherford on radioactive decay at the British Association meeting in Southport, close to Liverpool.

Lodge still kept his ties with the university and the city, though his visits gradually became less frequent. On September 27, 1906, the Victoria Monument had its official opening; on July 13, 1907, he received an honorary DSc at the college's Silver

Figure 2.2. (*From right to left*) Oliver Lodge, James Chadwick, and Lionel Wilberforce, 1935. Special Collections and Archives, University of Liverpool.

Jubilee; on March 12, 1908, he gave a memorial lecture to Ferdinand Hurter, once president of the Liverpool Physical Society, in the university's Chemistry Theatre on the structure of the atom; on the following day, he appeared for the last time at the Liverpool Physical Society to give a memorial lecture to Lord Kelvin, in the Arts Theatre in front of nearly one thousand auditors. His next visit does not appear to be until March 12, 1912, when he became an honorary member of the Literary and Philosophical Society. On March 19, 1914, he addressed the annual meeting of the charity known as the Victoria Settlement on life after death. Then, in the postwar period, after he had retired from Birmingham, he returned for another Liverpool meeting of the British Association, in September 1923, to speak on the significance of the ether. On September 16, 1923, he spoke at Sefton Park Presbyterian Church on "man and the universe." On November 11, 1927, he opened the extension to University Hall at Fairfield. Three days later he was at the Literary and Philosophical Society to give the Roscoe Lecture on energy.

Lodge was then seventy-six and this may well have been his last public lecture in Liverpool. He had, however, made such a massive impression during his time there that his contribution was remembered for many more years up to and beyond his death in 1940. He remained a vice-president of the Liverpool Physical Society for life, and his birthday was regularly celebrated in the *Liverpool Daily Post*, sometimes with an anecdote, probably supplied by himself. When James Chadwick was appointed professor of physics at Liverpool in 1935, he went with his immediate predecessor, Lionel Wilberforce, to Lodge's retirement home at Normanton House, near Salisbury, where a remarkable picture was taken of three professors who, between them, held the Liverpool chair for a total of sixty-seven years, from 1881 to 1948 (fig. 2.2). The *Liverpool Daily Post* commented that "it was a reunion which must be almost unique in academic annals."[44]

After Lodge's death at Normanton House on August 22, 1940, the *Post* also had a final word on his importance to the creation of the University of Liverpool and the university's importance in his own development: "His appointment at the early age of thirty was a brilliant choice by the founders of the old University College, and it meant much to the University that its own years of growth were the period of some of Lodge's greatest scientific discoveries. The young professor, indeed, soon became not only a national but an international figure."[45] The Liverpool years were the pinnacle of Lodge's career as a scientist and teacher, but he would later vastly extend the work he had begun there as a popular educator and public figure.

THREE

LODGE IN BIRMINGHAM

PURE AND APPLIED SCIENCE IN THE NEW UNIVERSITY, 1900–1914

Di Drummond

AT THE University of Birmingham during the days of its founding and consolidation between 1900 and the First World War, the first principal of the university, Oliver Lodge, a formidable popularizer of science, met with one of the leading promoters of the new, provincial industrial culture in Britain, Joseph Chamberlain. This changed the nature of universities in Britain, making way for the cultures of science to become paramount.

Lodge would play a critical part in forming the nature of the new University of Birmingham. He had engaged throughout his career with key institutions promoting physics, and his experience as a professor at University College Liverpool stood him in good stead to take the helm. It was his personal ideals, however—educational, political, religious—together with his reputation as an experimental physicist and popularizer of science, through public lectures, demonstrations, and publications, that would inform Lodge's conception of what a university could be. In Lodge, Chamberlain found not only a formidable promoter of the value of science but also someone who recognized the importance of the other disciplines. Crucially, however, while Lodge defended the place of pure science in the curriculum, he was also deeply interested in applying science to real-world problems.

Lodge's views of science thus corresponded closely with Chamberlain's vision for his new university. When Chamberlain, former mayor of Birmingham, by then serving as secretary of state for the colonies, launched a campaign to found the University of Birmingham in July 1898, he declared that the proposed university should be "redolent of the soil, and inspired by the associations within which it exists."[1] This was to be a new type of university, where teaching and research would be based upon such eminently practical subjects as "railway engineering, electric lighting, railway management and every large trade of the town," underpinned by pure science, a combination that lay at the heart of Lodge's work.[2]

Since 1877 Chamberlain had also been arguing that "the power of the provincial cities" now demanded that they should be given "opportunities for the highest culture."[3] In other words, the University of Birmingham, along with other, as yet unformed civic universities, should become sites not only for embedding traditional forms of learning into the industrial provinces but also for generating new forms of university scholarship. There were stark tensions between the industrial leaders of the Midlands and the new university, but Lodge and Chamberlain often found in their shared aims and vision—as well as Chamberlain's adept handling of the region's liberal elite—a means through which to win.

Both Lodge and Chamberlain, I argue, were of key importance in founding the university, and of redefining ideas of "the university" in England in the process; however, it was Lodge who set out the place of science within this new university, determining the type of science that would be taught, and, finally, the relationship between science and other disciplines.

PURE AND APPLIED SCIENCE IN THE NEW CIVIC UNIVERSITIES

The nature of university scholarship changed rapidly and dramatically during the nineteenth century, and with it the character and function of the university. In early nineteenth-century England universities were few (Oxford and Cambridge, plus University College, London [1826] and Durham [1832]) and the range of subjects taught was narrow, consisting of "traditional" subjects such as classics, mathematics, theology, law, and medicine.[4] In regard to the sciences, a handful were taught at Oxford and Cambridge, including astronomy, mineralogy, and anatomy. Scientific research was carried out in various national and local learned societies rather than in the established universities. The aim of the English university during the early part of the nineteenth century was to provide an education for the established professions (the church, medicine, the law) or to the "liberal elites." The traditional culture and values of the universities were often pitted against that of the new, rising middle classes of the United Kingdom, including those of Birmingham and its Midland region.

However, the character of the established English universities was beginning to be seriously questioned as Continental universities, especially in Germany, began their explorations in the sciences, mathematics, and theology. T. H. Huxley, in an address delivered at the opening of Mason Science College in 1880, contrasted the utility of these Continental institutions with the lack of practical application to be found in the English university.[5] By this point, not only had the established universities in England gone through a transformation in their structure together with the range of subjects researched and taught but a host of new civic colleges were formed in provincial cities such as Manchester (Owen's College, 1851), Southampton (Hartley College, 1860), Leeds (Yorkshire College of Science, 1874), Sheffield (Firth College, 1879), Liverpool (University College, 1881), and Reading (University College, 1892).[6]

Alongside the expansion in the number of universities and colleges together with extension of the subjects they taught was a debate about the nature of scientific instruction. Stathis Arapostathis and Graeme Gooday consider Lodge's understanding of the shifting relationship between the academic disciplines of physics and electrical engineering in Britain. Arguing that there was a shared and common culture between the two pure and applied disciplines, as arbitrated through the patent system, "physicists such as Oliver Lodge" were attempting to "decouple . . . electrical researches from technological matters" in order "to pursue an agenda of pure science."[7] Lodge did not demand a complete separation between the two; rather, he held that a continuum between them might be established. Arapostathis and Gooday also argue that Lodge, not only in his many roles in various national bodies representing physics and wider science but also in his university career, was important in calling for central government support of pure scientific research.[8]

THE FIRST ENGLISH CIVIC UNIVERSITY

Since the 1870s, Chamberlain had been planning a university "redolent" of the Midland soil that would both represent Birmingham's industries and serve its peoples. The new university, when founded in 1900, did just that, not only representing new developments in this area of education but also continuing the nature and role of the two institutions that preceded it, Mason College (1881) and then its successor, Mason University College (1897).

There was a great deal of continuity in subjects taught at the college and the university, and in the staff engaged; however, as a consideration of the professoriate of the university in 1910 reveals, the applied sciences were by no means predominant.[9] The university was founded with four faculties: Science, Medicine, Arts, and Commerce. The applied sciences were represented by mining (Sir Richard Redmayne, superseded by John Cadman); brewing (Adrian J. Brown); and civil (Stephen Dixon), mechanical (Frederick William Burstall), and electrical engineering (Gisbert Kapp); and a chair had been established in metallurgy, as a subsidiary of chemistry, at Mason Scientific College in 1881 (the professor of metallurgy in 1910 was Ted Turner). But while the numbers of professors of pure science were fewer than those of applied (George West, botany; Percy Frankland, chemistry; Charles Lapworth, geology; and J. Henry Poynting, physics), there were also a further eight professors across Arts and Commerce. The University of Birmingham's professoriate had a similar number of "medical men," a reflection of both the needs of the Midland region and the legacy of the two earlier medical schools that merged with the university.

A similar story can be told of the numbers of students taking the various courses at the new university. Nearly half of those attending were taking arts, social sciences, and educational subjects, with these particularly being favored for the daughters of the Midlands either seeking future careers in teaching or pursuing recreational subjects.[10] Again, the latter factor was a continuation of habits developed at Mason College. Medicine and science were most usually the preserve of male students, but some of the applied subjects were thinly supported, with mining having only ten students in the early 1900s.[11]

Birmingham certainly gained a reputation for its applied sciences at the expense of arts and social sciences. Lodge complained in 1902 that "there is an unfortunate impression abroad that Birmingham either does not possess or does not encourage a faculty of arts."[12] Possibly this miscomprehension was a result of the large sums of money raised for the various departments that taught applied science and the publicity that followed them. For instance, the School of Malting and Brewing opened in 1900 (at the Mason College building in Edmund Street), with a gift of £28,000 from the Midland Association of Brewers.[13] A "Practical Mining School" had begun at the university in 1902–1903 after an earlier version of this endeavor had failed at Mason College.[14] Another form of applied education, this time outside of the sciences, established at Birmingham as a result of a generous gift, was commerce. Chamberlain found the Birmingham Chamber of Commerce, of which he was a leading protagonist, willing but unable to support the initiative financially; instead, Lord Strathcona, the Canadian high commissioner, gave £50,000 to start the enterprise.[15]

Some photographs of the various installations provided for the applied sciences, sited first at the existing Edmund Street buildings of Mason College in the center of Birmingham and then, as they were constructed and opened between 1909 and 1911, in the new Aston Webb buildings on the Edgbaston campus, demonstrate some of the sheer practicality of the various applied science courses. There were foundries, machine shops, and "model mines" where safety and rescue drills could be carried out by students who were destined to become mine overseers and managers. Even more unusual, and not at all redolent of the Midlands, were the drilling derricks of the Petroleum Mining Department, erected during the 1930s. But alongside these facilities necessary for the new applied sciences, often within the departments themselves, were laboratories for pure science, such as chemistry, and museums for samples and exhibits. The professors of both the pure and applied sciences also, and quite logically, liaised with the Aston Webb firm of architects in determining the layout of their various wings of the new building. All came under the auspices of the University Building Committee established in February 1901 and Lodge, as principal and scientist, worked alongside them.

One thing Lodge was particularly concerned about was the possibility that as the sciences became established at the new campus, the arts and humanities would be left in the old buildings in the city center. However, he was also concerned that the design of Aston Webb might lead to the separation of pure and applied science. Arapostathis and Gooday quote Lodge to this effect: "To divorce pure science from its applications would be deadly: but to treat pure science as solely concerned with applications, and to eliminate it from general education and from the training of teachers, would be no less deadly. Something must be done to avoid both of these evils."[16] Lodge's concern for the balance of subjects on the new campus can be seen in the various plans for the Aston Webb building as they developed. The campus was made possible by a grant of £50,000 by Andrew Carnegie, the Scottish industrialist who made his massive wealth in the iron and steel industries of the United States.[17] A number of plans for the layout of the building were produced by Aston Webb for the university council, but the more fundamental requirements of the different departments to be housed in the building were forwarded by the professors of all subjects, under the direction of Lodge and command of Chamberlain.[18] Providing offices, teaching space, laboratories, museums, libraries, and various installations for practical training for the applied sciences, the building, based on a semicircular corridor with separate wings for each subject radiating out, was to include all commerce and the arts, as well as the applied and pure sciences. This was an indication of the intent of the university chancellor, principal, and the university council and senate, that all university courses and subjects were to be considered to have a form of equality, as each shared a part in providing a wider "liberal education" at the new university.[19]

THE FORM AND CHARACTER OF THE NEW UNIVERSITY OF BIRMINGHAM

The juxtaposition of work room and laboratory was necessary for the applied sciences as their curriculum, even in the Mason College days, had a strong element of pure science, including physics, chemistry and, later, electrical engineering. As a result, the mining and engineering wings had their own laboratories. The question remains, however, why science took this institutional position within the University of Birmingham as it developed. As its first principal, Lodge undoubtedly played an important role. He was a well-established physicist in his own right, active in a number of scientific organizations, a well-known popularizer of science, and, by this point in his career, an experienced university administrator. However, it was not just Lodge who had the final say. The university ultimately had to serve those who paid for it: the Midland middle-class industrial and commercial elite whose money was intended to provide the intelligent trained labor, managers, researchers, and technical advisors of the future. Lodge also had to work with Chamberlain, a national and international representative of that elite and its needs, who was key not only in translating those needs to reality in "his" university but also in taking the demands for higher education to government and nation. In doing so, Chamberlain helped to transform those values deemed characteristic of the Midlands and the provincial cities, "the industrial arts and sciences," into the highest form of British culture.

Chamberlain was also able to position the new university as an expression of civic pride, an example of the industrial spirit that had helped forge modern Britain. Along with other Birmingham notables such as George Dawson, Chamberlain had formulated and championed "the Civic Gospel," with Chamberlain seeing the University of Birmingham as its crowning achievement, taking its place in the context of the elementary school system that had been produced from such ideals, not just in the Midlands but nationally.[20] Chamberlain was also the center of a most important network of the industrial and commercial elite, not just in Birmingham and the Midlands but nationally, through his work with the Birmingham Chamber of Commerce together with his various national roles such as that of president of the Board of Trade (1880–1885) and secretary of state for the colonies (1895–1903).[21] This, too, was to play a very real role in fashioning the new university, but in this case in the areas of commerce and law rather than industry. However, Chamberlain's overwhelming influence declined quite quickly after he suffered a stroke in 1906, while Lodge's grew in importance and stature, not just in the university but also in his dealings with various government bodies as the new institution faced financial difficulties. The First World War made applied science subject to the demands of the national government, prompting a renewed emphasis upon it within the University of Birmingham, an emphasis of which Lodge did not approve.

The Midlands elite remained important, however. In making his address in 1898, appealing for financial support by promising a new university that was "redolent" of the Midlands, Chamberlain was very consciously considering the monetary needs of the new institution. The response was generous, but interestingly, few, if any, of the industrial donors made contributions to the development of departments that would serve their interests directly. Important gifts, such as for the Feeney Chair of Metallurgy and the Chance Chair of Civil Engineering, were given by prominent local men whose interests lay a long way from the subjects they sponsored: John Feeney was the editor of the *Birmingham Daily Post* and Chance a glass and lighthouse manufacturer, indicating that for these local men it was supporting the future of the industries for which the region was famous that was important.[22] The School of Mining, which was revived with the idea of a new university, was well supported through local gifts, but, probably more importantly, found consistent income in providing safety training courses in the model mine.[23]

It was in the social sciences where the local elites had the biggest say. Chamberlain, with first Mason University College and then the university, worked very closely with local chambers of commerce, formulating with the Birmingham Chamber a board of studies that worked at collectively producing new curricula.[24] Here, commerce, law, languages, geography, and education were of greatest importance; the applied social sciences and arts, rather than applied science, providing a further aspect of the University of Birmingham as an eminently practical and local university.

Most contributions, however, came without any particular requirement as to where the funds should go, even when, as in the case of the chemical company Albright and Wilson, there might have been good reason to be more focused in their giving. In all, 55 percent of the money given to the prospective university between 1898 and 1900 came from the industrial and commercial elite of the Midlands, many providing small sums of money.[25]

There were three reasons why this elite was not so focused and discriminating in their giving. First of all, even if some of the key industrialists of the region contemplated an endowment dedicated to a particular subject, they were at a loss as to how this might be accomplished. Chamberlain set up an advisory subcommittee, suggesting that an applied science course in an appropriate subject would be ideal for their purposes and seeking the wisdom of local industrialists in formulating the curriculum of the applied subjects at the new university.[26] Some did participate, but others declined, either failing to see the necessity of such higher training or indeed how such technical education might be considered a university education. In a few other instances members of the Birmingham professoriate, newly empowered by their position in the senate, resisted some of the more difficult curriculum developments forwarded by members of the local industrial elite, although it should be noted that key members of the professoriate in the applied sciences were also very important figures in regional organizations representing the various industries (Turner for metallurgy, for instance, and Redmayne for mining).[27]

Secondly, some firms stated that they provided their own training. For instance, in his 1898 address Chamberlain had advocated railway engineering; however, local railway companies—notably the Midland and London and Northwestern Railways—refused his overtures, and so Chamberlain wrote to Lodge, then newly appointed, asking him to produce a paper that might be used to convince them of their need for the University of Birmingham. Lodge's work advocating a school of railway engineering looked most promising, with ideas for test beds for running in locomotives, but still the railway companies did not respond to the idea that they would be better served by sending their premium apprentices—future managers and design apprentices—to the university. Possibly part of the problem lay with Chamberlain, who, in his letter to Lodge, displayed little understanding of the key differences in training and expectations of ordinary and premium apprentices in the rail engineering industry.[28]

The difficulty in recognizing, defining, and conveying the difference between those who were to receive a university education, and so become the future managerial/technical elite of industry and business, and those who were to be operatives was at the heart of the problem of defining how the university might serve local industry. The problem defining this critical division of labor was even captured in the ceramic friezes that adorn the Aston Webb building. Designed by Robert Anning Bell, who provided friezes in the House of Commons, for the University Building Committee, they were to "take colour from the university's environment . . . rooting the university in the soil," by depicting the industries that made the region and the subjects that were to be pursued in the new institution.[29] However, a consideration of most of these friezes shows a series of images where industrial equipment and activities such as "pipe-laying, forging, pattern-making, cable-laying, bridges and colliers," along with the laboring operatives performing these tasks, are prominent, and the educated managerial overseers, the university's future graduates, are marginalized and represented only at the edges of these processes.[30]

Finally, there were those like the jewelers of Birmingham, who were simply baffled by what might be included in a university curriculum, or those, like the Midlands chemical industry, who found it impossible to draw up a course that would cover the extensive and often contradictory requirements of their particular area.[31] Part of the uncertainty for such groups lay with the lack of a clearer lead from Chamberlain on how local industries might be best served by the emerging university. In time, and in certain areas of study, Lodge provided that lead.

The unevenness of the university's funding—well endowed in some areas but lacking in others—can be attributed to Chamberlain himself. There is much evidence to suggest that many who gave were Chamberlain's long-term political supporters. Indeed, the fundraising for the planned university functioned in a very similar way to the Birmingham Caucus that lay at the center of Chamberlain's power, initially in local municipal politics during the 1860s through to 1874, when he was elected a Member of Parliament for Birmingham, and then after, as he made his often highly controversial way, first as a Liberal and then a Liberal Unionist, as a member of the cabinet and a statesman. The City of Birmingham and other local authorities levied a rate to support the new university; in return, Chamberlain formulated its system of government so as to represent the Midland industrial elite in the university council, as well as the professors in the senate.

Chamberlain was an important figure in the formation, finance, and management of the University of Birmingham, but over the course of his career he was a strong advocate for extending a "liberal" education system to all. This can be seen in his contribution to the work of the National Education League. This national organization developed from the Birmingham Education League, founded in 1867 by MP George Dixon and other local Liberal activists, including Chamberlain. The National Educational League became important in running the Liberal party more broadly, influencing parliamentary policy and leading to the creation of an elementary school system in Britain in 1870. Chamberlain also become concerned that the deeply flawed education system that existed before this time was undermining workers' abilities to deal with new advances in technology, with Britain losing out in competition with other industrial nations such as the United States, France, and especially Germany. As a result, Chamberlain envisaged the future University of Birmingham as the crowning glory in his and other local Liberals' endeavors to provide a more effective system of education, in Birmingham as well as nationally.[32]

In 1905, the highly practical journal *Engineering* voiced the opinion that "it is clear that the Chamberlain ideal for the Midland University has always been a school of general culture, specialising in the faculties for training applied scientists."[33] Chamberlain foresaw in "his" university an institution that would represent this wider and liberal form of learning for all, including for those sons of industry and future industrialists and men of commerce who were to populate the applied and pure science courses that he felt were vital to the Midlands. There was a need, in other words, to present a balanced form of liberal education at the new University of Birmingham. Chamberlain found an example of such institutions in the United States, especially in Cornell University, in New York State, with Carnegie funding a visit there for some of the Birmingham professors in 1899.[34]

However, Chamberlain's liberal values concerning education were shared by many in Birmingham and the Midland region. A good example of this, important to the future University of Birmingham and to Lodge's understanding of the place of science within it, was Huxley's speech at the opening of Mason Scientific College (1880). Huxley was, at this time, a leading advocate of the culture of science, and his speech made a forceful case for scientific education at university level; however, he also argued that there was a need to value advanced knowledge of whatever form.[35] For Huxley, physical scientists were the representatives of a new age (that both the endower of the new college, Josiah Mason, and the city of Birmingham also represented); as such, physical science was as able to make a "criticism of life" and so constitute "culture" just like other, better-established disciplines such as classics and modern literature. He also argued that "men of business"—"the rule of thumb men," as he dubbed them—were as dangerous to this new endeavor of formulating the culture of science as any classicist trying to defend their monopoly over liberal education.[36]

In forwarding this ideal that the physical sciences could define culture, Huxley was advocating a new form of university. Lodge, as one of the best-known scientists of his day—a reputation garnered through research and as an energetic popularizer—was the perfect embodiment of this new culture of science. Chamberlain had first approached two other academics of the time for the position of principal: William Tilden, then professor of chemistry at Mason University College, and Samuel Butcher, professor of Greek at the University of Edinburgh. However, it was, apparently, Chamberlain rather than Edward Sonnenschein, also of Mason University College, who fell upon Lodge as a further candidate once the first two candidates had refused. Initially Chamberlain simply asked for Lodge's views as to who should have the principalship, summoning him to London from Deal, where Lodge was playing golf. Not only was Lodge already very well known but he also had the requisite administrative experience from his time at Liverpool. Lodge suggested the Cambridge physiologist Michael Foster and the physicist Arthur Rücker, at the Royal College of Science. After these two men also declined, Chamberlain offered the post to Lodge. Lodge's initial response was to refuse the post—he felt he was too well established in Liverpool—but he consulted those he trusted: Edward Talbot, then Bishop of Southwark, and Arthur Balfour, who was visiting at the time. Balfour told Lodge to consider it like a cabinet appointment—something not to be turned down—and suggested some conditions.[37] After visiting Birmingham and meeting some of his future colleagues, Lodge told Chamberlain he would accept as long as Robert Heath, the principal of Mason University College, would continue to do the day-to-day administrative work as vice-principal of the new university, and that Lodge could continue his research, physical as well as psychical.[38]

This separation of administrative duties from both the formulation of policy and Lodge's wider role as spokesperson was undoubtedly key to his—and the university's—success. It was also of great importance in defining the nature of this new type of civic university. Throughout, Lodge was keen that the arts and humanities be represented alongside the sciences, something illustrated by the debates around the entrance portal to the Great Hall in the Aston Webb building. In a paper to the building committee of the new university, Lodge wrote that it would be "an advantage to select Englishmen, to a great extent, and men connected with the Midland District of England, in its widest sense. We must further remember that the University in the future will include all branches of learning, and not merely the more technical branches which are in special evidence today; and we may hope that among the Humanities, thus dealt with, some form of Arts will not be absent."[39] Various key figures were suggested to represent the industrial arts and science, but it proved more difficult to find those to represent the arts and humanities. The only people suggested who fit the two criteria were Shakespeare and possibly Dr. Johnson. Finally, a committee decided upon Darwin, Faraday, Watt, Newton, Shakespeare, Plato, Michelangelo, Virgil, and Beethoven.[40]

Lodge also fostered the important relationship between Carnegie and the new university. He and his wife, Mary, had already visited the Carnegies at their luxurious home in Scotland, Skibo, during his Liverpool days.[41] The official history of the University of Birmingham highlights Chamberlain as the architect of this new civic institution because of his national and international connections. While Chamberlain was of fundamental and paramount significance, this neglects the importance of Lodge and the national and international networks with which he was engaged.

However, the work that Chamberlain and Lodge did together was also important, especially in terms of their shared understanding of higher education. As Arapostathis and Gooday note, there were two key elements of Lodge's vision. First, much of what Lodge argued about the purity of pure science and its relationship to applied science was predicated on his concerns that pure scientists such as himself should find enough remuneration in their research. He argued that overreliance on producing applied, saleable, outcomes for research often did not result in adequate forms of research, pure science being proven to be more (albeit serendipitously) effective. Lodge also argued that government funding that was too dependent on the need to produce applied science (as he would later experience as he ran the university during the First World War), was highly detrimental to the necessary balance between pure and applied science, and higher learning in general. Lodge's solution was that pure science should be funded either by central government or, more in line with his Fabianism, from municipal funding. This latter point provided particularly rich common ground for Lodge and Chamberlain, and the University of Birmingham, in many ways, offered an example of municipal funding for the sciences.

Finally, while Chamberlain and Lodge were of different political parties and ideologies, they shared longstanding ideals regarding the nature and form of higher education in England, and this vision of the municipal sponsorship of science was common to them both. Chamberlain was, after all, the architect of the city of Birmingham's earlier form of municipal socialism, while Lodge, in his Fabianism, provided a newer, and more socialist version of this. Together, then, Chamberlain and Lodge, with their liberal ideals of higher education, were equally important in forming the University of Birmingham. While Chamberlain embodied the new middle-class elite, Lodge embodied scientific culture. Both made key, and often highly complementary, contributions to the future of the new university.

PART TWO

SCIENCE AND COMMUNICATION

IT IS a core theme of this volume that Oliver Lodge's scientific activities were deeply enmeshed with his commitments to communication. Far from the stereotype of the unworldly natural philosopher—notwithstanding the manufactured image of such that might be discerned in some parts of *Past Years*—Lodge was deeply interested in securing work that was both useful and remunerative. This commitment ran through all his career, first in London, as he attempted to earn enough to marry, then in Liverpool as he sought funding for his work and an income with which to support Mary and their increasing family.

Thus we see in what follows that Lodge did not simply build his grand theories of the electromagnetic ether out of a curiosity-driven cerebral encounter with James Clerk Maxwell's *Treatise on Electricity and Magnetism* of 1873. Moreover, the work that led him both to this all-encompassing ether narrative and to his practical techniques for wireless telegraphy and syntony (tuning) did not start with his commemorative lectures on Heinrich Hertz's death in 1894, as Lodge himself often seemed to suggest. Instead, they arose from a deeply practical commission six years earlier from the Royal Society of Arts in London to undertake a course of lectures on how best to protect buildings by the installation of lightning conductors. From this we see a long-running dialectic in Lodge's work between the utility of material models for understanding electricity and the more abstract varieties congenial to fellow Maxwellians trained in higher mathematics such as George Francis FitzGerald. In understanding Lodge's relationship with mathematics we necessarily see how he relied on his brother Alfred for assistance with some of the more technical aspects of mathematical physics. We also see Lodge's involvement in a broader movement to reform mathematics education to speak more closely to the engineering concerns of practical audiences for technical training than was feasible by reference to the ancient work of Euclid or the higher-level skills of calculus. At the same time, we can recognize the role that this approach to mathematics played in shaping Lodge's resistance to the ether-free metaphysics of Einstein's elegant relativity theory that attracted so many during the interwar period.

Lodge's position in relation to the so-called new physics of space and time—free of all mechanical theorizing—can, however, all too easily be misunderstood as a reactionary move that ended his authority in the discipline. In fact, Lodge was a valued interlocutor in the debates about the new physics, his critiques welcomed both within the specialist domain of professional physics and in discussions with broader audiences. The high level of regard for Lodge's work owed much, of course, to his great skills as a "popularizer" of science. While some looked to the mystical algebraic symbols of relativity theory and the new quantum mechanics for enlightenment, many more saw Lodge as a well-established and reassuring narrator of physics. And it was that comfortable relationship with his many readers that shaped not only Lodge's public profile but also the way that his ether philosophy evolved and the forms in which it came to be published.

FOUR

THE ALTERNATIVE PATH

OLIVER LODGE'S LIGHTNING LECTURES AND THE DISCOVERY OF ELECTROMAGNETIC WAVES

BRUCE J. HUNT

EARLY IN 1888, the Society of Arts invited Oliver Lodge to deliver a pair of public lectures on how best to protect buildings from lightning. In preparation for the lectures, the young Liverpool physicist set about performing a series of experiments on Leyden jar discharges and electrical oscillations along wires. These experiments led him very close to Heinrich Hertz's discovery, announced that same year, of the production and detection of electromagnetic waves in free space: a discovery that was widely hailed as confirmation of James Clerk Maxwell's theory of the electromagnetic field. In fact, as Lodge often later pointed out, the waves that he observed passing along wires were essentially the same as those Heinrich Hertz had detected in free space. In both cases, the waves were really located in the surrounding space, with Lodge's wires serving merely as guides for disturbances in the electromagnetic field. Lodge acknowledged, however, that Hertz's experiments were beautifully done and that they demonstrated the existence of waves in the field, rather than just in conductors, more clearly than did his own.[1]

Lodge's experiments had important technological consequences, leading within a few years to his own work and that of others on wireless telegraphy, and particularly to the development of tuning, or what Lodge called "syntony." Achieving this precise resonance between transmitting and receiving circuits was the key to using the electromagnetic spectrum for the innumerable distinct channels of wireless communication that we all rely on today.[2] In many ways, Lodge's work on electromagnetic waves looks like a classic piece of "applied science," with pure laboratory experimentation on electrical discharges, combined with even purer mathematical field theory, leading to scientific insights and thence to world-changing practical technologies. Nor is such a view altogether wrong; certainly, Lodge was guided in important ways by his understanding of Maxwellian field theory, and he interpreted his experiments in that light.[3]

The picture changes, however, when we look more closely at the concrete roots of Lodge's experimental work. As we shall see, those roots in fact lay in a very practical concern with how to protect buildings from lightning, as well as in Lodge's desire to find a spectacular way to display lightning bolts—or something like them—for his lecture audiences.[4] Picking up on a phrase Lodge used to describe some of his most striking experiments on electrical discharges, we will examine the "alternative path" that led him to his work on electromagnetic waves, and will see the light this case can shed on the different ways that science and technology can interact—in particular, the ways practical concerns and the demands of public presentation can sometimes lead by winding paths both to advances in scientific knowledge and to useful new technologies.

The story of Lodge's work on lightning protection goes back to R. J. Mann, an English physician and popular writer on scientific subjects. Mann had taken an interest in electrical phenomena as early as 1857, when he wrote an anonymous pamphlet promoting the first Atlantic cable project and particularly the experiments of its chief electrician, Wildman Whitehouse.[5] Mann spent the next ten years, from the late 1850s to the late 1860s, in Natal, South Africa, organizing educational efforts for the local Anglican bishop, John Colenso, and later for the Natal government. There he developed an interest in meteorology and especially in thunderstorms and lightning protection. On returning to England, Mann became active in the Meteorological Society, serving as its president in the 1870s, and also in the Society of Arts, whose African and later colonial sections he headed for many years.[6] Following his death in 1886, his widow gave the Society of Arts a sum of money to fund a pair of lectures in Mann's memory on a subject dear to his heart: the protection of buildings from lightning. Trueman Wood, the secretary of the Society of Arts, invited Lodge to deliver the two Mann Lectures, which he did in London on March 10 and 17, 1888.[7]

Why did Wood choose Lodge? The young Liverpool physics professor had not previously been much involved with lightning protection, but he was well known as an electrical theorist and experimenter, and particularly as a popular expositor of what he called "modern views of electricity," a phrase that also served as the title of a long series of articles he was then writing for *Nature* and that were later published as a very successful book.[8] Perhaps more importantly from Wood's point of view, Lodge had gained a reputation as an impressive public speaker, known, in W. P. Jolly's words, for his "great success and popularity with mixed audiences of scientists and interested laymen."[9] "Dust," a lecture on the electrodeposition of dust and fog that he had delivered at the Montreal meeting of the British Association in 1884, had won particular acclaim; Lodge himself later said it was "a tremendous success," and Wood no doubt hoped that Lodge would turn the Mann Lectures into a similar triumph.[10]

Lodge was happy to accept Wood's invitation; energetic and ambitious, he was always on the lookout for ways to raise his public profile. He could use the extra money, too; by 1888 he and his wife, Mary, already had six children (the other six were still to come). Having promised to lecture on a subject of which he knew relatively little, Lodge quickly set about learning all he could about lightning and lightning protection. His starting point was the *Report of the Lightning Rod Conference*, an authoritative volume published in 1882.[11] The conference, convened in 1878, had brought together leading engineers, physicists, meteorologists, and architects to lay down standards for best practice in the protection of buildings from lightning. In its summary recommendations, the report emphasizes the use of stout conductors (preferably copper) and good earth connections, asserting that all reported cases of lightning rods and conductors failing to protect buildings could be traced to poor grounding, bad joints, or other errors in installation. The report focuses almost entirely on how well rods and conductors conduct steady currents and notoriously declares that "a man may with perfect impunity clasp a copper rod an inch in diameter, the bottom of which is well connected with moist earth, while the top of it receives a violent flash of lightning" (2). Although the report notes that there was reason to believe that conditions during a sudden discharge might differ from those in a steady flow, its recommendations tend to downplay such concerns and focus on the conduction of steady currents.

Lodge read the report and also many accounts of buildings (mostly church steeples, it seems) that had been damaged by lightning, despite being "protected" by lightning rods and conductors. More importantly, he set out to study the phenomenon experimentally; he in fact performed most of his experiments only a few weeks before delivering the lectures, so that his results were certainly fresh.[12] Here, of course, he faced a problem: bolts of lightning are not very common in England, and in any case they do not strike conveniently when and where an experimenter might want them—and if they did, it would be far too dangerous to work on them directly.

Lodge therefore set out to model lightning on a convenient scale in his laboratory.[13] The choices he made in selecting and constructing such a model were crucial and would largely determine the validity of his scientific conclusions. He of course had no direct way to check whether his laboratory model of lightning really captured all or even most of the important features of the real thing. The best he could do was to construct a proxy, something that seemed in some ways to be *like* lightning, and then argue that it was close enough. The gap between the proxy and the thing itself is a perennial problem in scientific and technological research, and one that would particularly bedevil Lodge's work on lightning.

Like many others before him, back to the time of Benjamin Franklin, Lodge chose an electrical discharge, or spark—specifically the discharge of a large capacitor, or Leyden jar—as the best available proxy for a bolt of lightning. Aided by A. P. Chattock, his junior colleague at University College Liverpool, and using a set of large and well-made Leyden jars he had bought for the college on a visit to Chemnitz, Germany, several years before, Lodge performed a wide variety of experiments designed to represent different kinds of lightning.[14] His aim throughout was to produce material for the Mann Lectures that would be both practically useful for purposes of lightning protection and impressive to a lecture audience. Lodge's pursuit of spectacular effects no doubt influenced his choice of apparatus and the form many of his experiments took.

The practice of modeling phenomena has a long history in both science and technology, and it is useful to compare Lodge's work with some earlier examples. In 1775 Henry Cavendish fashioned an artificial "torpedo," or electric fish, out of pieces of leather and pewter and showed that, when connected to a set of charged Leyden jars, it could produce shocks that were sensibly the same as those delivered by a real torpedo.[15] Cavendish's aim was to shed light on a mysterious natural phenomenon by imitating it with something he had built out of better-understood components. In this way he could obtain a kind of "maker's knowledge"—not, it must be admitted, of the torpedo itself, but of an artificial version that he could argue was similar to it in important ways. Lodge's aim was very similar, though with the important difference that he could not as directly compare his laboratory Leyden jar discharges with the real natural phenomenon of lightning. Whereas Cavendish could perform parallel experiments with a live torpedo, taking shocks from it and then from his leather and pewter imitation, Lodge was forced to rely on others' reports of the effects of actual lightning bolts.

Another set of examples comes from technological testing. In the 1870s, William Froude began building models of ships' hulls and dragging them through the water in a large test tank, measuring their resistance and assessing their stability. By testing different configurations on a small scale, and by varying and comparing them, he sought to determine which performed best, so that shipbuilders could then scale them up to full size.[16] Much the same was done later, of course, with aircraft models in wind tunnels, and similar testing and variation of small-scale models has long been an important part of engineering practice. Again, we can see parallels in the way Lodge subjected different arrangements of points and knobs to discharges from his Leyden jars as he strove, through a process of variation and testing, to improve the design of lightning rods and conductors. There were, however, limits to how confident he could be that his miniature version of a lightning flash was a faithful representation of a real one, and how well, or how far, he would have to scale things up to match the real thing.

A final and especially apt example of scientific modeling dates from the late 1500s. William Gilbert constructed a "terrella," or little earth, out of a spherical lodestone to enable him to explore on a small scale his idea that the earth is itself a great magnet.[17] Here the smaller version was, in its inner nature, just as mysterious as was the great magnet of the earth, but since Gilbert could manipulate the little terrella and see how it affected magnetized needles and the like, he could use his model to gain insights if not into the nature or cause of magnetism itself, at least into how magnetism could be expected to affect observable terrestrial phenomena. The parallel to some of Lodge's experiments is quite close, in that his model of lightning was built up from components (charges and currents) whose inner nature remained inaccessible to him, but whose behavior under varied conditions he could examine in some detail. It is worth emphasizing that Lodge's experiments with Leyden jar discharges could be of substantial scientific value even if they were not especially good representations of the natural phenomenon they were intended to model. Lodge's motive might have been to improve the protection of buildings from lightning bolts, but even if we put aside any value they may have had in that regard, his experiments could still shed considerable light on the effects of electrical discharges.

Lodge tried many variants as he connected his electrostatic generator (a Voss machine) to different arrangements of Leyden jars and then discharged them in various ways.[18] Many of his experiments involved offering what he called an "alternative path" for such a discharge and then seeing which route it actually took. Some of Lodge's most striking findings concerned the difference between instances of what he called "steady strain" and "impulsive rush."[19] "Steady strain" was the most familiar case. Using two flat metal plates (tea trays, he said) to serve as "clouds," Lodge connected the top one to the inner knob of a Leyden jar and the bottom one to its outer coating (51, 57, for tea trays as clouds 359). Then as he (or Chattock) turned the crank of the Voss machine and charged up the Leyden jar, the potential difference across the space between the tea trays gradually rose, putting the air between them under increasing electrostatic strain. A sharp point placed on the lower tray would produce a "silent discharge" (or, if the rate of charging was high enough, would audibly fizz), relieving some of the strain and, unless the charging rate was fairly high, preventing

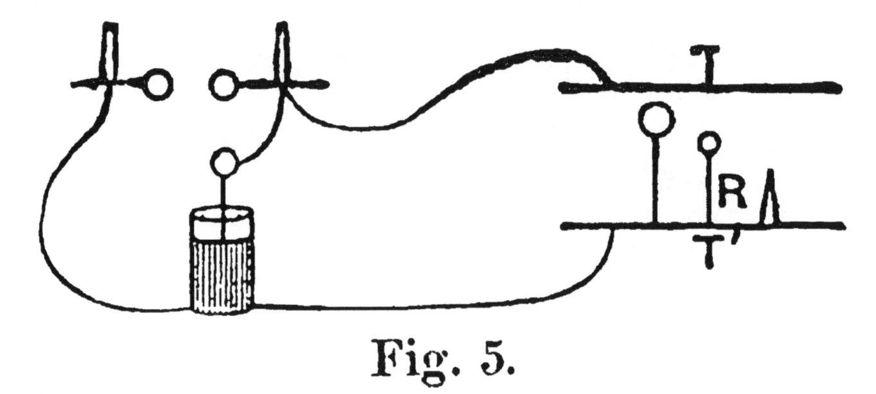

FIGURE 4.1. "Steady strain." Lodge attached an electrostatic machine (not shown) to a Leyden jar through the double leads at the top left. As he turned the machine and charged the jar, electrostatic strain gradually built up between the two tea trays (T and T′) connected to the inner and outer surfaces of the jar. A sharp point placed on the lower tray tended to dissipate the strain in a "silent discharge"; when the point was removed, the strain steadily grew until a spark, corresponding to a lightning bolt, jumped between the trays, preferentially striking the conductor topped by the smaller knob. Figure 5 from Lodge, *Lightning Conductors*, 55.

any sudden discharge. The strain simply dissipated without a flash ever jumping from the "cloud" to the point.

Removing the sharp point, Lodge next placed conductors topped with large or small knobs on the lower tray. He now found there was little or no "silent discharge"; instead, the strain built up until a flash—a miniature lightning bolt—leapt from the "cloud" to the knob. To his surprise, Lodge observed that the flash preferentially struck a smaller knob, even when it stood much lower than a larger one. Such cases of "steady strain" corresponded to ordinary cloud-to-ground lightning, Lodge said, and behaved generally in line with what the *Report of the Lightning Rod Conference* said one should expect—except that Lodge found that the conductivity of the rods did not matter much, and the smallness of the knob counted for more than its height. Even when the rod supporting it had a resistance of a million ohms, a small knob was still more likely to be struck than a larger and higher knob supported on a rod of very low resistance. "The flash actually prefers to jump three times as much air, and encounter a megohm resistance," Lodge said, "rather than take the short direct path offered by the bigger knob" (56). Why? According to Lodge, the steadily rising electrostatic strain gradually prepared a path for the discharge, breaking down the resistance of the air (ionizing it, we would now say) first around the small knob, where, because of its greater curvature, the electrical tension was the most intense. Since the resistance of the air gap—high if it was not ionized, relatively low if it was partially ionized—accounted for most of the resistance of the entire discharge path, the conductivity of the rod did not matter much, and there was no need to use a thick piece of copper. In fact, Lodge said, the discharge seemed to be quieter and less violent with a relatively poor conductor, suggesting that for such cases the usual insistence on using a well-earthed copper conductor might be misplaced (46–49).[20]

Lodge next turned to his second kind of discharge, the "impulsive rush." Here he used two Leyden jars, connecting the inner knobs of each to the Voss machine and their outer coats to the top and bottom tea trays. Then as he and Chattock cranked the Voss machine and charged up the two jars, their outer coats (which simply rested on a wooden table, and so were effectively grounded) remained at zero potential. There was thus no buildup of "steady strain" between the plates, which remained at the same zero potential throughout. But when the inner knobs of the two oppositely charged jars were allowed to spark into one another, neutralizing their charges, the difference in potential between their outer coats suddenly went from zero to thousands of volts, and the opposite charges that had accumulated on them flew together with an "impulsive rush."[21] When the jars were discharged into each other, a huge spark jumped between the plates, and unlike in the earlier case, did so without having had an ionized path prepared for it by any previous "steady strain." The charge, Lodge said, was forced to go in a hurry by any route it could find, or often by several at once; it could not be counted on to take the path that would have the lowest resistance for a steady current. This kind of "impulsive rush" corresponded, Lodge suggested, to a cloud-to-ground lightning bolt induced by a nearby discharge from one cloud to another. Such cases could produce side flashes, he said, as part of the bolt split off from the main conductor to strike other conductors in its vicinity. One might be safe enough grasping a well-grounded copper conductor in a thunderstorm in a case of "steady strain," as the *Report of the Lightning Rod Conference* had suggested, but if the conductor was struck by a bolt that instead resulted from an "impulsive rush," doing so might well be fatal.

Why do lightning bolts sometimes not follow the path of least electrical resistance? The short answer, Lodge said, was self-induction. For steady currents, all that matters is ordinary ohmic resistance, and if a stout copper rod of low resistance is available and is properly grounded, virtually all of the current will flow along it. But for a sudden jolt, in which the current is changing very rapidly, the self-induction (or more strictly the total impedance) will trump ordinary resistance, and the current may take quite different paths, jumping from a fat copper conductor to whatever else may be handy. In his Mann Lectures, Lodge cited many examples of buildings damaged by lightning bolts that did not behave the way the *Report of the Lightning Rod Conference* would have predicted, jumping in side flashes from conductors to rain gutters and the like, because the report had not taken the effects of self-induction into account.[22]

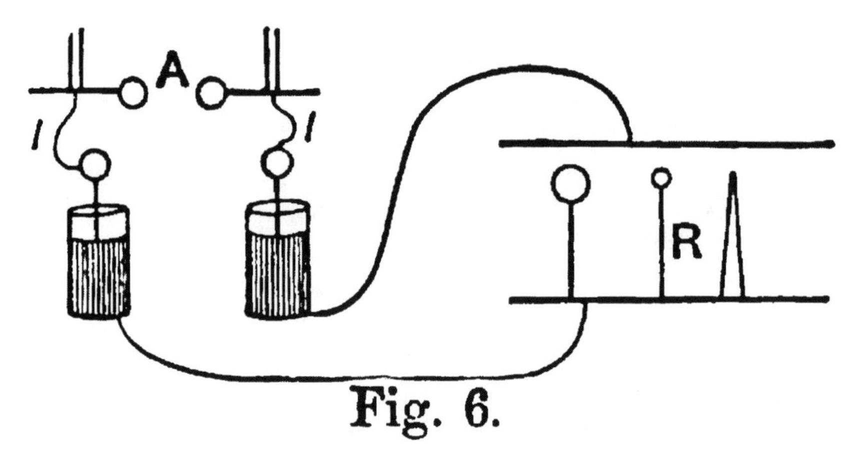

FIGURE 4.2. "Impulsive rush." Lodge connected the inner knobs of two Leyden jars to an electrostatic machine (not shown) and their grounded outer coats to two tea trays. As he turned the machine, giving the inner knobs of the jars large opposite charges, the potential difference between the trays remained zero; no strain built up between them. When he allowed the inner knobs of the jars to spark into each other at A, their opposite charges were suddenly neutralized while the difference in potential between their outer coats, and so between the tea trays, abruptly rose from zero to thousands of volts. This caused a large spark to jump in an "impulsive rush" from one tray to the other; it took any path available, Lodge said, often bypassing conductors that would have offered little resistance to a steady current. Figure 6 from Lodge, *Lightning Conductors*, 57.

Moreover, Lodge's thinking was imbued with Maxwellian ideas about fields and the energy they contain. The cardinal point of the Maxwellian view was that far more was going on outside a current-carrying conductor than inside it; the real action, according to the Maxwellians, was in the surrounding field. In his *Modern Views of Electricity*, written at just this time, Lodge vividly pictures the electromagnetic field as an array of whirling machinery in the all-pervading ether; a changing current, he says, has to work against the inertia of this whirling machinery.[23] An electric discharge, whether of a Leyden jar or a thundercloud, was not just a matter of neutralizing opposite charges, but more importantly of dissipating the *energy* stored in the surrounding field. "Attention is now directed," he says, "not so much to the opposing charges in cloud and earth, but to the great store of energy in the strained dielectric between."[24] A fat copper conductor might look like the quickest and best way to convey a quantity of charge from one place to another, but that did not mean it offered the best and safest way to dissipate the energy associated with that charge. "Given a store of energy in an illicit nitro-glycerine factory," Lodge goes on, "it could be dissipated in an instant by the blow of a hammer, but a sane person would prefer to cart it away piecemeal and set it on fire in a more leisurely and less impulsive manner. So also with the electrical energy beneath a thundercloud" (367, see also 39). In particular, a sudden discharge that struck even a well-grounded copper conductor was likely to splash around in dangerous ways, Lodge said, until all of its energy was dissipated.

A closely related consideration led Lodge to some of his most important scientific findings, although these turned out to have less relevance to lightning protection than he initially believed. If the energy of a thundercloud was not all dissipated by an initial flash of lightning, the current might overshoot and bounce back, Lodge suggested, producing an *oscillatory discharge*. That a Leyden jar discharge could oscillate, especially when it passed along a conductor of relatively low resistance, was not at all a new discovery in 1888; the phenomenon had been observed decades before by Félix Savary and studied closely by Joseph Henry and others.[25] But Lodge pursued the subject very thoroughly, and argued that such oscillations could occur with lightning bolts as well Leyden jar discharges. Making some very rough estimates of the effective dimensions of thunderclouds, he suggested that some lightning bolts oscillated millions of times per second, dissipating their energy over many cycles.[26]

As studies of electrical discharges, Lodge's experiments were valid and valuable. As models of real lightning they left much to be desired, as Lodge himself later admitted.[27] It turns out that thunderclouds are not much like tea trays; rather than forming single connected conductors, they are complicated collections of free electrons and rapidly moving and highly charged water droplets. Moreover, the usual discharge path—basically a channel of ionized air—is not much like a metal wire. The upshot is that lightning bolts do not oscillate in the way Lodge suggested, though their extreme suddenness and intermittency can sometimes lead to similar effects, particularly in cases of paired discharges that resemble Lodge's "impulsive rush."

In his second Mann Lecture, Lodge said some harsh things about ordinary lightning protection practices, and sometimes made it sound as if following the advice of the *Report of the Lightning Rod Conference* could do actual harm.[28] Press coverage tended to play up this conflict and emphasize Lodge's more dramatic criticisms. Lodge later toned down such statements and said that the usual practices undoubtedly did much good; in fact, other than suggesting that multiple small rods and conductors be used instead of single tall ones, in the end his practical advice did not differ much from the standard guidelines on lightning protection.[29] Perhaps his main contribution was to steer people away from elaborate and expensive measures (tall and sharply pointed copper rods, thick copper conductors, elaborate earthing arrangements) that were not really worth the trouble and expense; ordinary galvanized iron rods and conductors worked just as well, he said, and often better. Moreover, by pointing to the effects of self-induction, Lodge helped explain why even seemingly good rods and conductors did not always provide the promised protection from bolts of lightning.[30]

Lodge's Mann Lectures had several important consequences, both for Lodge himself and more broadly. First, they raised his public profile and burnished his reputation as a scientific lecturer, as he punctuated his remarks with dramatic flashes from his Leyden jars. When he followed them the next year with a lecture at the Royal Institution on Leyden jar discharges, he drew a huge crowd, and Sir Arthur Rücker told him that "it isn't often a young man comes up from the provinces and sets all scientific London agog."[31]

The Mann Lectures also led to Lodge's first significant contact with Oliver Heaviside, who soon became his friend and ally in the "Maxwellian" group.[32] Heaviside, an eccentric self-taught mathematical physicist and former telegraph engineer, was already engaged in a bitter dispute with William Henry Preece, the head of the British Post Office telegraph engineers, over the role self-induction played in telegraph and telephone transmission.[33] The year before, Preece had blocked the publication of an important paper Heaviside had written on the subject with his brother Arthur, a post office engineer, and Heaviside was intent on finding a way to strike back at Preece and again get his ideas before the scientific public.[34] After reading press accounts of the Mann Lectures, he wrote to Lodge in early June 1888 to ask for more details, particularly on what he had said about self-induction.[35] Lodge obliged by sending along copies of the lectures when they were published later that month.[36] Lodge had already read some of Heaviside's papers in the *Philosophical Magazine*, and in his second lecture had said, "I must take the opportunity to remark what a singular insight into the intricacies of the subject, and what a masterly grasp of a most difficult theory, are to be found among the eccentric, and in some respects repellent, writings of Mr. Oliver Heaviside."[37] Lodge's reservations about his style notwithstanding, Heaviside was delighted to receive this kind of public recognition; he later told Lodge that he looked on the second Mann Lecture as "a kind of special providence."[38] Significantly, Heaviside did not wholly endorse Lodge's claim that his Leyden jar experiments provided a good model of real lightning, but he emphasized the value of Lodge's experiments on electrical oscillations along wires and the evidence they provided in support of Maxwellian theory and of Heaviside's own results concerning the effects of self-induction, particularly the way the "skin effect" confined rapidly varying currents to the outer layer of a conductor.[39]

The Mann Lectures also landed Lodge in his own confrontation with Preece. Preece had taken an active part in the work of the Lighting Rod Conference and in 1888 he emerged as the main public defender of its report. All of this got tied up in complicated ways with Preece's ongoing dispute with Heaviside over the effects of self-induction. Preece was president of Section G (Engineering) at the 1888 meeting

FIGURE 4.3. "Waiting for the Verdict (with apologies to M. Gérome)," *Electrical Plant* 1 (1888): 5. Derived from the 1872 painting *Pollice Verso* (*Thumbs Down*), by Jean-Léon Gérôme, this cartoon depicts William Henry Preece as a gladiator standing with his foot on the throat of a vanquished Oliver Lodge. Though most other accounts said that Lodge was regarded as having had the better of it at the debate, opinions evidently differed.

of the British Association, held that September in Bath, and he used his opening address to pour scorn on the pretensions and speculations of scientists, including Lodge's claims about lightning and self-induction.[40]

Preece and Lodge were both known as lively public speakers, and the organizers of the Bath meeting arranged to spice things up by pitting them against each other in a special three-hour "joint discussion" on lightning conductors, held on September 11.[41] The two men tossed various charges and countercharges back and forth, along with a lot of good-natured chaffing. Preece was repeatedly dismissive of "the way in which lately self-induction had been brought in to account for every unknown phenomenon," and was sharply critical of mathematicians telling "practical men" how to go about their business.[42] A report in the *Times* said that "it was generally agreed that Professor Lodge had the better of it" in the debate, but not everyone agreed, and the journal *Electrical Plant* ran a cartoon showing Preece as a gladiator representing "Experience" standing with his foot on the throat of the defeated Lodge, labeled "Experiment."[43]

Lodge was far from being a mere ivory tower experimenter, however, and in 1889 he put his Mann Lecture results to practical use by filing for UK and US patents on a new form of "lightning arrester" (or surge protector) designed to protect telegraph cables, power lines, and similar equipment from damage by surges of current. He developed these "Lodge Patent Lightning Protectors" together with Alexander Muirhead, whose firm manufactured them, and this collaboration on lightning protectors helped pave the way for their later partnership in the wireless telegraphy business, the Lodge-Muirhead Syndicate, in 1901.[44]

This brings us to the most important result of the Mann Lectures: Lodge's work on electromagnetic waves. As Lodge experimented on "alternative paths" for Leyden jar discharges, letting them "choose," as he put it, between various short and long wires, he noticed a surprising phenomenon: when he connected a pair of parallel wires to his Leyden jars and then discharged the jars, he often got much longer sparks at the far end of the wires than he did closer to the jars. He traced this effect to what he called the "recoil kick": "The electricity in the long wires is surging to and fro," he said, "like water in a bath when it has been tilted; and the long spark at the far end of the wires is due to the recoil impulse or kick at the reflection of the wave."[45]

As he adjusted the length of his wires to match a half-wavelength of the oscillations produced by the discharge, Lodge found clear signs of resonance effects. It was Chattock, Lodge later said, who insisted that in these recoil kicks they were seeing evidence of true electromagnetic waves, and that these waves were really traveling in the space surrounding the wires, with the conducting wires serving simply as guides for disturbances that subsisted in the surrounding field, or ether.[46] This was very much a Maxwellian view of the phenomenon, and it was in line with a search Lodge had been pursuing off and on since 1879, when he first looked into how he might produce electromagnetic oscillations of such high frequency that, in accordance with Maxwell's electromagnetic theory of light, they would be visible to our eyes.[47] Lodge's friend George Francis FitzGerald had initially put him off this track, argu-

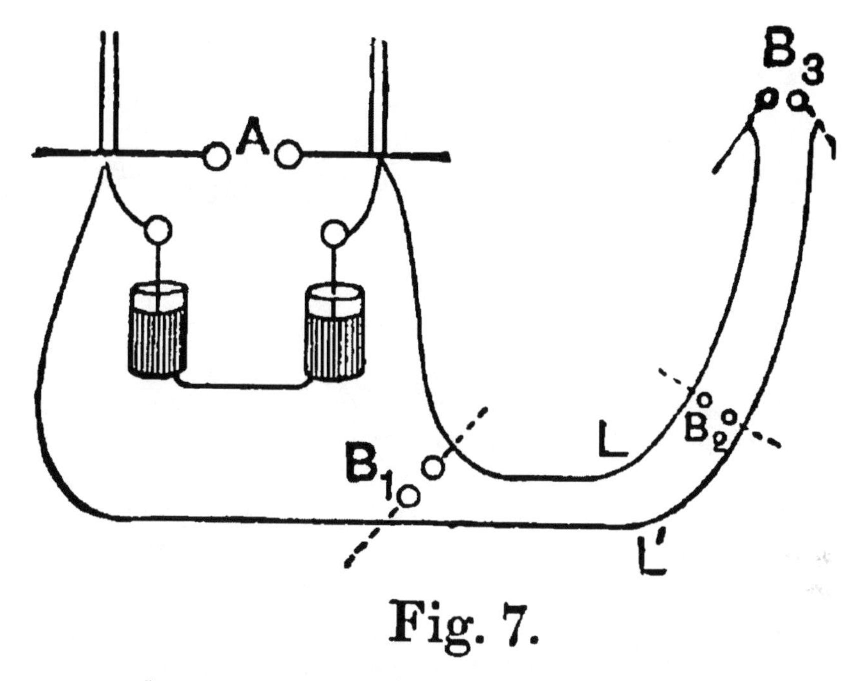

FIGURE 4.4. "Recoil kick." Lodge found that when he connected long wires to the inner knobs of two connected Leyden jars, charged the jars, and then allowed them to spark together, he could draw longer sparks from the wires (at points B_1 and B_2) than from the jars themselves, with the longest sparks appearing at the far end of the wires (B_3). This showed, he said, that discharging the jars produced electromagnetic oscillations along the wires, marked by a "recoil kick" as the surging waves were reflected back from the ends of the wires. Figure 7 from Lodge, *Lightning Conductors*, 60.

ing, based on a misreading of a subtle point in Maxwell's *Treatise*, that purely electromagnetic forces could never produce waves that would break off and propagate through the ether. FitzGerald found his error two years later and went on to show that oscillating currents should indeed produce such electromagnetic waves; at the 1883 meeting of the British Association, he even declared that by discharging a condenser through a small resistance, "it would be possible to produce waves of as little as 10 metres wavelength, or even less."[48] The only problem, as he told J. J. Thomson in 1885, was finding a way "to *feel* these rapidly alternating currents."[49] It was a challenge one might have expected Lodge to take up, but, busy with his new post at Liverpool and juggling a thousand other tasks, he did not return to the search for electromagnetic waves until his preparations for the Mann Lectures practically dropped them in his lap. Although Lodge did not make electromagnetic waves the focus of his lectures, he mentioned them several times, notably when he calculated that some of his Leyden jar discharges were producing waves about five meters long. In a footnote added just before the lectures were published in June, Lodge said that he had recently found clear experimental evidence of waves three yards long, and "I expect to get them still shorter," as he soon did.[50]

Lodge and Chattock carried out an increasingly quantitative program of experimentation after the Mann Lectures, and in early July 1888, Lodge wrote up a paper, "On the Theory of Lightning Conductors," for the *Philosophical Magazine*, with a substantial section at the end detailing the new work on electromagnetic waves along wires, including measurements of their wavelengths. Lodge sent the paper off just before setting off on a hiking holiday in the Tyrolean Alps with his friend A. C. Bradley, the Liverpool professor of literature. Lodge had high hopes that his work would be the hit of that September's British Association meeting at Bath, but on the train out of Liverpool he happened to read the latest issue of *Wiedemann's Annalen*, the German physics journal, and there came across Heinrich Hertz's account of the experiments he had recently conducted at the Technische Hochshule in Karlsruhe on electromagnetic waves both along wires and in air. Writing from Cortina in the Tyrol on July 24, 1888, Lodge added a postscript to his *Philosophical Magazine* paper in which he notes Hertz's experiments and adds that "the whole subject of electrical radiation seems working itself out splendidly."[51] Splendid or not, it was clear that Lodge had been scooped and that his own work on electromagnetic waves would likely be overshadowed by Hertz's.

FitzGerald, Lodge's friend and fellow Maxwellian, was president of the mathematics and physics section of the British Association that year, and he used his opening address on September 6, 1888, to take up the question of whether electromagnetic forces acted directly at a distance, as Wilhelm Weber and most other German physicists had long held, or were exerted through an intervening medium, as Maxwell had argued. "The year 1888 will be ever memorable," FitzGerald declared, "as the year in which this great question has been experimentally decided by Hertz in Germany, and, I hope, by others in England," confirming Maxwell's theory.[52] At the meeting, Lodge gave a brief account of his own experiments on waves along wires, but said they were now of interest mainly as supplements to Hertz's more dramatic experiments.[53] Although Lodge said that he was happy to see the subject in such good hands (he had met Hertz on a visit to Germany in 1881 and thought well of him), he had the sense that a German had stepped in ahead of him and carried off a prize he had hoped to win himself—though he consoled himself by remarking in a March 1889 lecture at the Royal Institution that Hertz was "no ordinary German."[54] Congratulating the Scottish physicist Alfred Ewing in 1890 on his work on magnetism, Lodge wrote, "I know the feeling of the 'Deutsches im strasse' slipping in front of one; glad you have safely landed the discovery on the British Isles, even if it is north of the Tweed!"[55] But Lodge liked Hertz personally and always had high praise for his work; he translated several of Hertz's papers for publication in *Nature*, dined with him when Hertz came to London in 1890 to receive a medal from the Royal Society, and carried on a friendly though not voluminous correspondence with him.[56] Lodge also conducted his own further experiments on electromagnetic waves along wires and in air, focusing in particular on producing resonance effects. By carefully adjusting the capacitance and inductance of the discharging and receiving circuits, Lodge was able to bring them into what he called "syntony."[57] As the foundation of tuning, this was perhaps his most important contribution to what would become radio technology, for it made it possible to select waves of particular frequencies from the vast electromagnetic spectrum. After Hertz's early death in January 1894, Lodge delivered a lecture that June at the Royal Institution titled "The Work of Hertz and Some of His Successors" that cemented public recognition of Hertz's achievements and also played an important part in sparking the development of wireless telegraphy.[58]

Hertz had been led to his discovery by an interest in deciding experimentally between Maxwell's field theory of electromagnetism and Weber's theory of action at a distance, and by a chance observation of some unexpected sparks in coils at his Karlsruhe laboratory. Though he had a strong background in engineering and at Karlsruhe was in fact teaching at an engineering school, the path Hertz followed to his discovery of electromagnetic waves was almost purely scientific; it certainly was not motivated by any thought of producing a wireless communication system. There were, however, "alternative paths" to the discovery of electromagnetic waves, and we have good grounds for believing that, had Hertz not gotten there first, Lodge would have detected electromagnetic waves in air for himself very soon, probably by the end of 1888. Had he done so, perhaps we would now speak of "Lodgian waves" and measure frequencies in kilolodges and megalodges—or perhaps not. And had Lodge been the first to discover such waves, we might view the relationship between pure science and practical pursuits in this case somewhat differently.

Lodge's work on electromagnetic waves emerged, like Hertz's, from a physics laboratory and from a combination of experimental results and theoretical analysis. But Lodge was led to embark on his investigation of electrical discharges and electromagnetic oscillations not by simple scientific curiosity or a desire to test a particular theory but by a request from the Society of Arts that he deliver a pair of public lectures on a very practical problem: the protection of buildings from lightning. Lodge did not hit on evidence for the existence of electromagnetic waves in the course of "blue sky," undirected scientific research, nor did he find it as part of a deliberate effort to devise a way to signal across space without wires. His path toward the discovery of electromagnetic waves instead exemplifies a different way science and technology sometimes interact, in which the pursuit of a deeper understanding of phenomena of practical concern, and the effort to display those phenomena to a public audience, may lead to scientific results whose importance extends far beyond the circumstances that first prompted the investigation.[59] In this case Lodge's attempt to model lightning in the laboratory—and on the lecture stage—turned out to be flawed in some basic ways, and his findings about electrical oscillations thus had far less relevance to lightning protection than he had hoped. They nonetheless proved to be of great scientific value, and eventually of great practical value as well, as Lodge followed an "alternative path" to results that went far beyond those he had originally set out to pursue.

FIVE

LODGE AND MATHEMATICS

COUNTING BEANS, THE MEANING OF SYMBOLS, AND EINSTEIN'S BLINDFOLD

MATTHEW STANLEY

MANY PEOPLE in the twenty-first century encounter Oliver Lodge initially as the last bastion of scientific resistance to Einstein's special theory of relativity (1905). Lodge's critiques were sharp. It was an "unphysical" theory, he said, and he was amazed by how Einstein's allies were "untrammeled by a sense of physical reality."[1] He wondered how Einstein's mathematical sleight-of-hand had fooled great minds into embracing bizarre ideas such as the bending of space-time. Bitingly, he suggested that the popularity of relativity was "surely a tribute to the beauty and complexity of the mathematical scheme which can temporarily so warp the judgment of even the most competent."[2] These comments, combined with his well-known predilection for mechanical models, seem to paint a clear picture: Lodge was a skeptic of the value of mathematics. Past historical work on Lodge has only reinforced this. W. P. Jolly's classic biography mentions mathematics briefly in passing; Bruce Hunt's *The Maxwellians* refers to mathematics only insofar as Lodge avoided it in favor of models.[3] For Lodge, physics was about hands-on laboratory work and modest theory based closely on tangible physical concepts, not symbols.

However, Lodge's story was more complicated. He was certainly no Pythagorean. But he *was* willing to embrace the value of symbolic reasoning, and actually developed a nuanced understanding of mathematics—from counting beans to vector algebra. Much of his writing on mathematics was concerned with pedagogy. He wondered about the right way to train both ordinary students and future physicists. More deeply, though, his approach to mathematics was centered on issues of practice and theory. How did he think mathematics could, or should, change the way a scientist worked?

LODGE'S MATHEMATICAL EDUCATION

Lodge's ideas about these questions were shaped by a profound tension in his embrace of Maxwell's electromagnetic theory. He was a full ally of Maxwell's *ideas*, but he was never part of Maxwell's *tradition*. He explicitly described himself as following in Maxwell's footsteps by (as he saw it) creating a physically meaningful ether theory, though he did not imagine that he had the same set of skills. To explain his own thoughts on the relationship between math and physics, Lodge described hearing Maxwell's 1870 address to the British Association for the Advancement of Science (BAAS) on the same subject. The Scot's lecture split men of science into two groups. One could grapple with purely abstract quantities, the other needed to place physical meaning into them. To the latter group, "momentum, energy, mass, are not mere abstract expressions of the results of scientific inquiry. They are words of power, which stir their souls like the memories of childhood." Lodge recognized himself immediately.[4] So he saw both Maxwell and himself as demanding physical meaning behind symbolic expressions and resistant to purely abstract mathematical reasoning.

Unfortunately for Lodge, Maxwell's theory demanded a great deal of skill in abstract mathematical reasoning. His *Treatise* was an extremely difficult book to learn from. It had complicated mathematics, was highly idiosyncratic, and was full of errors (at least for the first two editions). Andrew Warwick has shown how making sense of it was a collective enterprise that relied on the whole pedagogical infrastructure of Cambridge. W. D. Niven's class on the *Treatise* was as much exploration as instruction.[5] Unfortunately, this Cantabrigian instruction was unavailable to Lodge. He did what he could on his own, but it is extremely difficult to learn mathematical cultures of this sort from the outside. Lodge noted that he "never took kindly to the Cambridge textbooks."[6] This disconnect effectively barred him from fully grasping the mathematics of Maxwell's theory. Lodge was keenly aware of his lack of sophisticated mathematical training and how it made him an outsider. He wrote, "I . . . always regretted that I didn't go through the Cambridge grind; for I am somewhat isolated from all those who did" (88). While he was a student he felt "ashamed of [his] ignorance" of higher mathematics, and worked hard to rectify it.[7]

This is not to say that Lodge had no training in mathematics. He studied briefly with William Grylls Adams at King's College London and, more thoroughly, with W. K. Clifford and Olaus Henrici at University College, London (UCL).[8] He described Clifford as brilliant, though not a very successful teacher (85). Henrici seems to have been the more lasting influence. Henrici is not a very well-known figure to historians, though he was well respected among his contemporaries (he served on the Royal Society Council and was president of the London Mathematical Society). His approach to mathematics was somewhat unusual for the time, and it is worth a moment of description. He studied mathematics in Karlsruhe, Heidelberg, and Berlin after working as an apprentice engineer. After having trouble finding a job he came to London, where he taught pure math at UCL for ten years, and switched to the applied mathematics chair after Clifford died. He introduced projective geometry, vector analysis, and graphical methods into the UCL curriculum, which was seen as a significant difference from the previously taught Cambridge-style analysis. Lodge noted that Henrici did not care for Cambridge methods and explicitly taught in a German style. After UCL Henrici moved on to South Kensington, where he set up a mechanical laboratory. He was known for building models, machines, and apparatus, especially calculating machines (he built one of the first mechanical differential analyzers). In addition to his tangible creations, he was prominent for his skill in projective geometry and visual methods.[9]

When Lodge joined Henrici's course, he had missed some preparatory classes and "was plunged into the theory of equations without the necessary preliminary." They studied many parts of higher mathematics, including Riemann surfaces. Oddly, Lodge described Henrici as having "an engineering outlook" but not being very good at teaching actual applied mathematics.[10] So it might be correct to say that Lodge's major mathematical influence was someone who approached the subject through a visual and (literally) mechanical style.

By the beginning of his career, Lodge had come to accept that he did not have the skills to delve seriously into mathematical problems. There is no question, however, that he admired higher mathematics and the people who practiced it: "My instinct seemed to be more abstract, rejoicing in hidden forces, atomic occurrences, and other things which can never be seen. . . . If my training had been other, I believe I could have been a mathematician. Mathematics always fascinated me, though I never acquired the power of rapid manipulation of symbols, such as my brother Alfred possessed. . . . I do strongly admire the achievements of high mathematicians, even when I am unable completely to follow them. I have had to be mainly an experimentalist" (111). The reference to his brother Alfred is very important. Oliver's younger brother was professor of pure mathematics at Cooper's Hill Royal Engineering College. He was a mathematician of some ability: he calculated the tables of Bessel functions for the British Association for the Advancement of Science. And more significantly for this essay, he regularly assisted his brother: in *Past Years*, Lodge noted that he "was always ready to help me with any problem requiring special mathematical knowledge or ability, and I owe a great deal to his assistance" (31). They coauthored a few pieces on mathematics, and Alfred was an important influence on some of Oliver's ideas, as we will see later.

Despite his brother's help, Lodge's background and education placed him firmly outside the tradition that produced James Clerk Maxwell. Fortunately, he was able to find colleagues in a similar situation of mathematical exile. George Francis FitzGerald and Oliver Heaviside were also distant from the British mathematical establishment, which allowed the Maxwellians to develop their own systems of interpreting the *Treatise*, as Hunt and others have shown.[11]

MATHEMATICS VS. MODELS

Lodge's self-identification as someone who demanded physical meaning behind abstract symbols leads us naturally to think about his famous tendency toward model building. Like his idol Maxwell, Lodge spent a significant amount of energy trying to establish the proper relationship between mathematics and models.

A good first stop for understanding his views is his papers on "modern views of electricity." There he discussed how far one could push the analogy between electricity and fluids. He wrote that we were "impelled" to follow the analogy as long as the laws seem to apply and there was no discrepancy. If we resist using this analogy, he said, there are two courses: "either we must become first-rate mathematicians, able to live wholly among symbols and dispensing with pictorial images and such adventitious aid; or we must remain in hazy ignorance of the stages which have been reached, and of the present knowledge of electricity so far as it goes."[12] Here he seems to contrast what mathematicians do and what he thinks physicists should do. Working purely with symbols was only for those who rejected analogies and models. And he definitely discouraged that strategy. Models provided more than one could get from just the symbols. He warned that it is "unwise" to use only "hard and rigid" equations to do physics. Sometimes it can be done by certain people: "Few, however, are the minds strong enough thus to dispense with all but the most formal and severe of mental aides; and none, I believe, to whom some mental picture of the actual processes would not be a help if it were safely available."[13] As he described it, it is just possible to work only with mathematics—but perhaps only geniuses can do it right. First among those was Maxwell, whom Lodge described as dropping his mechanical model and still remaining "satisfied with his more abstract equations."[14] So it was not necessarily that models were better than mathematics, or vice versa, but that models were more helpful to more people. Working only with equations demanded a special sort of person, which, apparently, did not include Lodge himself.

WHAT CAN MATHEMATICS DO?

The next question, then, is what benefits Lodge thought these mathematical explorations could bring. In rare cases he conceded that almost completely mathematical work could provide for leaps forward in physics. He praised both J. J. Thomson and FitzGerald for real progress based on mathematical reasoning.[15] However, it seems that Lodge wanted to contrast this kind of work with a true physical discovery. Thomson and FitzGerald had done great service. Their results, though, were already hidden in Maxwell's physical insights: mathematics, "like all other reasoning, never gives us anything really fresh; it only brings out explicitly what is already contained implicitly in the *physical* statements which we subject to reasoning. The physical statements must be the results of the observation of nature, which is the only way of arriving at fundamentally new truths. Mathematical reasoning will, however, serve to bring out and make manifest what was really involved in the statements themselves when put together, if only we had sufficient insight to perceive it."[16] Equations could take us nowhere without the firm foundation of observation and models. It could provide only a secondary kind of progress.

This is not to say that Lodge wanted to discourage physicists from learning mathematics. He thought it could enrich one's appreciation and understanding of science. Put simply, physics was "more interesting when you know mathematics than when you don't."[17] This applied to ordinary people as well as scientists. Their ignorance of mathematics was "responsible for [their] incapacity to understand the truths of physical science." Higher mathematics was the "alphabet" that allowed understanding of science. Unfortunately, most simply did not know their ABCs: "The mathematical ignorance of the average educated person has always been complete and shameless."[18] He believed this ignorance was largely due to poor teaching.[19] The Royal Commissions on education had also shown that many elementary schools simply did not teach mathematics—over a third did not even instruct in arithmetics.[20]

This meant that Lodge could not assume that the readers of his popular writings would be able to handle even the simplest mathematical expressions. In some of his work he apologized for the resulting vagueness, blaming the difficulty of conveying the essence of physics without the language of mathematics.[21] Interestingly, he said that this meant it was the *aesthetics* of science that was lost:

> Only those however who are familiar with mathematics can appreciate the beauty and intricacy of the reasoning of the great masters of science; who seek to apprehend the intimate working of such laws of nature as have so far been disclosed, who express the strains, velocities, and forces in operation, by means of recondite equations, and

> who then solve the equations in accordance with fixed rules, and deduce from them a multitude of previously unsuspected consequences. The intellect required for such mastery over natural processes seems almost a different order from that common to the sons of men.[22]

Lodge wrote this near the end of his life, in a popular science book. Perhaps reflecting on his own career, he again invoked the idea that skilled mathematicians were very different sorts of people than the rest of us. We still could (and should) acquire enough mathematics to appreciate what we were seeing.

One of the great contributions that mathematics could make to scientific work was to generalize and link together different truths. For example, only the wave equation could provide an "all inclusive" description of waves, and he said that a physicist would be "entirely justified" in refusing to discuss waves in any other way.[23] The conservation of energy was one of his favorite examples for this. Once it was established by experiment, it "led to a great mathematical superstructure, and the domain of physics was illumined by a fresh light."[24] Mathematical formulation allowed the idea to be applied to areas that were not at all obvious. An equation's departure from direct observation, sometimes described by Lodge as a danger, was here presented as a benefit: "A proof means destroying the isolation of an observed fact or experience, by linking it on with all pre-existent knowledge; it means . . . bringing it into its place in the system of knowledge; and it affords the same sort of gratification as finding the right place for a queer-shaped piece in a puzzle-map."[25] All of physics was universal, of course, but mathematics helped make that universality clear. The number *e*, for instance, "must be regarded as typical of the way in which general facts in Physics are simplified, summarized, and compactly treated, by aid of more or less easy mathematics."[26]

This claim was also, unsurprisingly, used to explain Maxwell's great achievement. Lodge criticized Michael Faraday's mathematical unsophistication for preventing him from expressing his ideas properly. "Then comes Maxwell, with his keen penetration and great grasp of thought combined with mathematical subtlety and power of expression; he assimilated the facts, sympathises with the philosophic but untutored modes of expression invented by Faraday, links the theorems of Green and Stokes and Thomson to the facts of Faraday, and from the union there arises the young modern science of electricity."[27] So mathematics allowed for a process of generalization and assimilation. It did not discover these results, it simply described them in a form that allowed for new conceptions and research.

One benefit of this abstraction was the ability to address situations that were, for whatever reason, inaccessible to direct investigation. As Lodge put it, mathematics proved useful when "we want to express harder and at present unknown ideas."[28] For example, Joseph Larmor's equations were a bridge between macroscopic observations and conclusions about microscopic states.[29] No one really knew what electricity was, but mathematics allowed us to describe its characteristics. Lodge thought that numerical results (as opposed to symbolic expressions) were particularly useful for this. He admitted that the mathematics in Maxwell's *Treatise* was "so difficult as to be almost unintelligible."[30] Regardless of this opacity, however, "the ratio between the velocity of light and the inverted square root of the electric and magnetic constants was found by Clerk Maxwell to be 1; and a new volume of physics was by that discovery opened." He argued that whenever you see an integer like that in a calculation, "there is necessarily a noteworthy circumstance involved in the fact, and it means something quite definite and ultimately ascertainable." Similarly, numerical relations among spectral lines suggest a theory of atomic vibration. Mathematical patterns could indicate reality, at least at some basic level.[31]

MATHEMATICAL EDUCATION

Lodge did think mathematics could do useful things for science. One of his major concerns was that math tended to be *taught* in the wrong way, and thus students came to either hate it or misunderstand what it was for. He disliked that geometry tended to be taught "systematically"—that is, as a series of abstract theorems and proofs, rather than letting its principles emerge naturally from one's experience. To this end, he advocated letting children play with blocks before getting any instruction in geometry.[32] Even arithmetic should not be taught formally. Instead, children should be allowed to experiment with "handled things" like counters or beans. This was cheap, easy, and "very instructive" (34).

Lodge's arguments were very similar to those of the "Perry movement" pushing for mathematics teaching reform around the turn of the twentieth century. This movement emphasized "practical mathematics" for utility rather than mental training. These reformers called for mathematics education that involved the hand and eye, rather than just the mind. They hoped that this kind of hands-on learning would prepare students better for the scientific laboratory and the engineering workshop.[33] This stress on tactile and visual methods was very similar to Henrici's approach, which perhaps accounts for Lodge's receptiveness to the Perry ideas. Lodge stressed that this was an exploratory mode of learning mathematics. "Subjective discoveries" could be made this way and excite interest among students. Tedious matters such as multiplication tables should be discovered by investigation and experiment.[34] The students should then be encouraged to express these subjective discoveries in whatever crude way. If they create bad versions of the laws of arithmetic first, they will better appreciate the correct versions later (36).

These teaching reform issues were widely discussed at the British Association, which seemed to be a natural site at which to construct a new national mathematics education scheme.[35] Lodge was pleased to see this exploratory mode defended in J. J. Sylvester's 1869 address on the nature of pure mathematics. Sylvester was contesting T. H. Huxley's critique that pure math knew nothing of observation, experiment, or induction. In reply, he described a process of mathematical exploration.[36] Lodge agreed that this idea should be the basis of mathematics. Instead of reading Euclid, draw two lines and see what happens.[37] At that meeting the association appointed a committee to study the state of geometry teaching in general and the suitability of Euclid in particular.[38]

Really, Euclid seemed to be the heart of the problem. Lodge argued that simply memorizing the perfect way to solve a problem was "training in classics, not in geometry."[39] Sylvester recalled his own encounter with the classics: "The early study of Euclid made me a hater of geometry."[40] This was a contentious issue in mathematics teaching reform. Instructors were deeply divided over whether geometry should be taught via Euclid's deductive methods or via practical techniques (often involving physical instruments).[41] The former was the traditional method in British education, and was seen as primarily beneficial in terms of intellectual training. The latter was better for solving the problems students would actually encounter in their careers. It was not obvious that utility was the standard against which a mathematics education should be judged.

Lodge worried that not every student could appreciate the true beauty of Euclid, so forcing everyone to follow that path was foolish. It was better to teach by direct experience. Geometrical drawing, experimental trigonometry, and surveying would let a boy outstrip his classical colleague in "real and intrinsic" knowledge of the subject. The student who learned his mathematics through direct experience "will know things in a blunt, utilitarian, practical-engineering sort of way . . . his culture will be deficient." But he will still be useful and make a mark on the world.[42] It is hard not to imagine Lodge considering his own education as he wrote this. Indeed, the Perry reformers explicitly placed themselves against the Cambridge analysis tradition from which Lodge felt so excluded. He defended the value of the incomplete knowledge that came along with this approach. He noted that in his own experience at scientific meetings he would often see speakers fill blackboards with complicated equations. And even though he could not always follow everything, he "got a general notion of the methods used, and was stimulated to learn more about the detailed machinery of the process."[43]

Indeed, a completely thorough mathematical education could actually be counterproductive. Insisting that a student fully grasp every "microscopic quantity of geometry and algebra" would destroy any interest in either. Imagine, he said, forbidding study of biology until every part and function of one animal were known. But this was exactly what was expected with mathematics.[44] It was not *bad* to learn mathematics in detail, it was simply not worth the time and energy: "True thoroughness in the rudiments is only possible when progress has been made in the parts further on; the higher parts react upon the lower. The full meaning and beauty of the base of a pillar cannot be perceived until the column is erected" (113). He called this "overteaching" a subject. Lodge wanted to make the teaching of mathematics more lively and interesting, and this chiefly meant avoiding wearisome overpractice. When teaching science students, there was no point in drilling them on large groups of sums, which rarely came up in real scientific work. Simply doing physics provided plenty of practice in calculation.[45]

Lodge was certainly not advocating a one-size-fits-all approach to learning mathematics. He suggested that one went through several stages while learning them properly. First was adding and subtracting quickly, "like a bank clerk." Second came amusing oneself by solving puzzles with algebra. Third, ordinary formulae used in elementary textbooks need not be skipped. Fourth, ordinary treatises on physics using calculus can be read. Fifth, math itself becomes a tool for investigation, so "that discoveries can be made by its aid." Sixth, the highest branches of math are reached, and progress in math itself can be appreciated. Seventh, creating new regions in mathematics, like Joseph-Louis Lagrange or Isaac Newton.[46] Not everyone needed to achieve every level. It was perfectly fine to ascend to the appropriate level for oneself. For example, Lodge did not think that all students needed to be proficient in fractions: "It is not worth while to exaggerate this practice, because the resulting art is not an accomplishment capable of giving pleasure to other people."[47]

Presumably physicists were expected to advance to the fifth or sixth level. Once there, they had to grapple with the profound issue of the meaning of mathematical symbols. Lodge worried that many teachers introduced symbolic reasoning before students had full command of mathematical operations: "to supply the label and withhold the object, to lecture about daisies or stars or numbers before they have been seen, is, let us politely say, unwise" (2, 4). Symbols could provide "both power and clearness." This was particularly true in physics, where "the symbolic treatment of unknown quantities is essential, and the sooner children are accustomed to it the better. . . . The *x* is to be thought of as a kind of crutch: but sometimes it is like a leaping-pole and enables heights to be surmounted which without it would be impossible" (109). But he recommended that this not be the end of the process. Once *X* is used to solve an equation, one should go back and try to express it without that assistance. It was critical to be able to grasp it directly, without the opaque processes of algebraic reasoning. Those tools could make problem solving so quick that a student could miss the essential features. Once the answer was known, there could be a productive discussion of what it meant (111–12). And how was one to do this?

Lodge's recommendation was, essentially, to go back to counting beans: "It is highly desirable that arithmetical practice should be gained in connexion with laboratory work, for then the sums acquire a reality, and interest is preserved. It is absolutely essential that all concrete subject-matter be based upon first-hand experience, for unless that can be appealed to, abstractions have no basis, but are floating unsupported in air" (193). Once this concrete experience was established, abstraction could follow.

THE MEANING OF SYMBOLS

In the 1890s Lodge engaged in an extended discussion in *Nature* that helps illustrate his particular concerns about the use of symbols. The battle was with Professor (later Sir) A. G. Greenhill over whether algebraic symbols should be thought of as representing physical entities or simply numbers related to those entities. Lodge said that he had been considering this question for some time, and credited his brother Alfred for first pointing it out to him (430). Greenhill had been Second Wrangler at Cambridge, won the De Morgan Medal, and at one point was a professor in the same department at Cooper's Hill as Alfred. He wrote to *Nature* in 1891 regarding Lodge's "whirling machine" (a device intended to measure a kind of ether drag). As a self-identified "practical man"—he developed formulae for casting bullets and artillery shells—he complained at length about how Lodge's formulae did not include specifications about what units were to be used. Was length to be measured in feet or inches? Was the mass in tons? Without clear specifications of units, the symbols could have no definite numerical values, and therefore the formulae were "meaningless."[48] Lodge responded by declaring Greenhill to be naïve and showing what straits a "practical man" is put to when "he wishes to interpret the simplest general formula." He went on to describe how the difficulties came from a poor mathematical upbringing: "I have always held, not in sarcasm but in sorrow, that students brought up on the system of specialized and limited numerical formulae used by Prof. Greenhill and some other Professors of Engineering in this country, must necessarily go through the tentative trial-and-error sort of process which he so graphically describes, whenever they have to obtain a numerical result from anything not already arithmetical. In other words they are unable to deal with complete algebraic symbols or concrete quantities."[49] Lodge's position was that engineers such as Greenhill thought that a symbol like *V* represented a number, and that it could only represent a physical quantity once you add units—literally writing "feet" next to "23." They were being *too* concrete.

Lodge suggested that it would surely be better to invest the symbol with the full aspect of the thing in itself. He completely agreed with Greenhill's assertion that "all expressions should be complete and capable of immediate practical numerical interpretation" (513). But even given this, he argued, a formula such as $T=\rho\cdot v^2$ is "perfectly complete, and that it is true and immediately interpretable in every consistent system of units" (513). There is no need to say any more than identify the concrete meaning of each symbol. No properly taught student, he wrote, ought to have the slightest difficult in obtaining a numerical result directly in any system of conventional units that may be offered him.

Lodge concluded that the problem seemed to be that engineers had separated three intertwined entities: the symbol itself; the physical meaning of the symbol; and the particular quantity involved. They literally separated the numbers from the concept, and therefore made their lives much more difficult: "It is the frequent recurrence of such ghastly parodies of formulae as

$$T = \frac{62\cdot 4}{2240 \times 144} \cdot \frac{pv^2}{g}$$

in many engineering treatises which makes them such dismal reading. It is a standing wonder to physicists how a man of Prof. Greenhill's power can fail to see the inadequacy and tediousness of expressions which are only true in one particular system of units, and which to be true even in that require the special statement of every unit employed" (513). Engineers had to put all the units into the formula and then "wearisomely" take them out again. Physicists knew that it was better to have a "concrete interpretation of algebraic terms (wherein each symbol is taken to represent the quantity *itself*, and not merely a numerical specification of it)." The physicist's procedure had "extreme simplicity and reasonableness" (513).

Lodge explicitly wrote that he hoped to persuade Greenhill to his way of thinking, because that conversion would bring along many other teachers. Alfred, his mathematician brother, was credited with teaching Oliver this important lesson: "Ever since my brother showed me the advantage of consciously interpreting algebraical symbols as standing for concrete quantities, and not merely for abstract numbers, the advantage of doing so has presented itself to me with cumulative force. Most physicists are, I think, now of a similar opinion" (513). Lodge asserted that Greenhill was almost alone in his error, though perhaps "he has some of the pure mathematicians with him for company."[50] So this mistake—separating physical concepts, algebraic symbols, and numerical reasoning—was made by *both* engineers and mathematicians. Either end of the spectrum of abstraction led to the same problem. Engineers were obsessed with the numbers, mathematicians with the symbols. Lodge summarized his view nicely a few years later when he engaged in a similar argument: "Mr. Cumming says that "2 R $h = d^2$ is an algebraic equation, and as such its symbols express numbers, not things." Whereas I say truly it is an algebraic equation, and as such its symbols may express things and not numbers."[51]

THE PROBLEMS WITH MATHEMATICS

We have established what Lodge thought mathematics could be useful for, and what benefits it brought. But he did have serious critiques and warnings as well. The basic danger was the temptation to "hide ignorance under the cover of a mathematical formula."[52] This danger emerged in several different ways. One was that mathematics could sometimes give nonphysical results. He commented that the results of calculations "may be real and useful, or non-existent and meaningless, according to circumstances."[53] Physicists sometimes do things such as multiplying a volume by a length, giving a four-dimensional entity, "which so far as we know has no existence. Nevertheless such operations are often performed." These quantities can sometimes "land us in difficulties; though these can be tackled by those specially competent" (61–62). Similarly, students with a "budding aptitude for science" may find it "strange and rather uncanny, unexpected and perhaps rather disappointing" that irrational numbers exist.[54] These nonphysical results could be useful, but they could also lead one astray.

He warned that it was dangerously easy to become naively attached to the results of calculations. This could be numerical—someone thinking that the 3 in an astronomical measurement of 17.4673 represented an "absolute fact." Any physicist would recognize that as an impossibly accurate measurement, but a mathematician might not (194). Or it could be conceptual. In 1889 he worried that some chemists were "permitting themselves to be run away with by a smattering of quasi-mathematics and an over-pressing of empirical formulae."[55] This was within the contemporary discussion of whether certain chemical properties were continuous or discontinuous. Lodge caricatured chemists as assuming "some elementary form of empirical expression for the function, say a quadratic expression with three arbitrary coefficients, and they determine these coefficients to suit three points on the curve" (273). They then plot the derivative of this function and conclude that discontinuities exist. "Now, were it not that eminent persons appear to lend their names to this kind of process, one would be inclined to stigmatize this performance as juggling with experimental results in order to extract from them, under the garb of chemistry, some very rudimentary and commonplace mathematical truths" (273). This crude mathematics was simply recovering the original assumptions brought to the calculation. Lodge warned that "no juggling with feeble empirical expressions, and no appeal to the mysteries of elementary mathematics, can legitimately make experimental results any more really discontinuous than they themselves are able to declare themselves to be when properly plotted" (273). The chemists' equations were only a fig leaf covering their complete ignorance of the actual physical situation.

EINSTEIN'S BLINDFOLD

These kinds of criticisms bring us back to where we began, with Lodge's concerns about relativity. He described Einstein as having taken force, matter—virtually all the explanatory categories of physics—and "reducing it all to pure mathematics."[56] According to Lodge, relativity (he did not typically distinguish between the special and general) was a most complicated theory "elaborated by a singular genius with the aid of innumerable symbols and difficult reasoning."[57] The crux of his worry was this: "That theory gives us equations, but leaves us in the dark as to mechanism; it dispenses with mechanism; being that respect rather like the second law of thermodynamics and the principle of Least Action" (1200–1201). He was certainly not alone in this concern, and many British physicists worried about formidable mathematics at the heart of Einstein's work.

The complicated mathematics was not necessarily wrong. Rather, it concealed what was happening. The physical meaning of the symbols had become opaque, and therefore the calculations involved could be dangerously misleading. In *Past Years* he marked modern physics as having the peculiar property of trying to deal with phenomena "without contemplating their detailed machinery": "Much of the treatment of modern physics is of this highly abstract and non-pictorial character, though the reasoning is obscure and difficult to follow. Most of our progress in the past has been due to men who tried to form clear conceptions of what was happening, and who sought the aid of analogies and what have been called 'models.'"[58] That is, modern physics was defined by its lack of recourse to the models that underlay the great contributions of Faraday, Lord Kelvin, Maxwell, and Heinrich Hertz: "Their mathematics embodied and were dominated by physical ideas," Lodge wrote. "They never tried to transcend those ideas or take refuge in vagueness or uncertainty" (350). He warned that when students of modern physics encounter an opaque quantum equation, they should ask for the meaning behind it. "It is only fair to add that the greatest mathematicians confess to the same wish. Their symbols are stimulating and exciting rather than satisfying."[59] The equations of relativity and quantum mechanics were appetizers, not main courses.

Lodge hoped that this tendency to abstraction was simply a fashionable phase of physics: "It is being conducted through the temporary haze with great ability; but in time, I believe, it will emerge on the other side and become intelligible once more."[60] And since the mechanisms were concealed, he said, it was not at all clear that relativity had overthrown anything. The equations were so abstract and mysterious that they were compatible with a wide range of physical interpretation. In a letter to the *Times* in 1922, Lodge complained that Einstein's "pure mathematical equations, invented and developed in order to express mental imagery, may be utilized and interpreted otherwise, and can be applied to actual things, which to all appearance are of a different order, and perhaps really are essentially different except in so far as they are susceptible of similar mathematical treatment."[61] That is, the equations of hypergeometry could easily be erroneously thought to apply to all sorts of physical situations. But this could only be "done by ignoring and treating as unnecessary certain ancient and superficially obvious axioms" such as causality or the reliability of experience (13). If one took Einstein's equations too seriously, one could even think that life and mind could someday be treated mathematically "by a still further suppression of commonplace notions, or what some might call common sense" (13). Relativity could only be interpreted Einstein-style if one applied the equations incorrectly.

This position gave rise to one of Lodge's chief antirelativity strategies: whatever truth was contained in Einstein's equations was supplementary to existing knowledge, not contradictory. Consider the Michelson-Morley experiment: "A mathematical doctrine of relativity may be based upon this experimental result, and may be convenient for reasoning purposes, but no such doctrine is required by the facts. The facts are patient of the doctrine; they do not compel it, nor do they justify it. Any comprehensive mathematical expression is liable to permit other modes of interpretation, as well as the simplest and truest or the one most directly applicable to the problem in hand."[62] All the strange claims of relativity were based on such extended interpretations. Lodge wondered, then, why should we place a metaphysical structure on these equations (326)? In fact, he contended, for all practical purposes Newtonian physics still reigned.[63] Indeed, the equations of space-time were actually describing the behavior of the ether.

Lodge wanted what he called a "full blooded" universe, by which he meant a universe of physical sensations and conceptions based on ordinary experience, rather than solely on "complex mathematical machinery."[64] And this was in contrast to Einstein, whom he saw as attempting to "geometrise physics, and to reduce sensible things like weight and inertia to a modification of space and time. The work of great Geometers has been pressed into the service, and a differential-invariant scheme of expression has been utilised to do for physics in general, and especially for gravitation, what Maxwell's equations did for electric and magnetic forces. . . . The beauty and ingenuity of this scheme, and of the reasoning associated with it, are apt to overpower the judgment at times."[65] So aesthetic beauty, which Lodge earlier listed as one of the great benefits of learning mathematics, here became a danger. It had developed into a siren song that had lured the most skilled mathematicians away from the safe world of physical models into shoal-filled waters.

Even worse, the beauty of Einstein's equations seemed to have convinced everyone that they must be *real*. In something of a contrast to his earlier argument with

Greenhill, Lodge warned against taking too seriously the physical meaning of every element of a calculation: "It is scarcely wise to seek to interpret physically every link in a chain of mathematical reasoning. If the chain is coherent and if the terminal hooks are firm—that is, if the end results are intelligible and verifiable—no more need be expected from any system of equations" (796). His concern, apparently, was that the inner workings of differential geometry were so abstract that they could not seriously be connected to any of our direct categories of physical experience. The relativists misunderstood the relationship of physics and mathematics by assuming that every symbol must have a true physical meaning. He was apparently trying to refine his earlier position by specifying that only the beginning and end of a calculation needed to have direct physical corollaries. Along these lines he announced that "mathematically considered, relativity is a splendid instrument of investigation, a curiously blindfold but powerful *method* of attaining results without really understanding them."[66]

The innards of a calculation might have all sorts of unphysical entities—e.g., the square root of negative one—but no one should feel compelled to assert that they were *real*. "It is undeniable that mathematicians . . . can thus attain remarkable criteria, and are able to anticipate definite results; but we need not seek to engraft their modes of expression on the real world of physics. We need not consider realities superseded, because a system of pure space and time can be devised which can formulate, and be consistent with, the movements observed by ordinary men and animals."[67] He admitted that Einstein's calculations could replicate observed phenomena—but that did not mean that mathematicians should be able to impose their categories onto physicists. Lodge denied that everyday conceptions would "readily evaporate into a geometrical modification of empty space" (796). Time dilation should only be thought of as "a useful mathematical fiction," not reality.[68] At the end, our direct experience of the world had to triumph: "In the concrete reality of things there is something more, and physics is richer than geometry."[69]

He warned that following the mathematicians' path without any physical insight would never get science anywhere. Geometrizing physics would "complicate" it. Even if the relativity equations are used "a physical explanation can still be looked for, and our knowledge of the universe will not be complete until it is found. We cannot be for ever satisfied with a blindfold mathematical method of arriving at results. We can utilise the clues so given, and admire the ingenuity which has provided them, but that is not the end; it is only beginning. The explanation is still to seek; and when we really know the properties of the ether we shall perceive why it is that things happen as they do" (366). Lodge frequently returned to the metaphor of mathematics as Einstein's blindfold. He did not deny that the equations could be used, so the image is perhaps one of a craftsman working in the dark—through long practice, his hands could complete the desired tasks. But he could not see what he was doing—he did not truly understand what the parts looked like, and how they fit together. The blindfolded craftsman had deprived himself of his most important sense, just as Einstein had deprived himself of the ability to think about cause and explanation.

Abstraction thus became a dangerous practice. Once you were used to passing over physical meaning, you began to forget why it was there in the first place. In a letter to Arthur Stanley Eddington in 1929, Lodge wrote: "The danger is that from long habit of omitting things, one tends first to ignore them and then perhaps to deny them, that is to extend the notions appropriate to an abstraction and treat it as if it satisfied the whole philosophical outlook. It may be said that this is what scientific treatment has done for the Deity. Constant attention to transcendental realities is unnecessary; so they are legitimately ignored, and then in due course illegitimately denied."[70] Einstein's abstractions, if accepted, could imperil the whole future of physics.

THE VARIETIES OF MATHEMATICS

Lodge's opinions on mathematics were complicated. Partly that is because they certainly evolved over time. We can see how he responded to Victorian debates about mathematics education, and how Einstein's radicalism pushed him to reframe what it meant for equations to have "meaning." But he also acknowledged that the umbrella term *mathematics* applied to a wide spectrum of ideas and practices, from purely practical arithmetic to purely abstract symbology. So exactly what the relationship was between physics and mathematics depended on what kind of mathematics one was talking about, and how it was being used. To articulate these different kinds of relationships, it is helpful to use one of Lodge's own favorite comparisons: pure and applied.[71]

Lodge noted that of course pure mathematics could be of value to physics. He listed Maxwell's mathematical theory as the root of all the applications of ether waves.[72] He reported that H. J. S. Smith's address on the value of pure mathematics was what convinced him to buy a copy of Maxwell's *Treatise*.[73] But he also followed Smith in asserting that physics could inspire pure mathematics. The mathematical depth of the *Treatise* should hold great interest for pure investigators.

Lodge sometimes said that he pursued physics calculations because the results might be "of some little mathematical interest."[74] On the problem of electric oscillations in a compound body, he accepted that a full solution was beyond his skills. A complete result "must be left, I imagine, to the time when some pure mathematician may devote his attention to this particular shape of conductor, if the case appears to him of sufficient interest." Lodge was satisfied with a general notion of what was happening.[75]

Some kinds of math were dangerous, or at least uninteresting. He warned that it could sometimes be mechanical, in the sense of simply repeating some trivial operation over and over. Counting beans was a good start, though a man of science needed to move beyond. Anything that a calculating machine could do should not be part of a scientific mindset: "Nothing that can be performed by turning a handle can be considered an element in a liberal education: it can only be a practical and useful art."[76] It was actually a waste, Lodge argued, for smart people to spend their time on these simple operations. It took genius to design a machine, but nothing special to work it (234). "Men of application" had to grapple with "an immense mass of detail" from which men of pure science were free. But too much freedom was a problem, too. Algebra may have come from pure mathematicians playing with abstract problems, but physicists were interested in "things" and wanted symbols "to represent concrete quantities."[77] Drifting away from these concrete quantities was what happened if you put on Einstein's blindfold—mistaking the myriad possibilities of mathematics for the singular truths of the physical world.

So physics needed to have the right balance of pure and applied mathematics. No one should be surprised that Lodge held up Maxwell as the exemplar of the correct mix of physical understanding and symbolic power. Faraday did not have enough pure math; Einstein had too much. Einstein had been entranced by aesthetic beauty as a mathematical method, rather than as something that was found at the end of a well-established theory. Models were the touchstone that allowed physicists to set up reliable equations while also preventing unchecked mathematical adventuring. Some beings of extraordinary ability could move beyond their models—as when Maxwell developed his more abstract system. But according to Lodge, such people were few and far between—and included neither Einstein nor himself.

SIX

THE RETIRING POPULARIZER

LODGE, COSMIC EVOLUTION, AND THE NEW PHYSICS

Bernard Lightman

AS HE was looking back on his life in 1931, Lodge declared that his main interest had always been physics. "I perceive now," Lodge wrote, "that even in my childhood I was keenly interested in all the information that I could gather in connexion with that subject, though it was extremely little." He remembered "devouring Brewer's *Child's Guide to Knowledge,* or some catechetical book of that kind, which had accidentally come into my hands, and which contained simple information about why the fire-irons should be bright and the kettle black, and baby things of that sort; so that even in those days the connexion between matter and ether, illustrated by the elementary facts of radiation and absorption, were of fascinating interest."[1] Lodge's reference to Brewer's *Guide,* and his assertion that it contained information on the connection between matter and ether, are important keys to understanding the last two decades of his life. The Reverend Ebenezer Cobham Brewer's *A Guide to the Scientific Knowledge of Things Familiar* (1847) was quite possibly the nineteenth century's bestselling science book for a popular audience.[2] Like Brewer, Lodge became a prolific popularizer, devoting much of his time near the end of his life to writing books for a general audience that spelled out the religious implications of science. However, Brewer's *Guide* did not contain the term *ether*. Lodge's notion that Brewer was actually dealing with issues relevant to the connection of matter and the ether illustrates his tendency to interpret all scientific ideas "in light" of the ether. For

Lodge, the religious implications of the science of the early twentieth century, particularly the new physics, could not be fathomed without focusing on the role of the ether in a cosmic evolutionary process. Those implications, he thought, pointed to the reality of life after death.

An interest in psychical research would not have been encouraged by the scientists who were his heroes when he was a young man. As previous chapters of this volume have set out, his attendance, at the age of fifteen, at a course of Royal Institution lectures by John Tyndall on heat in 1866–1867 inspired him to pursue a career in science.[3] After attending the lectures, Lodge recalled that Tyndall became "one of my heroes, and I ever afterwards attended every Friday evening, when he was holding forth at the Royal Institution."[4] In the winter of 1873 he took a course for teachers at the Royal College of Science in South Kensington run by two of Tyndall's close friends, the biologist T. H. Huxley and the chemist Edward Frankland. In January 1874 he went to University College, London, where he took mathematics classes with W. K. Clifford and Olaus Henrici.[5] Unsurprisingly, Lodge was, as he put it, "under the influence of Huxley, Tyndall, and W. K. Clifford" during his time in London up to the early 1880s.[6] Huxley, Tyndall, and Clifford were all scientific naturalists deeply critical of psychical research. Lodge's interest in psychic research began in 1882 when he conducted experiments on thought transference.[7] In 1884 he joined the Society for Psychical Research, as he believed that psychical phenomena should be scientifically investigated.[8] Psychical research became a major preoccupation for the rest of his life, and by the early twentieth century Lodge had become an important critic of scientific naturalism.[9]

Lodge retired as principal of Birmingham University in 1919. W. P. Jolly, Lodge's biographer, asserts that the physicist intended this last part of his life to be a time when he could concentrate on extending the scientific and philosophical ideas that he had been developing for many years, free from the duties of university administrator.[10] During the last two decades of his life Lodge wrote a series of books dealing with the larger significance of modern scientific ideas, especially as they related to psychical research. Some of Lodge's books—for example, *Making of Man* (1924) and *Evolution and Creation* (1926)—did not go beyond a first edition. But others were more successful. *Atoms and Rays* (1924) reached a fourth edition by 1931, *Ether and Reality* (1925) an eighth edition by 1927, and *Relativity: A Very Elementary Exposition* (1925) a third edition in 1926. In the 1920s and early 1930s readers would have been familiar with Lodge as one of the most productive popularizers in England. During this period he wrote over fifteen books. It is perhaps no coincidence that Lodge's output diminished considerably after the death of his wife, Mary, in 1929. His last book, *My Philosophy*, was published in 1933.

Although a study of Lodge in retirement would give us an understanding of his mature position on the larger meaning of science, this period is given very little attention, even by Lodge himself. *Past Years* basically ends when he retires from Birmingham. Jolly devotes only one of fourteen chapters to the post-Birmingham period. Substantial scholarly research tends to focus on earlier periods. Bruce J. Hunt, for example, who treats Lodge as one of the core members of the group of Maxwellians, discusses how Lodge was among the first to work on describing how to generate electromagnetic waves (besides those of light) between 1879 and 1883.[11] Peter J. Bowler has briefly, and usefully, discussed Lodge's activities as a popularizer, and his role in attempts to reconcile science and religion in the early twentieth century.[12] Scholars have only begun to tackle the issue of popularization relatively recently. Since the publication of Roger Cooter and Stephen Pumfrey's groundbreaking article in 1994, popularization has been taken far more seriously by historians of science as an important cultural product in its own right.[13] But Lodge has yet to have his in-depth treatment as an important popularizer in the early twentieth century.

While Brewer, Lodge's predecessor from the nineteenth century, wrote his accessible *Guide to Knowledge* for a young audience, in his more demanding books Lodge wrote mainly for "the non-expert," as he referred to his audience in the preface to *Ether and Reality*. Lodge hoped that his "elementary treatment" of the topic would "prepare the way for a larger and more technical volume."[14] Similarly, in *Atoms and Rays*, Lodge characterized his book as being "introductory to more advanced treatises." However, some sections of the book required a basic knowledge of science. He warned the "general reader, and student desirous of information about modern progress in physics" that they would find the "more technical chapters worthy of sustained attention" provided that they were acquainted with the fundamental principles of mechanics and simple geometry.[15]

Lodge's main concern in his books was to educate his target audience about the larger metaphysical and religious meaning of contemporary scientific theories. This meant explaining the new developments in physics, especially relativity theory and quantum theory, and how they related to evolutionary theory. Since Lodge's popularization of science was deeply tied to his personal philosophical beliefs, educating the public also involved incorporating the new developments in physics into the metaphysical and religious framework he had previously developed, with its emphasis on the concept of the ether within a broad-ranging natural philosophy.[16] Lodge updated his previous framework by incorporating relativity theory and quantum theory into it, just as Christian intellectuals of the second half of the nineteenth century had updated natural theology in response to evolutionary theory. Lodge came to believe that the new physics helped him to articulate an even more convincing and gripping vision of the ether as the fundamental essence of the natural world, which provided scientific evidence for life after death. The supposed confirmation of

Einstein's general theory of relativity in 1919—the year Lodge retired—forced Lodge to come to terms with Einstein. Lodge's final book, *My Philosophy*, was published in 1933. These two dates, then, mark the bounds of a period in Lodge's life when he devoted himself to interpreting the larger significance of complex new ideas in early twentieth-century physics for the general reader.

LODGE AS POPULARIZER, EVOLUTIONIST, AND SPIRITUALIST

In his popularizations of science published after his retirement in 1919 as university principal, Lodge presented a new synthesis that combined concepts drawn from evolutionary biology and ether physics with humanism, psychical research, and Christian modes of thought. In many of his postretirement books Lodge started by discussing the world of matter as described by chemistry and physics. Matter was composed of particles held together by the ether, a substance responsible for their cohesion. The ether was a "cementing substance" or a "welding medium," and the first "building stone of the universe"; electrons and protons were merely "specialised specks" in it. Electrons and protons combined into atomic matter, then they combined to form chemical molecules, and then they aggregated into the visible bodies that appeared to our senses. In a further step, matter aggregated under gravitation into planets and suns.[17] The ether was not only the physical basis of matter, it was also "responsible" for "gravitation, and for electricity and magnetism and light, for elasticity also and all strain; it is also responsible for cohesion."[18] Taken as a whole, the physical universe, Lodge maintained, was like a "reversible engine: its operations are cyclical; that is, they go round and round in a cycle, with no advances, no progression, but with constant and eternal repetition" (122). Stimulated by the radiation of the sun, molecular aggregates could form into living, organic materials. But there was a fundamental difference between the physical universe and life. Life was not characterized by reversibility. Rather, we could see "real progress, development, evolution" in life (123).

Although Lodge was primarily a physicist, evolutionary theory was essential for his synthesis. Here he differed significantly from the nineteenth-century physicists that he admired. Lord Kelvin and James Clerk Maxwell had been resolutely against evolution, since scientific naturalists had seized upon Darwin's theory as a weapon to undermine Christian belief.[19] Lodge had pointed to the connection between evolution and evidence for immortality as early as 1908,[20] but his first major integration of evolution into his synthesis was presented in *Making of Man: A Study in Evolution* (1924). Evolution began where the laws of physics and chemistry ended. When the "supplementary laws of Biology" took over, "the vitalised molecules were no longer beaten about by every random force; they began in some dim way to control those forces, to form themselves into cells or communities."[21] When freedom was introduced into the world by evolution, "an element of risk and even of pain" was inevitable, so although the prospect for the future was "one of infinite progress," achieving progress demanded effort on the part of biological organisms. Humanity was in an "early stage of evolution, having but recently arisen from an animal ancestry," and was still therefore grappling with the gift of "conscious freedom" (vii–viii). We were intended for better things, Lodge declared, and "we are still far below the ideal. We are an unfinished article" (141). Writing just after the devastating chaos that resulted from World War I, Lodge stressed that the evolutionary process did have a direction. "The course of evolution," he wrote, "is not blind or unguided; the reasonable character of the result could not be accounted for in that way" (141). Lodge's synthesis was based on a complex combination of the concept of the ether with evolutionary theory.

Two years later, in *Evolution and Creation* (1926), Lodge developed the religious dimensions of his synthesis of evolution and ether. "My thesis," Lodge declared, "is that there is no essential opposition between Creation and Evolution. One is the method of the other." When Lodge referred to evolution as the method of Creation, he meant that it was through the evolutionary process that "things change and improve and come into existence." Humanity had an important role to play as "the co-operator in the Divine process of Evolution," since the result "depends partly on our own exertion." This insight had been obscured by the early evolutionists who had rallied to Darwin's defense in the wake of the publication of *On the Origin of Species* in 1859. Contrary to the bleak materialism of his former heroes, Herbert Spencer, Huxley, and Clifford, Lodge proclaimed, "Evolution is a discovery full of hope."[22] Raia has labeled Lodge's synthesis "ether theology." However, it would be more accurate to refer to it as an *evolutionary* ether theology.[23]

Lodge presented an evolutionary epic similar to Spencer's that treats the ether as an integral part of the developmental process. In fact, chapter four of *Evolution and Creation* is titled "Cosmic Evolution," a term used by some of Spencer's disciples, such as John Fiske and Grant Allen, to refer to Spencer's grand evolutionary scheme based on the notion of development from the homogeneous to the heterogeneous. The process, Lodge asserted, begins with an "undifferentiated all extensive substance, the raw material out of which everything composed, and which we call the Ether of Space." Then the ether knots into specks of two kinds, protons and electrons, which group into ninety-two different atomic patterns, followed by the formation of molecules from these atoms, later aggregated into larger masses due to gravity, eventually formed into a nebulous mass out of which stars and then planets form.[24]

Lodge pointed out that, strictly speaking, evolution was not a process in time that had a beginning or an end. "We see things in all stages of evolution," Lodge asserted, "not one after the other, but concurrently. We see the great spherical nebulae as an early state; we see the Companion of Sirius as a very late stage" (87–88). The fact that these stages were concurrent led Lodge to dismiss a simplistic interpretation of the second law of thermodynamics, which seemed to dictate the complete dissipation of energy. Instead, Lodge endorsed a cyclical cosmology, in line with his notion of nature as cyclical. Then, when he began to discuss spiritual evolution, he could contrast the repetitiveness of nature with the "growth, development, increase of value, [and] rise in status" of life. Lodge insisted that "the evolution of spiritual things has no necessary regress. They can advance continually through higher and higher stages towards perfection" (96). This led Lodge to declare that if death was not the end of humanity, "there may be infinite progress in store." The infinite progress of spiritual evolution was, for Lodge, "the real meaning of Evolution" and the reason why the physical world exists. "This, surely," Lodge affirmed, "is the real aim and purpose of the ultimate and infinite term 'God'" (98–99). Lodge could put forward this etherealized and Christianized form of Spencer's evolutionary epic and still be taken seriously in the 1920s because Darwin's theory of natural selection had not yet won the day. The modern evolutionary synthesis, which led to a consensus among biologists that Darwin's theory was valid, was not complete until after the 1930s.[25] Lodge exploited that lack of consensus. In *Evolution and Creation* he argued that "whatever may be the history and fate of the Darwinistic theory in its narrower sense, there is no reasonable doubt that evolution represents the method of Creation."[26]

Lodge's concept of cosmic evolution as the method of Creation was very much in step with the efforts of many nineteenth-century popularizers of science to perpetuate the natural theology tradition in an updated form. From the publication of *Making of Man* on he did not hesitate to draw on a discourse of design to describe the order in nature that was a manifestation of the guided nature of the evolutionary process. *Making of Man* climaxes with the statement that in the directed course of evolution there was "an evident effort to carry out a design."[27] In *Science and Human Progress,* Lodge declares that there is "evidence of Design and Planning" in the heavens as well as in the interstices of the atom.[28] Lodge closed *Modern Scientific Ideas* with the notion that the scientific evidence for the "unity of design running through the whole universe" led to the affirmation of the existence of "an all-controlling and all-designed Mind."[29]

In his popularizations of science produced after his retirement from Birmingham, Lodge developed a grand synthesis of scientific knowledge that drew far more on evolutionary theory than had his earlier writings. Lodge's evolutionary ether theology joined the biological and physical sciences into a coherent unity.[30] Like Robert Chambers and Spencer, he offered the British public a gripping evolutionary epic. Unlike Chambers and Spencer, he used two "glues" to hold this synthesis together. While Chambers and Spencer relied primarily on evolution to connect the various bodies of scientific knowledge, Lodge used the ether as a second unifying factor.

THE CHALLENGE OF RELATIVITY THEORY

Lodge met Albert Einstein when the German physicist was visiting Oxford in June 1933 to deliver the Herbert Spencer lecture. During the meeting Lodge took detailed notes on what Einstein had to say about the implications of relativity theory for the existence of the ether. According to Lodge, Einstein told him that he had gone through three stages in his thinking about the ether. First, he had believed in the old dynamical theory of the ether; second, he no longer believed in the ether; and third—this was his current position—he believed that the ether enters everything, although Einstein did not admit that the ether has motion.[31] Whatever the qualifications, Lodge felt vindicated by Einstein's comments. New developments in physics did not necessitate a rejection of the ether. Lodge had defended this position in his presidential address to the British Association in 1913.[32] However, he did not believe it was necessary to present a detailed response to those who argued that relativity theory did away with the ether. After 1919, when the general theory of relativity had been confirmed by two British eclipse expeditions, and Einstein became a world celebrity, Lodge could no longer avoid a thorough engagement with relativity theory. Lodge's meeting with Einstein in 1933 came after more than a decade of reflection, and writing, on the relationship between relativity theory and the concept of the ether.

* * *

In her chapter in this volume, Imogen Clarke argues that Lodge's role as an expositor of modern physics was firmly established and that he was not considered by contemporary physicists to be an embarrassing dinosaur due to his support for the ether.[33] Clarke is undoubtedly right. Lodge's defense of the scientific validity of the ether after 1919, despite the increasing acceptance of relativity theory, was both robust and detailed. Moreover, he could point to support from eminent physicists for his position. Lodge took the debate one step further. Not only did he maintain that relativity theory provided evidence for immortality, he also contended that the concept of the ether reinforced key elements in relativity theory.

Lodge's first detailed responses to relativity theory came shortly after the announcement of the results of the solar eclipse expedition, in publications that appeared from 1919 to 1921. All three of his publications were geared toward a popular audience. Initially, he was on the defensive and emphasized that relativity theory left the question of the ether to the side. In 1919 Lodge acknowledged in an article in *Nineteenth Century and After* that the eclipse expeditions had confirmed that the sun's gravitation field gave the deflection of light predicted by Einstein's general theory of relativity. Although this appeared to confirm his theory and to "substantiate all the implicated details which are associated with the main theory," which included the sounding of the "death-knell of the Ether," Lodge cautioned that this went "too far." The real significance of Einstein's achievement had nothing to do with "Space or Time as such, nor is it any true denial of the Ether." The achievement "is that Gravitation has been related to other forces for the first time," which to Lodge meant that Einstein had shown that gravity interacted with "the other properties of the Ether of Space." In sum, Lodge insisted that "for all practical purposes Galilean and Newtonian dynamics still reign, perhaps no longer supreme, but as a limited monarchy." Einstein's ideas supplemented the "old physics," rather than replacing it.[34]

In 1921 in the *Fortnightly Review* Lodge began by expressing his mistrust of many of the "popular interpretations which have come into vogue since the eclipse of May, 1919." To those writers commenting on the special theory of relativity who asserted that "Einstein's theory destroys the ether and shows that it has no existence," Lodge replied that that was "not the true interpretation." Rather, he insisted, "the equations make no explicit assertion as to the existence or non-existence of ether." Similarly, the general theory of relativity had replaced the notion of force with "complex mathematical machinery." But this was a merely abstract treatment of the issue that ignored the dynamical reality in nature. The "relativity equations can be used, but a physical explanation can still be looked for, and our knowledge of the universe will not be complete until it is found." A full physical—and philosophical—explanation, Lodge declared, would only be achieved when "we really know the properties of the ether."[35]

About a month after the *Fortnightly Review* article appeared, Lodge gave a lecture to the Literary and Philosophical Society at Liverpool on October 31, 1921, on relativity.[36] In this lecture Lodge rejected "full-blown" relativism, by which he meant the position that there were no absolutes in nature (9). Lodge believed there was an absolute standard of rest against which to measure motion; that is, the ether of space. However, since the relativists "differ among themselves" on the issue of the ether, Lodge claimed that they "do not say much about it" (19). But Lodge pointed out that neither Eddington nor Einstein had said that relativity theory "abolished the ether." Eddington had told Lodge that when he asked Einstein in Berlin about the ether he had replied that he had no objection to it, and that his system was independent of it (19–20). There was another reason Lodge was not a relativist. Relativity theory actually showed that there was a second absolute, in addition to the ether, the maximum velocity of light (29). Full-blown relativism was impossible. Lodge made similar points in a series of articles published in *Nature*, a venue less hospitable to the general reading audience.[37]

After 1921, when Lodge had more time to reflect, he was able to offer a more nuanced discussion of relativity theory in a number of his books. He could answer his critics, which he labeled "the Relativists," more effectively. He depicted them as closed-minded. Accepting relativity wholesale prematurely shut the door on possible future test experiments to ascertain the speed of matter through the ether.[38] Acknowledging that physicists were among his critics, he became more aggressive in responding. He considered his chief opponent to be Sir James Jeans. To counter Jeans, Lodge pointed to eminent physicists, such as Larmor and Eddington, who sided with him on the issue of the existence of the ether. But Lodge's ace in the hole was actually Einstein. In his *Ether and Reality* (1925) Lodge quoted at length from an address titled "Ether and the Theory of Relativity," delivered by Einstein in 1920, and translated into English and published in 1922.[39] In this lecture Einstein declared that the hypothesis of the ether in itself is not in conflict with the special theory of relativity, though it was necessary to guard against ascribing a state of motion to the ether. There was, Einstein argued, "a weighty argument to be adduced in favour of the ether hypothesis," as to "deny the ether is ultimately to assume that empty space has no physical qualities whatever." Next, when Einstein examined the relationship between the general theory of relativity and the idea of the ether, he again rejected the notion that there was a contradiction. "According to the general theory of relativity," Einstein affirmed, "space without ether is unthinkable; for in such space there not only would be no propagation of light, but also no possibility of existence for standards of space and time (meaning—rods and clocks), nor therefore any space-time intervals in the physical sense."[40]

Once he was aware of Einstein's lecture, Lodge became more confident that he could integrate relativity theory into his evolutionary ether theology.[41] Seemingly with Einstein's stamp of approval, he made a series of bold pronouncements in *Ether and Reality*, and in subsequent books. In *Ether and Reality* he used the ether to help his readers understand Einstein's conception of gravity in the general theory. Einstein, Lodge explained, replaced action at a distance by a "contact effect," or "a strain in the ether we should wish to say." "Einstein does not say as much as that, but he recognizes that it must be something directly in contact with the moon that is curving its path, even if it be only a warp in space." Lodge suggested that it was acceptable to say that "a large body like the earth warps the ether all round it, thus making other bodies fall towards it as if they felt its attraction."[42] As for the special theory, Lodge basically conceived of Einstein's notion of the constant velocity of light as a statement about the ether. The vibration in the ether was "systematic and orderly" since it always traveled at the same speed, "the one absolute speed in the Universe, which we have measured as the velocity of light" (158).

Later, Lodge began to insist that the concept of the ether was integral to both special and general relativity theory. Contrary to what the Relativists believed, the theory of relativity was "a quest for Absolutes. The theory proclaims a new and surprisingly small list of absolutes, in the midst of a wilderness of humanly interpreted and mainly relative phenomena, dislodged from their late pre-eminence."[43] Of course, Lodge had always emphasized that the ether was an absolute. In his final book, *My Philosophy* (1933), he argues, "The velocity c, and the ether to which it belongs, are fundamental realities, which can in no way be ignored." The inclusion of *c* in relativity equations was actually "a tacit recognition of the fact that the bodies are moving through a fundamental medium to which this velocity c intrinsically belongs." The ether, Lodge was maintaining, is built into relativity theory.[44]

Although it was "fashionable in a few quarters to doubt the existence of the ether of space, and to suppose that Einstein has exploded it," in reality "no great authority on Relativity really supposes that; certainly Einstein does not himself, nor does Eddington." For "Einstein's discovery, linking gravitation with light for the first time, strengthens the position of the ether" (173). Light had already been known to be an "etherial phenomenon" and now Einstein had connected gravitation with the ether (173–74). Einstein had also helped bolster the scientific validity of the ether by showing that the existence of something as elusive as a gravitation field around a large object could be revealed through intricate experimentation. In order for the general theory to be confirmed in 1919, "the entire mass of the sun had to be pressed into service; and the light had to come from a very great distance" (140). If a gravitational field could be detected, then there was hope that eventually complicated experiments could also be devised to detect the ether. Lodge wrote, "By exceptional experimenting such as that, the ether itself may possibly be constrained to yield an observable effect" (140).

Not only did Lodge argue that relativity theory offered more scientific evidence that the ether existed; he also maintained that it offered compelling reasons for rejecting materialism. In *Beyond Physics* Lodge declared that one outcome of the "Theory of Relativity is that there is no absolute distinction to be drawn between matter and other forms of energy."[45] If matter was actually energy, then a fundamental basis of materialism was no longer scientifically credible. Similarly, Lodge believed that relativity theory had raised questions about the ability of the human sensory apparatus to perceive reality. Materialism depended heavily on the notion that the senses provide access to reality. But, he argued in *Phantom Walls*, "modern physics is insisting that most of our mundane experience is illusory, that even space and time if taken separately are abstract frames dependent on our limitations, and that we are surrounded by phantasmal appearances through which our sense cannot penetrate. Matter is what we primarily apprehend though the senses; but the nature of matter is mysterious." Einstein and Eddington did not contend that matter produced the curvature of space but that "it *is* that curvature."[46] Although our senses tell us that we live in a world composed of discrete and disconnected objects, Lodge believed that relativity theory pointed to an underlying, nonmaterial unity in nature. "Meanwhile," Lodge affirmed, "matter and energy have merged into one another; both are treated geometrically, as if they were properties of space, or rather of the greater generalisation called space-time; and there is beginning a great unification which, in spite of present complexity, seems likely to lead to an ultimate simplification" (132). Relativity pointed toward a synthesis similar to the one Lodge had developed in his evolutionary ether theology.

After being on the defensive at first, by the middle of the 1920s Lodge was ready to confront those who claimed that relativity theory spelled the end of the ether. The turning point for Lodge must have been his reading of Einstein's lecture "Ether and the Theory of Relativity," in which the German physicist stated his acceptance of the ether as a valid scientific concept that would continue to play an important role in modern physics. To Lodge, Einstein had given him his blessing to integrate relativity theory into the grand synthesis he was developing. Lodge argued that relativity theory required the existence of the ether while at the same time it provided new scientific evidence for it. Moreover, by undermining the presuppositions of materialism, relativity theory bolstered a belief in immortality. When Lodge finally met Einstein face-to-face in 1933, he thought that he had received confirmation of all that he had written since the middle of the previous decade about the relationship between the ether, relativity theory, and a belief in life after death.

ENCOUNTERING QUANTUM THEORY

Lodge expressed his views on quantum theory in *Atoms and Rays* (1924). He regarded quantum theory, largely, in positive terms. "There is nothing altogether novel and perturbing in the idea of physical discontinuities like quanta," he wrote. He believed that research on atomic particles and their strange behavior would be "helpful and instructive, and contributions to further knowledge to a remarkable degree," although there was much that was not understood about the phenomena.[47] Even in 1930, after the Copenhagen interpretation of quantum theory had become well known, he remained undisturbed by the new developments within the field. "To the quantum I have no sort of objection," he declared in *Beyond Physics*, "only I think it may be rationally accounted for, some day, by regular mechanics."[48] Lodge engaged more with relativity theory than with quantum theory in his writings. The latter seemed to him to be less of a threat to his evolutionary ether theology. He was, however, concerned that quantum theorists relied too heavily on mathematics to conceptualize the physical world, and the implications of the theory forced him to discuss the issue of discontinuity at the atomic level. But these problems were outweighed by the new insights into human agency and unity that could be drawn from the study of the quanta. These features allowed him to integrate quantum theory into his synthesis.

Though quantum theory was not yet complete, in Lodge's opinion, it nevertheless offered opportunities for drawing metaphysical conclusions that were favorable to his synthesis. Instead of relying on Bohr or any of the German quantum theorists to guide him in determining what those conclusions were, Lodge looked to Eddington. In *Beyond Physics* he discussed Eddington's *Nature of the Physical World* at length. He praised Eddington for dealing in a "vivid and illuminating manner with very difficult subjects" (76). He added that Eddington's assertions about the new physics, which might seem to some to be "careless nonsense," actually represented "the outcome of profound knowledge, such as is possessed by few other men." Eddington was "an example of one of the most thoughtful of the explorers into the innermost recesses of mathematical physics" (78). Lodge was particularly struck by the fact that Eddington had become an idealist as a result of his research on the new physics. In Lodge's case, his rejection of materialism had come out of his psychical research, an area that lay outside of Eddington's investigations. This made Eddington's conclusions "all the more valuable, because when people arrive at similar results by different paths, it is an indication that they are proceeding on the whole in a right direction" (82).

* * *

With Eddington at the back of his mind, Lodge discussed some of the metaphysical implications of quantum theory in *Beyond Physics*. He appreciated Eddington's position that physical science was limited to the domain of pointer readings. Lodge agreed with Eddington that "great tracts of direct human experience which are not metrical at all" were excluded from the world of science, but that didn't mean that these dimensions of human experience were nonexistent and without value. Eddington admitted that "our aesthetic and religious convictions and intuitions are equally real, equally valid" (70–72). Lodge drew attention to Eddington's assertion that quantum theory showed that consciousness was a part of nature. He pointed to Eddington's belief that modern physics had "opened a fresh avenue towards Indeterminism and Free Will, by its apparent overthrow of the strict causality in which hitherto it had been bound" (74). All of these insights helped Lodge to emphasize that a world of freedom existed apart from the cyclical, deterministic realm of physical nature and that human agency played a crucial role in the unfolding of evolution.

Other aspects of quantum theory appealed to Lodge. At first, quantum theory appeared to threaten his emphasis on the continuity in nature provided by the ether. The idea that discontinuity existed in the atomic world implied that the whole of the physical world was riven by discontinuity. In *Atoms and Rays* Lodge acknowledged that there was discontinuity when dealing with individual atoms, but he refused to accept the notion that ultimate continuity had to be discarded. Lodge also argued that "when Matter is dealt with in the mass, and when the Ether is dealt with in bulk, the quantum is not necessary or applicable; the ordinary considerations of dynamics and optics then serve sufficiently well, or indeed to all appearance perfectly."[49] Since quantum theory was limited to an understanding of the atomic world, it did not apply to nature as a whole.

Lodge argued that quantum theory actually strengthened the argument for continuity. Referring to Heisenberg's uncertainty principle in *Phantom Walls,* Lodge argued that its larger implications were that a greater unity existed in nature. The boundaries of the electron had become blurred and its locality indefinite. From one point of view the definite points of light seem to be merging into a sort of continuum, Lodge wrote, "while, from another, a continuous luminosity gathers itself together into discontinuous points. The gain of definiteness on the one hand is mingled with a loss of definiteness on the other." Whatever the outcome of further research would be, Lodge thought that "a greater unity is beginning to be discerned throughout the material cosmos; and that the initial stages of some comprehensive unification are of great interest for the present and of good augury for the future."[50]

One of the reasons for Lodge's optimism was that quantum theory, like relativity theory, raised questions about the nature of matter that seemed to undermine the materialism that had flourished in the nineteenth century. While relativity theory had demonstrated that matter was convertible into energy, quantum theory emphasized the idea that at the atomic level matter behaved both like a particle and a wave. In a discussion of Louis de Broglie and Erwin Schrödinger's concept of matter as "a small localised area of group waves," Lodge acknowledged that quantum theory would likely not be the "last word on the subject." But he praised the work of the quantum theorists, characterizing it as "very brilliant and quantitative, and well deserving of study." In the end, he concluded that "the wave theory seems to me specially hopeful and likely to be physically justified, however remote it be from ordinary perception" (viii–ix). In any case, the notion that matter was as much a wave as a particle seemed to Lodge to strengthen the scientific validity of the concept of unifying ether. He could account for discontinuity within a larger continuity provided by the ether. In *Beyond Physics* he argued that the discontinuity of matter was due to "some unknown rotational boundaries existing even in the ether." The electron's partial resolution into waves, he declared, "is one of the most hopeful signs of the resolution of matter into some form of ether vibration or circulation, some form of constitutional periodicity in space."[51]

Lodge assimilated quantum theory into his grand synthesis, just as he had integrated relativity theory. Although Lodge believed that quantum theory tended to be too abstract and mathematical, it nevertheless offered opportunities for drawing metaphysical conclusions that reinforced the synthesis. Lodge was sympathetic to the interpretation of quantum theory put forward by Eddington and other scientists with idealist tendencies. Bolstered by Eddington's authority, Lodge boldly asserted that quantum theory affirmed the reality of human agency, since it undermined a strict causal determinism. He could continue to talk about the important role of human agents in the realization of the divine evolutionary goal. Quantum theory also strengthened the argument for continuity since it pointed in the direction of a unified field theory, and it weakened the case for materialism based on the notion that nature was composed of material particles. To Lodge, this reinforced arguments for the existence of a unifying medium like the ether and for the validity of psychical research.

THE PUBLIC SUCCESS OF EVOLUTIONARY ETHER THEOLOGY

Lodge saw himself as working in the tradition of nineteenth-century physicists like Kelvin, Maxwell, Peter Tait, and Balfour Stewart, men who believed that science should be conceived of as being within a larger religious framework. However, although nineteenth-century physicists provided Lodge with the goal of providing science with a religious framework, they could not help him integrate them in light of the new physics. Lodge was highly tentative right up to 1921 about the new physics. Many of his contemporaries interpreted it as fatal to the continued validity of the ether. But with the help of more contemporary physicists like Einstein and Eddington, whose authority in these areas was unquestioned, Lodge could show that the new knowledge yielded by relativity theory and quantum theory was "helpful and confirmatory, or, at least not hostile" to Christianity and psychical research.[52] In *Science and Human Progress* he declared that "the tendency at the present day is for science in its philosophic mood to become less materialistic," as the material universe was found "to be suffused throughout with spiritual reality."[53]

By the mid-1920s Lodge was confident that materialism was on the decline, so confident that he described the materialist as a victim of a crippling mental disability. "Persons afflicted with what psychoanalysts of the more reasonable variety might call 'a materialistic complex' appear constitutionally unable to open their minds to evidence of any non-material or anti-materialistic kind," he explained in *Ether and Reality*. "They exhibit a curious kind of mental aberration or unconscious warp, and yet they are quite unconscious that their perceptive faculty is atrophied in one direction."[54] Due to their "mental aberration" the materialists were now an obstacle to the progress of science. Those physicists who did not accept the ether believed that the "only things worthy of attention are those which make appeal to the senses."[55] These condemnations of materialism are stronger than any comments he had made before World War I. Perhaps he felt confident enough to state them due to his growing authority as a popularizer and his status as a well-known scientific celebrity.[56]

Lodge's optimism about the future of the ether, of psychical research, and of his entire grand synthesis during the 1920s and early 1930s should be a major factor when interpreting both his activity late in life as well as his life as a whole. Lodge is often depicted as a scientist whose time had passed—as a Victorian who outlived his moment. Long ago David Wilson portrayed Lodge as "the old scientist in clear counterpoint to a new age."[57] Lodge was fully aware that many scientists were hostile to his support for psychical research and that some theologians were suspicious of his reinterpretation of Christianity. But some clergymen praised him for his positive religious views.[58] When it came to his scientific views, Lodge would not have appeared to his contemporaries to be as antiquated as is usually supposed. Orthogenetic evolution would not have been completely discredited until after his death, when the new synthesis enshrined Darwin's theory of natural selection at the center of biology. Lodge could claim the support of important physicists, such as Einstein and Eddington, with impeccable credentials for being au courant with the latest theories.

As far as the British reading audience was concerned, Lodge was still a reliable guide to the development of scientific thought. A competition held by the *Spectator* confirms how highly he was regarded by the British public. Readers of the journal were asked to select the five best brains in Great Britain in 1930. A prize of five guineas was offered for the competitor whose selections matched most closely with the majority verdict.[59] Lodge placed second with 183 votes, behind George Bernard Shaw, who had 214 votes. The next scientist on the list was J. J. Thomson, much further down at number fourteen, with only 28 votes. Then came three more scientists, J. B. Haldane (23 votes), Julian Huxley (20 votes), and Sir James Jeans (20 votes).[60] To the British public of the time, Lodge's evolutionary ether theology seemed entirely reasonable and scientifically credible. As Richard Noakes has asserted, "the success of Lodge's books, wireless broadcasts and lectures, testify to the fact that well into the twentieth century there remained considerable audiences for speculations on the ether's broader functions, irrespective of its vanishing presence in cultures of experimental and theoretical physics."[61] Rather than seeing the postretirement Lodge as a somewhat tragic figure, whose fate was inevitably tied to the declining fortunes of the ether, he can be viewed as a productive popularizer whose evolutionary ether theology was well received by members of the public.

SEVEN

THE FORGOTTEN CELEBRITY OF MODERN PHYSICS

IMOGEN CLARKE

IN HIS chapter in this volume, Bernard Lightman laments (and corrects) the dearth of historical attention paid to Oliver Lodge's postretirement years, an absence that he notes has thus far resulted in a limited understanding of Lodge's mature position on the larger meaning of science. I would, however, go further, and suggest that in neglecting Lodge we lose much more than an understanding of the man himself. Where Lightman's chapter analyzes Lodge's popular science of the 1920s, here I look at the reception of this work. Lodge's extremely visible place in early twentieth-century Britain had an impact on contemporary wider conversations about physics. A renewed focus on Lodge can thus provide us with a more comprehensive understanding of perceptions of physics in 1920s Britain, in particular "modern physics."

In a period conventionally characterized by the rise of general and special relativity and quantum theory as "modern" physics, the relevance of a champion of the ether is not immediately evident. The task of inserting Lodge into this history is made even more challenging by the enduring assumption of a dichotomy between classical and modern physics.[1] When assessing his work within the framework of a sharp divide between the old and the new, historians place Lodge firmly in the camp of the former. And if there really was such a dramatic shift in conceptions of the natural world, a decisive rejection of the past, then what purpose would it serve us to study the work of a "classical" physicist?

I see hope for Lodge, however, in the emergence of a more nuanced picture of the period. It is now evident that to apply the terms "classical" and "modern" to the 1910s or even much of the 1920s would be an anachronism.[2] Furthermore, our present-day use of these categories may have obscured our analyses of the past through undue emphasis on what we now perceive as modern.[3] Hence, many "reception studies" have assessed historical actors in terms of the success which with they adopted new ideas.[4] Andrew Warwick moved beyond this in his study of the reception of relativity theory in different Cambridge traditions from 1905 to 1911, considering instead how physicists viewed the theory in the context of their own work.[5] If we apply Warwick's methodology to Lodge in the 1920s, then we do indeed see him incorporating new ideas into his existing framework of ether physics, the old and the new working in constructive dialogue, rather than outright opposition. But did this have an effect on his contemporaries, or was he merely a lone researcher, with little impact on wider discussions?

Such discussions took place in a multitude of spaces. Cambridge University—particularly the Cavendish Laboratory—receives a significant amount of attention in the history of twentieth-century physics, as the home of groundbreaking ideas and as a training ground for a diaspora that would go on to set up "modern" research schools.[6] However, studies that have moved beyond the narrow but influential confines of Cavendish physics have shown how that which we now consider to be physics was often researched in other disciplinary spaces, particularly chemistry, and that research did not necessarily take priority over teaching.[7] Indeed, the primary responsibilities of many physicists lay in training electrical engineers, a community that far outnumbered academic physicists.[8] Lodge, as both a Fellow of the Royal Society and honorary member of the Institution of Electrical Engineers, spanned multiple professional networks.

It was not only "professionals" who formed Lodge's audience in the 1920s. The widely reported 1919 eclipse expedition created considerable appetite for accessible accounts of the "revolutionary" changes under way.[9] James Jeans and Arthur Stanley Eddington subsequently sold millions of copies of their popular physics books, explaining "modern" physics to a large audience.[10] Lodge was ever present in this popular science "boom," but his contributions are often dismissed or ignored. In Bowler's study of early twentieth-century popular science, Lodge appears as a well-known public figure, but attention is focused less on his influence than on his factual errors and inconsistency with contemporary scientific consensus.[11] With scholarship on popular physics in the period heavily skewed toward the dissemination of "modern" ideas, Lodge is often sidelined.[12] While studies of popularization would seem to be his natural home, he is largely absent from accounts because, as is so common with Lodge, he does not quite fit the historiographical mold.

LODGE THE PROFESSIONAL

By the 1920s, Lodge was editor of the *Philosophical Magazine*, had served as principal of Birmingham University for two decades, had been elected a Fellow of the Royal Society, and had received a knighthood. On paper, he was certainly not an outsider, but he has retrospectively been treated as one. By examining Lodge's treatment at the hands of his peers, I consider the space he occupied in physics during this decade and what authority he held.

While Lodge's professional credentials remained firmly in the mainstream, his scientific views were diverging from those of his peers. As Lodge's skepticism of many aspects of relativity theory persisted, the theory was becoming more firmly established. The 1919 eclipse expedition brought wider attention to the theory, while Eddington was also working to entrench relativity theory within Cambridge pedagogy. By the mid-1920s, his course on the subject had been recognized as an official component of the Mathematical Tripos, and his 1923 book on *The Mathematical Theory of Relativity* was becoming a standard textbook in the field.[13]

While Lodge did not reject relativity theory outright, he interpreted it within a worldview of physics that firmly placed the ether in the center. The retention of the ether was not, in itself, necessarily a controversial position to take, and both Eddington and Einstein maintained their support (although many did not).[14] However, Lodge did differ from the majority of his peers in the extent of his belief, and the nature of his ether. For Eddington, the ether was entirely nonmaterial, an alternative to an absolute void in space. Lodge, on the other hand, not only maintained the belief that we should be able to observe motion through the ether but insisted that a national laboratory should take responsibility for carrying out such an experiment.[15] For Lodge, the ether was a fundamental and central part of modern physics; for his contemporaries, it was (at their most generous) a topic for debate.

And yet Lodge's views often were presented without any obvious indication that they strayed significantly from the mainstream. In February 1921, in the wake of growing interest in Einstein's theory of general relativity, *Nature* published a special issue dedicated to the topic. Lodge was one of the "leading authorities" invited to contribute, alongside such eminent physicists as Eddington, Jeans, and Norman Campbell.[16] From our current vantage point, Lodge's position in a group of relativity experts seems incongruous. Indeed, his article was focused on the Michelson-Morley experiment that had sought, thirty-four years prior, in 1887, to detect relative motion through the ether.[17] Lodge's focus on an experiment that was then three and a half decades old is particularly notable considering that he was writing less than two years after the famed eclipse expeditions that had set out to provide observational confirmation of relativity theory. (Eddington, Frank Dyson, and Andrew

Crommelin, writing alongside Lodge in the *Nature* issue, had all been involved in these expeditions). Furthermore, Lodge's article argued that the results of the 1887 experiment could be explained within the framework of ether physics, and that the drastic changes proposed by relativity theorists were certainly not necessary. We have thus a fairly staunch opponent of relativity theory, discussing ether physics and a Victorian experiment, in the 1921 "Relativity" issue of *Nature*. Furthermore, it would seem that the editor was so pleased to have Lodge involved that he allowed him to write an article roughly twice the length of the others in the issue.[18] This is not the treatment of a scientist on the fringes.

Lodge was given a more public platform by his peers when he was appointed vice-chair of the Royal Society's organizing committee for a display of "pure science" at the British Empire Exhibition in 1925. The exhibition, intended to display the very best of British science and industry, was held in Wembley from April to November 1924, before being reopened for a further six months in 1925.[19] The "Pure Science" display had sections on various types of physics, including precision measurement, geophysics, and atmosphere physics, as well as biology. At its focus was a display of research into the structure and behavior of the tiny particles that make up matter, with contributions from the Cavendish Laboratory.[20] Lodge had no involvement with the initial run of the exhibition, but in 1925 was given the lofty position of vice-chair, despite making it clear on appointment that he would be far too busy to attend any meetings.[21] Lodge's role was to write an introduction for the publication *Phases of Modern Science*, which was to be produced concurrently with the exhibition.[22]

Here Lodge featured alongside even more prestigious names than he had in *Nature*, including Ernest Rutherford, William Bragg, and J. J. Thomson.[23] His contribution was the opening piece, a general survey of radiation, which, as was to be expected, discussed matter and energy within the framework of the ether. And, again, Lodge's contribution was considerably longer than anything else in the publication. To a contemporary observer, Lodge appeared as an authority on cutting-edge physics, verified by his peers.

This approval was not limited to physicists, and Lodge also appears to have had an ally in the eminent zoologist Peter Chalmers Mitchell, a Fellow of the Royal Society who also served as Secretary of the Zoological Society of London from 1903 to 1935.[24] In addition, Mitchell had a strong interest in science writing, and after several years of writing zoological articles for the *Times*, was made a special writer on science in 1919 and given his own weekly column in March 1921. "Progress of science" was intended to be both "intelligible to the educated public" and "not distasteful even to the specialist in his own subject."[25] Here, Mitchell discussed a variety of topics, covering biology, physics, astronomy, chemistry, and technology, as well as broader considerations of the funding and management of science.

Throughout the 1920s Mitchell corresponded frequently with Lodge, predominantly on the subject of spiritualism. He did not agree with Lodge's views on the subject, describing himself as a "materialist," but engaged in friendly debate.[26] As is evident from the frequency in which Lodge appeared as an authority in the Progress of Science column, Mitchell viewed him as a reliable authority on physics. In an article on the transmutation of metals, Mitchell recalled "a famous dictum of Sir Oliver Lodge" on the amount of energy contained within a piece of chalk.[27] Discussing low-temperature research, Mitchell referred to a recent suggestion by Lodge that the changes in properties of matter at very high temperatures might also occur at the opposite end of the scale.[28] When writing on discussions concerning hydrogen as the "primitive element," Mitchell used Lodge's explanation on the matter, which "brings in the most difficult theories in modern physics."[29] Despite having no research experience in these areas, Lodge appeared as an authority. Indeed, in a 1925 article on the radioactivity work under way in Cambridge on the disintegration of matter, Mitchell quoted Lodge's discussions of Rutherford's work, rather than going to the source itself.[30] Lodge was repeatedly treated as a general expert on "modern" physics.

Mitchell was a Fellow of the Royal Society, but with regards to physics he was something of an amateur. In a column on quantum theory, he wrote that, in an attempt to become acquainted with "the conceptions of modern physicists," he had been "reading once more two books written by two very distinguished physicists, Sir Oliver Lodge's *Atoms and Rays* and Mr. Bertrand Russell's *The ABC of Atoms*."[31] Neither Lodge nor Russell was a researcher into quantum theory, and indeed Russell was not really a physicist but rather a logician and mathematician. As a nonphysicist, Mitchell no doubt found Lodge's expositions more accessible than more in-depth offerings, such as Edward Neville da Costa Andrade's 1923 textbook, *The Structure of the Atom*.[32] Furthermore, his friendship with Lodge meant that he was also more inclined to include Lodge's work in the *Times*. In 1923 he informed Lodge that he had "succeeded in persuading the *Times* to put in a small adaptation of part of your lecture [to the Roentgen Society]."[33] One of Mitchell's columns in 1927 was devoted entirely to Lodge's address "A Century's Progress in Physics," which had been delivered as part of University College, London's centenary celebrations.[34] Through his relationship with Mitchell, Lodge was able to further establish his reputation as an authority on "modern" physics.

Throughout the 1920s, then, it would appear to most bystanders that Lodge was an active and respected member of the scientific community. He was repeatedly presented as an expert, in *Nature*, at the British Empire Exhibition, and in the pages of the *Times*. He appeared alongside eminent figures: researchers into radioactivity, champions of Einstein, and Cavendish experimentalists. In the case of the *Times*, he often appeared *instead* of such men, a single authority on physics. In such places, Lodge was situated not outside of or on the fringes of scientific consensus, but square in the middle. His views were thus not only heard but given implicit approval by those at the very top.

LODGE THE POPULARIZER

Against this backdrop, the popular books that Lodge wrote in the 1920s were already primed to be received as those of an expert authority. The letters that appeared next to his name in print were an additional stamp of approval. But an additional reason to take Lodge's writings seriously came again from his fellow physicists.

In 1924 Lodge published *Atoms and Rays*, an "introduction to modern views on atomic structure & radiation," with publisher Ernest Benn.[35] An anonymous reviewer for the *Yorkshire Post* read the whole book in almost a single sitting, remarking that "Sir Oliver Lodge could hardly have found a subject more congenial to his pen than the one he has attacked so fluently and understandably in his latest popular exposition."[36] This reviewer was particularly interested in the book's intersections between physical and psychical research, an aspect absent from the *Observer*'s review of the book. The *Observer* review was written by Andrade, a professor of physics at the Artillery College in Woolwich, and some thirty-six years Lodge's junior. Andrade was well placed to review the book, having published his own, less populist book, *The Structure of the Atom*, the previous year.[37] He also had considerable professional experience in this aspect of physics, having completed a stint at the Cavendish early in his career, before spending a brief period from the autumn of 1913 in Ernest Rutherford's Manchester laboratory, conducting research on the gamma rays emitted by radium.[38]

In the *Observer*, Andrade discussed at length the "New Physics," which he believed had to some extent done away with the certainty of the previous century. While much of this work served to "increase the scope of older conceptions and simplify older problems," some research had added knowledge that was causing "grave embarrassments" and paradoxes awaiting solution.[39] He described the ether, which was absent from *The Structure of the Atom*, as the "New Poor, gallantly trying to adapt an organization designed for an older society, governed by tried conventions and approved procedure, to fulfill the demands of a critical age, that will not be dissuaded from destruction by the consideration that it has nothing that functions so well to put in the place of the overthrown" (5). The ether, of course, features heavily in *Atoms and Rays*, and Lodge ended his book by noting that much of what he had discussed in the final two chapters belonged "more to a work on the Ether than to a book on Atoms and Rays, though the whole subject is so interlocked that discrimination is rather arbitrary."[40] Despite this, Lodge did not view his book as departing significantly from Andrade's in terms of subject matter, and in his preface explicitly described the work as an introductory volume to Andrade's more advanced study (vii).

After discussing the key features of the New Physics, and casting aspersions on the ether, Andrade's review discusses the structure of the atom, knowledge of which was "key to the problems of modern physics." This is followed by a lengthy commendation of Lodge's book: "It is in the power of making striking generalisations and seeing recent things in a proper perspective that Sir Oliver Lodge excels. His object in this book is to expound, in a way comprehensible to the average educated man, some of the essential features of the physics of to-day; and he brings to the task that freshness, charm, and polished simplicity of style which make his extempore speeches so delightful a feature of any scientific gathering."[41] While Andrade praised Lodge's skillful explanation of various aspects related to the quantum nature of the atom, he did have criticisms of the text; in particular, Lodge's reference to C. G. Barkla's experiments on the scattering of alpha particles, which Andrade suggested could "only be due to a printer's error of a complicated kind, as Barkla never worked on this subject" (5). In fact, as established in subsequent private correspondence between Andrade and Lodge, this was not a printer's error but rather Lodge's own confusion of X-rays with alpha particles.[42] Of course, this clarification did not reach the readers of the *Observer*, confined as it was to personal letters between the two physicists. Thus for the reader, Andrade's criticisms were, as he himself wrote, "minor points," which it would be "ungracious" to mention "without paying homage to the masterly way in which Sir Oliver generally puts the essence of a difficult problem before the reader."[43]

Andrade's main concern with Lodge was his "unorthodox" approach, of which he was keen to warn the *Observer* audience. In *Atoms and Rays*, Andrade notes, we see a "constant reference of everything back to the ether," whereas we "know practically nothing about the ether with any certainty, except, as Einstein has justly pointed out, that it has not got mechanical properties, which rather spoils its usefulness" (5). Thus Lodge's book contained many "controversial matters," which Andrade noted were certainly welcome, provided that readers were able to differentiate "the certain and the less certain" (5). The review was a mixed one, placing Lodge certainly outside of the scientific status quo, but not to the extent where he should be ignored. Andrade heaped praise on Lodge's writing, strongly recommending the book, albeit with several caveats. What emerges is a sign that perhaps the classical/modern physics divide was not as distinct, or important, as we might now think. As a result, Lodge's etherial unorthodoxies did not exclude him from a position among more "modern" physicists.

Indeed, a few years later both men contributed to Benn's "Sixpenny Library," a series of booklets covering both scientific and nonscientific topics, which the publishers intended to provide a "reference library to the best modern thought, written by the foremost authorities, at the price of sixpence a volume."[44] Lodge's *Modern Scientific Ideas* and Andrade's *The Atom* were reviewed alongside each other in *Nature* and the more populist science magazine *Discovery*, with no suggestion that either book was any lesser than the other.[45] In *Discovery*, V. E. Pullin, director of radiological research at the Royal Arsenal in Woolwich (an institutional neighbor of Andrade's Artillery College), described *Modern Scientific Ideas* as "an excellent preamble" to the more specialized books in the series. Pullin declared that to "acclaim Sir Oliver Lodge as an expounder of modern science would be to gild the lily."[46]

Lodge was entrusted to guide the public through developments in physics, providing for them a clear introductory overview, before handing them over to a more advanced study. He appeared in the prestigious pages of *Nature* as both a contributor to a serious scientific discussion on relativity and the recipient of praise for his popular book on atoms. While such approval from Andrade was offered only with caveats, it was approval nonetheless. As a popular writer, Lodge sought to present himself as a trusted expert, and he was assisted in this venture by the endorsement of his peers.

LODGE THE ELECTRICIAN

Lodge's audience spanned a wide range, from the educated layman to the academic physicist. He spoke to both the "public" and the "professional," but the lines between these two categories are often blurred, particularly in the case of electrical engineers. This community was eager to learn about changes in a discipline closely related to theirs, and the ether's connection with modern wireless technologies made it a popular topic. As Jaume Navarro has shown, wireless amateurs, electrical engineers, and inventors maintained a healthy interest in the ether throughout the 1920s and early 1930s. As the ether's relevance to academic physics was waning, the introduction of the radio receiver into British middle-class homes created a demand for accessible explanations of the mechanisms behind new wireless technologies, and the ether provided a suitable foundation for commonsense descriptions of these inner workings.[47] Thus, outside of the elite circles of academic physicists, Lodge found alternative audiences who were very responsive to his particular worldview.

In 1925 Lodge's *Ether and Reality* was published by Hodder and Stoughton.[48] The book, comprising a series of recent lectures delivered by Lodge over the wireless, received a glowing review in *Wireless World*.[49] The reviewer was already familiar with the contents, having heard the original lectures, and indeed doubted whether there was "any listener who did not appreciate and enjoy Sir Oliver's clearly expressed, thoughtful, and practical discourses," but nonetheless continued to "read and re-read this book." For, the more they read, the "more channels of thought it opens out. There are no learned technicalities, only to be comprehended by the initiated, but every phase of the mighty subject is expressed in plain language which should appeal alike to the man of average intelligence and to the physicist, philosopher, poet or theologian" (16).

Lodge's widely acknowledged powers of clear exposition made complicated physics palatable to an audience that extended far beyond the academy. But it was not merely his delivery but also his content that appealed to engineers, who had begun to feel alienated by the changes under way in physics. This is evident from the response, in the technical magazine *Engineer*, to Ernest Rutherford's 1932 Hawksley Lecture to the Institute of Mechanical Engineers. The magazine printed the lecture, "Atomic Projectiles and Applications," accompanying it with one of that issue's leading articles, "Modern Physics and the Engineer" (the other being on trade with New Zealand).[50] Here, the author decried the uncertainty of modern physics, the seeming lack of stability caused by abandoning Newton's previously fundamental truths. While they had heard much in Rutherford's lecture about the progress of physics, the author "listened in vain to hear Lord Rutherford tell us something about its fundamental difficulties." Would Rutherford assure his audience that, having discarded the luminiferous ether "because of the intolerable complexities which its supposed existence entailed," the physicist had now "reached the open sea and plain sailing?" Was there now a new physics that would work "consistently and smoothly without absurd contradictions and without the necessity for doing violence either to mathematics or common sense?" (485).

From the perspective of *Engineer*, Rutherford's lecture was lacking in this respect, leaving engineers to form an opinion on these points only from their "own interpretation of the statements made to us by physicists." And such an interpretation led the author to doubt that modern physics contained "the elements of a consistent and harmonious conception of the universe" (485). Instead, it appeared doubtful that the relativity formula could successfully apply in cases where the Newtonian was already successful, and there were "many other aspects of modern physics which seem to require explanation before they can carry conviction to the ordinary person." "If modern physics is to maintain any connection with practical life, if it is to avoid isolating itself as a department of science dealing purely with abstractions and speculations, it is essential that it should explain itself in terms comprehensible by the ordinary electrical and mechanical engineer. It is they who would be called upon to apply whatever discoveries it might make" (486). The overall tone of this account of Rutherford's lecture was one of mistrust and confusion. Here, Rutherford did not appear as the trusted expert he did elsewhere. Engineers seemed to feel they had been cut out of the conversation with the move from Newton to Einstein. They might one day be required to use the new physics in their work, and they did not feel comfortable doing so when they did not understand it. They wanted a clear explanation, and physicists such as Rutherford were not providing them with one.

Two years later, the same publication reported on James Jeans's presidential address to the British Association.[51] Again, this was preceded by a leading article, this one titled "The Old and the New Physics," that decried the growing separation between physicists and engineers.[52] For while the "new physics indubitably presents us with a fresh basis for the interpretation of life and Nature," in order for it to "influence progress in the physical sciences it must emerge from the interpretative stage and demonstrate that it possesses the qualifications of bold and inspiring leadership" (237). The author doubted it would be able to do so, believing that "its transcendentalism will deprive it of practical value": "The fluids, billiard balls, jellies and spinning tops of the old-fashioned physicist, the whole materialistic conception of the universe on which he rested his faith, may be illusory and false. But we have a feeling that their falseness is that of the falsework on which engineers erect actual structures and that the human mind, being as it is, will always require their aid to help it towards fresh achievements in the physical world" (237). Engineers were crying out not just for a clear exposition of "modern physics" but for one that spoke to them in their own terms, the language of tangible realities. How were they to apply the new physics of abstractions to the everyday world in which they worked? "To what discoveries will the new physics lead?" (237). If anybody were to answer this, it was to be a physicist who could speak the language of the engineers, and nobody was more suited to the task than Oliver Lodge. Furthermore, Lodge provided a more comforting version of events. The reports of both Jeans's and Rutherford's talks began with a hark back to the nineteenth century, and an observation of how alarmingly unfamiliar present-day physics would seem to a scientist from that era. But Lodge *was* a Victorian physicist, and his version of modern physics did not require such a revolutionary shift in beliefs as demanded by his contemporaries. He could serve as a reassuring connection between the past and the present.

THE LEGACY OF OLIVER LODGE

On August 23, 1940, the *Times* announced the death, the previous day, of "Sir Oliver Lodge, the famous scientist."[53] Three pages on, an obituary described him as "one of the most distinguished scientists of the age," listing his many awards and honors, his achievements in engineering and physics, his gifts as an expounder of knowledge, his success at the University of Birmingham, and (somewhat less enthusiastically) his work on spiritualism.[54] Similar dedications followed in the *Philosoph-*

ical Magazine, *Electrical Review*, and the obligatory *Obituary Notices of Fellows of the Royal Society*.[55] At the end of that year, Sir William Henry Bragg delivered the presidential address to the Royal Society. Before discussing his wider topic, "science and national welfare," Bragg began with a tribute to two recently deceased men of science, Oliver Lodge and J. J. Thomson, who had died a mere eight days apart: "The present generation is quick to honour the names of J. J. Thomson and Oliver Lodge; but they cannot remember, as we older men can, the brilliant years when these men and their contemporaries were writing the chapter's first pages. What they wrote was eagerly read, their lectures were heard with rapt attention; they were the pioneers, and the men of science of that time, nearly half a century ago, streamed after them."[56] In the years and decades following their deaths, the histories of Lodge and Thomson have diverged considerably, with the former taking on the character of a Victorian scientist out of touch with modern developments and the latter depicted as one of the great forerunners of modern physics. But in 1940, the similarities between the two men were still evident. They were of the same generation, and both maintained a belief in the ether, although Thomson was less outspoken in this regard.[57] With Thomson's death occurring a little over a week after Lodge's, joint accolades to the two of them were inevitable. Thomson's obituary in the *Times* lamented that his death "coming so soon after that of Sir Oliver Lodge—robs the country of another of its most famous scientists."[58]

Nonetheless, differences between the statuses of the two scientists were apparent in the weeks following their deaths. Lodge's funeral was held at Wilsford Church in Salisbury and attended by (among others) Alfred Egerton as a representative of the Royal Society and Royal Institution, and Allan Ferguson representing the British Association, the Physical Society, and the Institute of Physics. An impressive sendoff, but not quite the burial at Westminster Abbey afforded to Thomson. For Thomson's funeral, Bragg was a pallbearer and a representative for Winston Churchill was in attendance. This was perhaps fitting of a man whose institutional connections were more prestigious than Lodge's, having been Cavendish Professor of Physics at Cambridge from 1884 to 1919, and Master of Trinity College, Cambridge, for the last two decades of his life. A strong Cambridge presence provided Thomson with certain connections less readily available to Lodge: the day before Thomson's funeral, his wife received a message of sympathy from George VI, who remembered the recently deceased "well from my Cambridge days."[59] Furthermore, while both Lodge and Thomson were Fellows of the Royal Society, Thomson had also served as president, from 1915 to 1920.

But a more significant hint as to the posthumous legacies of the two men can be found in the opening words of their *Times* obituaries. Where Lodge's was titled "A Great Scientist," Thomson's tribute began with a more specific reference, "The Discovery of the Electron." By the time of his death, it was common knowledge that Thomson had "discovered" the electron in 1897, a narrative that persists to this day. The actual version of events is somewhat more complicated. Thomson did indeed find experimental evidence for "corpuscles," but he did not conceptually connect them to contemporary theories of the electron, nor was he motivated in his experimental work by such considerations.[60] As Isobel Falconer has shown, not everybody originally conceived of Thomson as pivotal to the history of the electron, with Walter Kauffman's 1901 lecture titled "The Development of the Electron Idea" affording him only a minor role. Fortunately for Thomson, an alternative account was produced by none other than Lodge. Originally delivered in a series of lectures to the Institution of Electrical Engineers in 1902, Lodge's thoughts, and Thomson's prominent contribution, were more permanently cemented in book form in 1906.[61] Falconer argues that it was this account that was responsible for the discovery narrative, subsequently spread by Thomson's students and physics textbooks.[62] By the time of Thomson's death, his history had already been written.

In Geoffrey Cantor's chapter in *Telling Lives in Science,* he considers how Michael Faraday's life has been rewritten to suit the agendas of biographers. He notes that while Faraday's life was originally valued for "either its Romantic connotations or as a paradigm example of self help," more recently the emphasis has been on his discoveries, and the technological and scientific progress to which they contributed, with his biography "constructed in the service of technology and the powerful electrical industry."[63] Thomson was thus well placed to be remembered: long before his death, his position in the history of physics had been established, within a clear framework of progress. He himself perpetuated this, with efforts to present himself as among the first of the modern physicists, not the last of the classical physicists.[64] He also used the electron as an example of the unexpected practical applications of academic research, thus situating his work in a line of both intellectual and technological progress (135). However, when a biography must connect the past to the present in such a way, Lodge becomes a difficult figure to write about.

That is not to say that it has not been attempted. In the 2014 *Biographical Encyclopedia of Astronomers,* the entry on Lodge describes how he "was able to show that the (then widely accepted) ether could not be carried along by moving bodies, and this can be seen as a step toward Albert Einstein's theory of special relativity."[65] This is, at best, a stretch. Furthermore, such teleological accounts obscure Lodge's actual significance in early twentieth-century physics, that of an educator and public authority. I would argue that we learn more about Lodge, and physics during this period, from his 1902 lectures to the Institution of Electrical Engineers than from his nineteenth-century work on the ether. Here, Lodge's skill as an orator, ever-increasing public presence, and ability to frame a topic in a manner relatable to an audience of engineers (through the familiar technology of a vacuum tube rather than abstract reasoning) contributed to the spread of an enduring tale.[66] In this respect, he had considerable power. If we are to argue for Lodge's importance in the history of physics, it is not on the evidence of discoveries or developments but rather on the basis of his work disseminating contemporary ideas. And Lodge's considerable position as a public educator deserves particular attention because of his—in the words of Andrade—"unorthodox" approach to the period. Lodge did not act as a conduit from the academic physics community to wider publics; he actively steered the conversation, in ways now forgotten.

If we try to frame Lodge within a narrative of progress from the past to the present, we quickly become hindered by the man himself. He argued for a continuing prominent place for the ether in modern theories, downplayed the potential effects on the discipline of relativity and quantum theories, and ultimately advised a large audience that everything would eventually "settle down," with business almost as usual. Lodge cannot be considered a champion of modern physics, not with respect to our current conception of the subject. Rather, Lodge is crucial *because* his account was so very different from that we now tell ourselves, and because that account was heard, and trusted, by so many.

* * *

Is there a significant place for Lodge in our histories of early twentieth-century physics? Or rather, have we been neglecting him to our detriment? This reassessment is about more than simply giving Lodge his "rightful" place in history. It is not an attempt to add another "great man" to the history books. It is instead about constructing a more comprehensive understanding of early twentieth-century physics, one that moves beyond the key actors who played a part in developing what we now see as "modern" physics. A history of early twentieth-century physics that omits Lodge presents a narrow and incomplete picture of the period.

Throughout the 1920s Lodge's views were widely heard, and endorsed by his peers. His depictions of the state of physics were disseminated not just through his own popular books but also in professionally "accredited" spaces. Thus he provides us with an insight into what physics looked like to the majority of observers during the period, and it is not necessarily the physics of Eddington or Jeans. Furthermore, his popularity suggests a large public appetite for what he was offering, a less "revolutionary" approach to changes in the discipline, a crucial connection to the past. Through Lodge we see an alternative "modern" physics, where the developments of the early twentieth century are incorporated into an etheric worldview. This is vital for how we approach early twentieth-century public and professional conversations about physics. The treatment of Lodge by his peers is evidence against a clear divide between "classical" and "modern" physics. It suggests a period of flux, with the old and new sitting side by side. Lodge was not dismissed as an out-of-touch Victorian clinging to the past. He was given a platform and welcomed in professional spaces. This observation extends to our studies of the popularization of physics. It should not be taken as given that a "classical" physicist such as Lodge is of little importance to a study of early twentieth-century popular relativity theory. Nor should relativity theory necessary take center stage over older ideas, such as the ether, which continued to occupy an important space in discussions.

Finally, the case of Oliver Lodge provides us with guidelines for how better to write our histories. We must consider the "losers" as well as the "winners." We must not allow our historical judgment to be clouded by present-day preconceptions. We must move beyond the academic elites in Cambridge, and look for the less prestigious voices. We must consider not just the practitioners, those who developed the theories and carried out the experiments, but also anybody playing a part in the conversation. These are things we already do, often, but it is easy to forget. And Lodge, out there somewhere in the ether, is our gentle reminder.

PART THREE

SCIENCE, SPIRITUALISM, AND THE SPACES IN BETWEEN

ALTHOUGH INITIALLY skeptical about spiritualism, Oliver Lodge became one of its most visible and respected adherents. He joined the Society for Psychical Research (SPR) in 1884, a year after its foundation, and became its president in 1901, stepping in after the death of his friend Frederic W. H. Myers. Its empirical methods appealed to him and he brought scientific rigor to his psychical investigations. However, as the following chapters demonstrate, for Lodge psychical research was an integral part of his broader project to understand what he called the "imponderables," those phenomena on the edge of the tangible universe necessary for its physical operation. The most significant of these was the ether, the perfect yet elusive medium that accounted for electromagnetic effects of various kinds while also providing a possible explanation for other, more occult modes of connection. As such, there were not just points of correspondence between his science and spiritualism but clear continuities.

Spiritualism opened up other worlds for Lodge. On the one hand, it gave him access to realms beyond the physical, allowing him to study the ways in which people were connected, whether living or dead. On the other, it allowed him to meet and collaborate with a much wider social set. His involvement with the SPR introduced him to Myers and the rest of the Cambridge group with whom he was associated. It was through his psychical work that he knew Percy and Madeline Wyndham, getting to know their family and spending Easters at their house, Clouds. The Wyndhams were at the heart of the aristocratic clique known as The Souls, and it was through this set that Lodge became friends with Arthur Balfour, the Conservative prime minister from 1902 to 1905. Lodge's interest in spiritualism also introduced him to J. Arthur Hill, a spiritualist from Bradford with whom he corresponded throughout much of his later life.

The chapters that follow set out the connections between Lodge's scientific work and his psychical research, and how they shaped his philosophy overall. They explore how Lodge established a reputation as a cultural authority, able to communicate the latest developments in each field and their wider significance. And finally they investigate how Lodge's work informed other areas, whether discussions of shell shock, the church's response to spiritualism, or the etheric ideology of the early British Broadcasting Corporation. Although Lodge was always conscious of how his psychical research would be received, whether by his scientific peers or the broader public, it was nevertheless integral to the way he understood the world.

EIGHT

GLORIFYING MECHANISM

OLIVER LODGE AND THE PROBLEMS OF ETHER, MIND, AND MATTER

RICHARD NOAKES

THE YEAR 1900 marked a turning point in Lodge's career. He resigned his professorship at University College Liverpool to take up the distinguished position of principal at the newly formed University of Birmingham. The administrative burden of the new position proved much heavier than Lodge had anticipated, and left him with little time for the experimental research on wireless telegraphy and psychical research that he insisted on pursuing as part of his agreement to come to Birmingham.[1] He may have had less time than he desired for new experimental work, but an analysis of his bibliography demonstrates that he found plenty of time for writing, and on a far broader range of subjects and for a greater diversity of readerships, than he had managed at Liverpool.[2] Now that he had established himself as one of the leading British scientists of the day, Lodge was an attractive proposition to editors and publishers who offered him a plethora of opportunities to engage more openly and extensively with the psychical, religious, and philosophical questions that had preoccupied him for over a decade.[3]

Central to these questions was his conception of the ether, the hypothetical, invisible, space-filling medium that had long been regarded as necessary to understanding the transmission of energy and force through space but whose elusive nature had been the subject of much scientific debate for decades.[4] By 1933 Lodge could claim the ether as his "life study" and a defining aspect of his "philosophy," but, as he elaborated both its nature and place within his thought, Lodge confused many of his critics.[5] On the one hand he insisted that the ether had inertia, density, and other properties associated with ordinary matter; on the other hand he proposed that it had to be radically different from ordinary matter and that this quasi-material, quasi-mechanical substance was, accordingly, a possible mediator between matter and the immaterial domains of mind and spirit. Yet for some this ether still seemed too "materialistic" and a contrast to the "anti-materialistic" point of view that Lodge supported from his psychical investigations.[6]

I argue that Lodge's ether-based philosophy was deterministic, but it was a determinism that embraced spirit and mind as well as matter in one seamless chain of cause and effect. He certainly had moved on from the *materialistic* determinism apparently propounded by the scientific naturalists who taught and impressed him as a young science student in the 1860s, but he never relinquished a belief in a fundamentally causal cosmos reinforced by most Victorian scientists, whether scientific naturalists or the devoutly Christian physicists who became his intellectual heroes from the 1870s. Indeed, in many ways Lodge's philosophy reflected the fact that, as Matthew Stanley has so eloquently shown, his scientific mentors and heroes shared more principles and values than their apparently different worldviews suggested: although scientific naturalists and "theistic physicists" disagreed profoundly on the relationships between religion and science, they agreed that the sciences were built on such principles as uniformity of nature, intellectual freedom, and a recognition of the limits of science.[7]

CHALLENGING MATERIALISM

Among the questions Lodge took up soon after arriving at Birmingham was the materialistic "tendency" of the scientific world of his youth, a tendency that in 1900 he believed "leading physicists" were now vigorously and successfully opposing with new, electrical, and etherial theories of matter.[8] Lodge's critical writings on this question constituted an important part of the framework within which he would try to explore the ways in which some kind of etherial mechanism could allow a material cosmos to be guided by purposeful immaterial agencies. Like so many scientists writing in the early twentieth century, Lodge believed that nineteenth-century materialism was best represented by the Anglo-Irish physicist John Tyndall's notorious address to the Belfast meeting of the British Association for the Advancement of Science in 1874. Few passages had caused more controversy than Tyndall's declaration that brute matter contained the "promise and potency of all terrestrial life" and defense of the universal domain of the doctrine of energy conservation, a doctrine "which 'binds nature fast in fate,' to an extent not hitherto recognized, exacting from every antecedent its equivalent consequent, from every consequent its equivalent antecedent, and bringing vital as well as physical phenomena under the domain of that law of causal connection."[9]

Tyndall's statement captured what many critics believed were the worst aspects of scientific naturalism. As Bernard Lightman has argued, scientific naturalism was the "English version of the cult of science" pervading late nineteenth-century European culture, which championed scientific methods, ideas, and theories of matter, energy and biological evolution as the ultimate basis for understanding the cosmos, including "humankind, nature and society."[10] Scientific naturalists fully accepted the reality of consciousness and volition but sought to regard them as attributes of life that were strictly limited to what psychophysiological laws permitted. For many, the heavy restrictions on free will implied by Tyndall's address were tantamount to materialistic determinism.[11]

Lodge was among Tyndall's auditors, and in 1931 he recalled the atmosphere becoming "more and more sulphurous as the materialistic utterances went on and on in that strongly Protestant atmosphere of Northern Ireland."[12] In 1874, however, Lodge may well have been more sympathetic to Tyndall's philosophical position than this retrospective view implies. His scientific education owed a good deal to the leading scientific naturalists: he attended Tyndall's lectures at the Royal Institution in the 1860s and early 1870s, took T. H. Huxley's lessons in biology at the Royal College of Science in 1872, and studied mathematics under W. K. Clifford at University College, London in 1874–1875, and he often remembered these formative scientific experiences as being inspirational.[13]

It was only after the 1880s, when he developed his lifelong interest in psychical research, that he developed a more critical perspective on his naturalistic scientific education. In 1884 he joined the Society for Psychical Research (SPR), a London-based organization that sought to apply the methods of the sciences to a host of obscure psychological and psychophysical phenomena, including thought reading, apparitions, mesmerism, spiritualist mediumship, and haunted houses.[14] Lodge's earliest original contributions to the SPR's swelling body of research persuaded him that the human mind could exist independently of the material brain. These included the positive results of investigations into thought reading, or what Frederic W. H. Myers in 1882 christened "telepathy" (the capacity of one individual to perceive images, words, and other impressions in the mind of another person independently of known sensory channels), and, in 1889, the results of his study of the trance utterances of the American spiritualist medium Leonora Piper. Piper had impressed no less acute an observer than the American psychologist and philosopher William James with her apparently genuine capacity to commune with invisible spirits of the dead.[15] During séances with her at Myers's house in Cambridge and his own house in Liverpool, Lodge spoke to disembodied personalities professing to be deceased relatives and who revealed information about themselves that, as far as Lodge was concerned, Piper could not have obtained even by fraudulent methods (such as researching Lodge's family tree or "fishing" for clues in his answers to questions) and that he later verified with those relatives' living descendants. Having convinced himself that he had taken every precaution against fraudulence and self-deception, Lodge declared publicly in 1889 that while some of Piper's revelations could be explained in terms of thought transference, others were achieved by "none of the ordinary methods known to Physical Science."[16] Lodge later revealed that the Piper séances started him on the path toward a belief in survival of the soul following bodily death and of the capacity of disembodied souls to communicate; but in 1889 he kept such convictions private, undoubtedly to keep his distance from those who exemplified this position—the spiritualists—and who were frequently lambasted by his professional scientific colleagues as dupes and fraudsters.[17]

In 1894, a few years after the Piper séances, Lodge investigated another class of psychic effect that further challenged the materialistic idea that the mind could not exist independently of the material brain. Christened "telekinesis" by Myers, this was the apparent capacity of conscious or unconscious will to move objects without the use of any known form of physical contact. Lodge gained his first and most substantial experiences of the effect during séances with the Italian medium Eusapia Palladino, and although he was convinced that some of her startling physical effects (including levitating and playing a musical box at a distance) suggested the operation of new power, damning evidence of her trickery produced by SPR colleagues in 1895 persuaded him to share his personal convictions only with fellow psychical researchers and close colleagues and friends.[18]

Looking back in his 1931 autobiography *Past Years,* Lodge claimed that the "tuition" of Myers, that dominant figure in the early SPR, was what proved so decisive in his rejection of the key materialistic belief that he had held during his student days: the inseparability of mind and body.[19] Lodge regarded Myers as "among the chief influences of my life" and as someone who, despite a lack of formal scientific training, showed "many of the faculties and instincts of a man of science, combined with such a mental grasp, vivid imagination, and power of expression, as would put most of us [scientists] to shame."[20] Myers and Lodge corresponded at least once a week for over twenty years and frequently met at SPR meetings and each other's homes.[21] The correspondence testifies to the enormous respect the two men had for each other: Myers for Lodge's scientific knowledge and judgement, and Lodge for Myers's scientific "instincts" and imagination.

By the early 1890s Myers had long been preoccupied by the extent to which the feats of inspiration shown by poets, saints, and seers revealed vast, complex, but hidden strata of human personality that had been shown more dramatically in psychical research's studies of hypnotism, telepathy, multiple personality, and spiritualist mediumship.[22] Having accepted by this period that psychical research lent strong support to the Christian ideas of prayer, duty, immortality, and God, Myers undoubtedly laid the foundations of Lodge's own attempts from the early 1900s onward to update, strengthen, and liberalize Christian doctrine in the light of psychical research, physics, and biology.[23]

In the years immediately after Myers's death in 1901, Lodge's reassessment of his naturalist mentors was, not coincidentally, at its most critical. In introductions to cheap new editions of Huxley's most popular books, Lodge praised the biologist for the battles he had successfully fought against "forces of obscurantism and of free and easy dogmatism," but criticized him for giving the impression that the "material side of things" was the "only side that mattered" or existed.[24] Ultimately, however, Huxley remained an important scientific icon for Lodge, not least because he had helped spread the idea of the uniformity of nature and because his materialism was only apparent: it was a pragmatic, successful methodological position rather than a "philosophy."[25]

In what proved to be a controversial entry on Tyndall for the tenth (1902–1903) edition of the *Encyclopaedia Britannica*, Lodge concluded that his former physics teacher had been far less able than Huxley at countering a materialistic approach in the laboratory with a broader-ranging cosmic philosophy. The simplicity, vividness, and "crude fearlessness" of Tyndall's "conception of physics" suppressed a frequently "reverent and religious spirit" and led him to make dogmatic statements about the efficacy of prayer and miracles and to the "materialistic position" of the Belfast Address, which failed as a "serious contribution to philosophy."[26] Lodge was arguably more sensitive to and critical of Tyndall because Tyndall posed a far greater threat and opportunity than did Huxley. Tyndall, who had died in 1893, had fulfilled many roles that Lodge now occupied—experimental physicist, physics popularizer, and contributor to philosophical and religious debates—but Lodge clearly felt that he could improve greatly on his mentor, not least in his overall "conception of physics."

EXTENDING THEISTIC PHYSICS

Lodge's rhetorical strategies for challenging Tyndall's "conception" drew heavily on the examples set from the 1860s by James Clerk Maxwell, William Thomson, George Gabriel Stokes, and other theistic physicists whom Lodge venerated. As Crosbie Smith and others have shown, these physicists were particularly hostile to Tyndall and scientific naturalism because scientific naturalism appeared to challenge their strongly held Christian beliefs regarding humanity and the cosmos: it seemed to be reducing spirit to matter and, in binding the fate of nature and humans to the conservation of energy, insisting that free will and divine agency were illusions.[27] Theistic physicists agreed with scientific naturalists that the laws of nature were uniform and that the cosmos was fundamentally causal, but vigorously denied that these effectively prohibited the agency of mind (whether human or divine) in the cosmos. In response, theistic physicists sought to show how the very physical laws and principles on which Tyndall rested his materialistic determinism left spaces for the possibility of mind and spirit.

A key argument of the older physicists was that energy conservation did not bind organic nature "fast in fate" and contradicted neither the common experience of free will nor the theistic belief in divine agency. It was possible for a material system to be subject to the directing or guiding action of an immaterial will or a soul without the violation of energy conservation in particular and the uniformity of nature in general.[28] To illustrate their argument, Maxwell and Stokes employed a fairly widely used metaphor of a railway pointsman. This compared the relationship of the human will and body to that of the intelligence of a railway pointsman or an engine driver to a locomotive, insofar as intelligence did not supply the motive force for changing the locomotive's direction (this force derived from coal), but it did start a psychophysiological chain of events resulting in this significant change. As Stanley has argued, the pointsman metaphor served a "critical" rather than "constructive" function, in that it warned men of science (and especially scientific naturalists) that with all the dynamical, thermodynamical, physiological, and other scientific knowledge about the pointsman and locomotive at one's disposal, it was still not possible to determine when and how the locomotive switched tracks.[29] Tyndall and other scientific naturalists had erred in extending the "dominion" of physical laws to psychological regions where they were not applicable.

In published essays in the late 1890s and early 1900s Lodge found himself in a similar position to the one held by Maxwell and Stokes thirty years earlier: he wanted to police unwarranted interpretations of physics with a view to safeguarding the domain of physical laws where they legitimately applied and to leave a space for life and mind. By showing that physical laws were not incompatible with the idea of purposeful guidance from immaterial agencies, Lodge was preparing the way for later speculations on what kind of cosmic mechanism might enable such an interaction. Many of these essays were presented to the Synthetic Society, a short-lived British forum of philosophical, religious, and intellectual debate where Lodge mixed with leading statesmen, philosophers, and theologians of the day.[30] In a contribution of 1903 he defined his position in opposition to two groups of thinkers who had radically different views on the relationship between physical laws and immaterial or "non-mechanical" agencies.[31] On the one hand were those who pressed the laws of physics to their extreme conclusion, and applied the conservation of energy "without ruth or hesitation," which resulted in the possibility of free will being negated.[32] On the other hand were many who, for reasons of faith, instinct, or experience, denied that volition and other nonmechanical agencies could be so excluded.

Lodge rejected both positions, which clearly belonged to the scientific naturalists and their theological and philosophical opponents. As far as he was concerned, the common experiences of volition, historical evidence of divine agency, and psychical research's evidence of disembodied minds defied the materialistic/deterministic position that many attributed to scientific naturalism. But as a professional physicist who had invested much time and effort into promulgating the foundations of the relatively young discipline, he was also critical of those, such as the statesman and philosopher Arthur Balfour and the psychologist James Ward, who attacked scientific naturalism in general, and the foundations of physics in particular, on the grounds that such laws failed to cover life and mind.[33]

Like Maxwell, Lodge's solution to the problem was to emphasize the modest scope of the laws that naturalists took too far and that opponents threatened to abandon in toto. In doing so, Lodge was also contributing to the wider debate about the limits of physics with an argument for the integrity of such laws.[34] Dynamical laws, he insisted, "do not exclusively cover the realm of Nature even on the material side."[35] By "material side" he meant biology and psychology, the regions where there appeared to be something vital or mental "operating consistently and legally no doubt, in no way upsetting dynamical laws, not in the least infringing the conservation of energy, but contriving interactions and initiating changes which in the region governed *solely* by dynamics would never had occurred."[36] Implicitly drawing on Maxwell's and Stokes's pointsman analogy, Lodge asked whether "solely dynamics" ruled the direction and departure of a steamboat, which was steered by a helmsman who was following orders from his captain, and who himself adjusted the power after consulting a timetable or thinking about the future trip.[37]

Lodge's argument that the physical universe (including mankind) could be guided or controlled by purposeful vital and mental agencies without interfering with dynamical and energy laws had profound religious implications. This is clear in his *Life and Mind* (1905), an attack on the first English edition of Ernst Haeckel's *Riddle of the Universe* (1900), in which Lodge savaged the eminent German biologist and philosopher for supporting a materialistic monism on what Lodge believed were misunderstandings of physics.[38] Haeckel had erred in assuming that life and mind were forms of energy inextricably associated with matter, and overlooked what physicists knew: that life and mind *guided* energy. Developing a well-known nineteenth-century analogy between human and divine will, Lodge argued that once we accepted that the human will could guide and control the "physical scheme" we could no longer deny the possibility of "such power and action to any higher being."[39]

Haeckel was not the only person who, in Lodge's opinion, allowed poor physics to determine a skeptical view of religious questions. Addressing the Synthetic Society in 1900, Lodge sought to "oppose as a physicist" what he saw as Tyndall "speaking as a theologian" on the thorny subject of prayer.[40] His target was an essay that Tyndall had written on prayer in 1861 and which was republished in a collection of essays that Lodge read when he was attending the older physicist's lectures.[41] Here, Tyndall anticipated much of the determinism that he would amplify in his Belfast Address, but tied it to an argument against the physical, as opposed to moral, efficacy of prayer. The robustness of energy conservation in particular and natural law in general made prayers for rain as likely to succeed as the "stoppage of an eclipse, or the rolling of the river Niagara up the Falls."[42] For Lodge, Tyndall's argument revealed a "theological" animus toward prayer that had blinded him to the fact that from the perspective of "pure physics" there was no essential difference between praying for rain and asking a gardener to fetch a watering can: in both cases there was a request for water to fall when and "where otherwise it might not."[43] The interference with energy conservation or dynamical laws was no greater in the case of prayer than in the banal but uncontroversial horticultural example. Prayers could not be dismissed as futile solely on the basis of physical laws that better-informed physicists agreed had a limited scope even on the "material side" of Nature.[44]

For all his confidence in judging Tyndall's physics by the higher standards of his own, Lodge was acutely aware that his argument against Tyndall's determinism was not as convincing as he would have liked it to have been. When, in a 1902 *Hibbert Journal* article, he attacked Tyndall for "needlessly" dragging energy conservation into the scientific appraisal of prayer, he insisted that this was because the principle had "nothing really to do with it."[45] The use of "really" veiled what many readers would have understood as the difficulty Lodge was having in demonstrating what he elsewhere characterized as his "contention" that the "superposition" of life and mind on the "scheme of Physics" upset physical and mechanical laws "no whit" despite being able to "profoundly affect the consequences resulting from those same laws."[46]

This difficulty was the subject of cordial debates in the correspondence pages of *Nature* in the 1890s and early 1900s.[47] One of the most incisive criticisms of Lodge's contention came from the British psychologist William McDougall, who in 1903 could not imagine mechanical guidance without momentum and energy conservation being violated. A guiding force deflecting a particle at right angles to its original direction of motion did in fact seem to do mechanical work because it led to a change in the particle's overall momentum.[48] At the time Lodge sought to deal with McDougall's criticism by asserting that since the kinetic energy of a body was independent of direction then no work was needed to change its direction, and that the example of a perfectly elastic rebound (as in the case of a comet passing near the sun) showed how change of direction did not always mean expenditure of energy.[49] Yet it is possible that McDougall's argument affected Lodge more than he was prepared to admit. In 1931 he revealed that the idea of mind changing the momentum of matter without imparting energy to it was a "hitch" that required a "mechanism" to solve it.[50] It was to the ether, and what Lodge believed to be its strange and "accommodating" properties, that he turned early in the 1900s for ways of resolving the hitch.[51]

SPIRITUALIZING THE ETHER

One of the most notorious responses of theistic physicists to philosophical materialism, and one to which Lodge would claim some intellectual debt, was *The Unseen Universe; or, Physical Speculations on a Future State.*[52] First published in 1875, the year after Tyndall's Belfast Address, it was the work of two of Maxwell's closest allies, Balfour Stewart and Peter Guthrie Tait, although the earliest editions of the book were anonymous. *The Unseen Universe* posed a scientific challenge to the "materialistic statements now-a-days freely made (often professedly in the name of science)" by arguing that well-established conceptions of energy, ether, and matter were not incompatible with Christian beliefs in the future state, the immortal soul, and such miraculous events as the Resurrection.[53] The universal dissipation of energy suggested that the visible universe was transient; but the principle of continuity, which the authors vaguely defined as that which guaranteed reconciliation between a cosmic state and its immediately antecedent one, suggested that this dissipated energy required a destination. This destination was, in the authors' view, an unseen, eternal, and immaterial universe connected to the visible by the ether of space.

By the 1870s most physicists accepted the need for some kind of subtle, transparent, imponderable, and universal medium that would explain the propagation of light waves across empty space. The nature of this etherial medium, however, was extremely puzzling, as it needed to be rigid enough to sustain the transverse vibrations constituting light but rarefied enough to allow material bodies to pass through it relatively unhindered. The hypothesis of William Thomson (first proposed in 1867) that the atomic constituents of matter were vortex rings in the ether only compounded the problem of representing the medium. Nevertheless, Thomson's friends and fellow Scotsmen Stewart and Tait held that the ether remained a key component of their physical theory of the afterlife. Central to their argument was evidence that the ether appeared to absorb light, which suggested that it could dissipate energy from the visible realm. Moreover, Stewart and Tait held that the thoughts and memories that constituted human identity could be interpreted as the energy associated with the movement of the atomic constituents of the brain, and that this energy was eventually dissipated in the super-rarefied and super-energetic realm that was the unseen universe. The transfer and permanent storage of this form of energy thereby lent credibility to the Christian ideas of the afterlife and immortality. Death was one of many dramatic cosmic events that breached neither energy conservation nor the authors' more general principle of continuity. The origin of life and the Resurrection could also be regarded as transfers of energy to and from the unseen.

Maxwell was one of several physicists who found Stewart and Tait's ideas unconvincing and inappropriate. Although he embraced an older argument that the ether's apparently perfect continuity and universality evoked the idea of an omnipotent and omniscient deity, Maxwell judged their attempt to extend the ether into the psychological and spiritual realms as something "far transcending the limits of physical speculation," and that questions regarding the fundamental nature and post-mortem existence of the human personality were beyond the limits of scientific inquiry.[54] But for Lodge, who knew about *Unseen Universe* at least as early as the 1890s, the fact that Maxwell had bothered to engage at all with Stewart and Tait's controversial text actually reinforced what Stewart and Tait had done: to give scientific legitimacy to speculations on the ether's possible psychical and spiritual functions.[55]

A measure of the growing psychic significance with which Lodge was prepared to invest the ether is evident from comparing comments he made in the early 1880s and early 1890s. In a public lecture on the ether in 1883 he merely hinted that it might be the medium that carried the mesmeric influence for which psychical research was amassing experimental and clinical evidence.[56] In an address to the British Association for the Advancement of Science in 1891, however, he was less restrained and proposed that the ether might provide an explanation of telepathy and telekinesis, as well as an answer to the more general question of the way that material will interacted with matter. One mind might be able to enter direct communion with another independently of the "ordinary channels of consciousness and the known organs of sense" via "some direct physical influence on the ether."[57] And the startling evidence of bodies being moved "without material contact by an act of will" could not be ruled out as a violation of the conservation of energy because of the possibility of a "novel mode of communicating energy, perhaps some more immediate action through the ether" (552). Exploring these possible functions of the ether constituted a "line of advance" for the "orthodox scheme of physics" because it had the potential to address the more general question of how the will interacted with the body (554). Two years after the British Association address, Lodge was further diversifying the possible psychic functions of the ether. In an article for a leading British journal of intellectual debate, the *Fortnightly Review*, he anticipated that the ether might eventually prove to be a "region of the universe which Science has never entered yet, but which has been sought from afar, and perhaps blindly apprehended, by painter or poet, by philosopher or saint."[58] The ether was a possible region from which humanity received inspiration and spiritual guidance.

There are at least three reasons for Lodge's greater confidence in the ether's psychic possibilities. The first is that Lodge, like many other physicists of the 1890s, was supremely confident that the ether was more than just a luminiferous medium. Heinrich Hertz's detection of electric waves in space in 1887 and Lodge's own generation of electric waves along wires the following year lent powerful support to Maxwell's theory that the ether was a medium of electrical and magnetic energy and that light was a form of vibration of this electromagnetic medium.[59] However, the ether's electromagnetic function only compounded the existing problem of its constitution. One of the most taxing problems for British physicists was devising a satisfactory mechanical model of the ether that would explain its extraordinary physical properties and functions in terms of force, motion, and energy, thus extending the tradition of dynamical explanation that had reigned in British physics for decades.[60] By the time he wrote his *Fortnightly Review* article, Lodge had produced experimental evidence that the ether was not dragged by matter moving rapidly past it, which not only contradicted Michelson and Morley's better-known evidence of an absence of an ether drifting past the earth but suggested that the ether was unlike any known material body and could not be "brought under the domain of simple mechanics."[61] But the ether's very elusiveness constituted a second reason why Lodge judged its psychic functions more plausible. It was the prospect of a new, ether-based type of mechanics that would enlarge the "foundations of physics" and help achieve one of Lodge's great ambitions: to help "annex vital or mental processes" to the science.[62] Already accepted by British physicists as the mysterious mechanism by which electromagnetic energy was generated and propagated, and by which the constituents of matter were formed, the ether's extension into the realms of life and mind would also extend into biology and psychology the mechanical intelligibility that had proved so successful in the physical sciences.

The third reason why Lodge was confident in the ether's possible psychic functions was undoubtedly Myers. Lodge's reference, in the *Fortnightly Review* article, to a possible etherial region from which artistic, philosophical, and religious inspiration derived owed much to his blossoming friendship with the psychical researcher. In late 1889, for example, Myers reacted to Lodge's expressed belief in the genuineness of the disembodied spirits speaking through Piper by revealing that while he was "not yet clear" on this specific case, he was "assured" that spirits could so commune with the living and that "the ether will bathe an endless evolution not of worlds only but of souls."[63] Here, Myers distilled ideas from esoteric, religious, philosophical, and scientific discourses relating to the capacity of some kind of invisible ether to be a repository of the dissipated energy generated in the cycles of planetary evolution and of the souls that once lived on such planets.[64] Myers's thinking about the ether provided the basis for his later speculation that within or "beyond" the ether was a "still more generalised aspect of the Cosmos" which was the "world of spiritual life" or "metetherial" environment.[65] It was a world possibly continuous with, but that ultimately transcended, the physical ether of space, and a world where telepathic impressions, souls of the departed, and other psychic phenomena operated.

While Lodge's correspondence with Myers does not reveal intense exchanges of ideas about the ether's possible religious or spiritual functions, Lodge certainly appreciated Myers's speculations in this direction. In a review of Myers's *Human Personality and the Survival of Bodily Death* (1903), a colossal work published two years after the author's death, Lodge explained that Myers's ultimately "optimistic" view of the future state owed much to his belief in the capacity of the human personality to retain all of the characteristics it had before death and to exist solely in the form of the "etherial or, as some would say, spiritual body" that it always had.[66] Lodge's reading of his much-lamented friend was that the physical concept of the ether and the Christian idea of the spiritual body were interchangeable, and this shaped his emerging view of the ether as the "instrument of Mind, the vehicle of Soul, the habitation of Spirit."[67] Until the 1910s, however, the legacy of Myers's ether thinking in the work Lodge presented to scientific and nonscientific audiences was muted, and Lodge was not prepared to do more than hint at the likely psychic and spiritual significances of the space-filling medium.[68] What undoubtedly held back such ether thinking was the ongoing problem of the ether's constitution. Lodge faced enough of a problem persuading scientific audiences of the ether's extraordinary physical and mechanical properties (notably, its colossal density) without introducing matters that would have made his approach seem even more speculative.[69]

The introduction of theories that threatened the ether's very existence—the special and general theories of relativity—would provide Lodge with a powerful incentive to assert the scientific, philosophical, and religious significances of the physical medium. His hostility to the special theory of relativity was well captured in his widely reported presidential address to the British Association for the Advancement of Science in 1913, in which he criticized relativity's abandonment of the ether as a modern example of the unwarranted and "dogmatic negation" in science that had made nineteenth-century materialism and agnosticism so unpalatable.[70] Lodge's response to the general theory of relativity was much more ambiguous but still provided him with an additional incentive to champion the ether and its wider possible functions. Although he criticized the theory for abandoning physical notions of ether for geometrical notions of space-time, it still constituted a "fuller realisation of the wide-spreading influence of a medium with finite properties, essentially pervading all space in which phenomena occur, and away from whose perfect but dominating uniformity we cannot escape."[71] For this reason, it was hardly surprising that Lodge should have helped himself to Einstein's own admission of 1921 that relativity demanded the existence of an ether insofar as it endowed empty space with "physical qualities," although Lodge disregarded the fact that Einstein's "ether" differed significantly from his own in being bereft of all mechanical qualities.[72] The success of general relativity in connecting gravitation to light and therefore etherial vibrations only strengthened Lodge's hope in the possibility that the ether would one day embrace those "other forms of existence" revealed in psychical research.[73]

IDEALIZING THE ETHERIAL MECHANISM

By the early 1900s Lodge had reached a philosophical position that he described as being part of a wider tendency in physics to refine or invert the materialism of the 1860s and 1870s.[74] The key driving forces had been the electrical and electron theories of matter of Joseph Larmor, Hendrik Antoon Lorentz, and others, which proposed that matter was built from microscopic electrified particles called "electrons," whose inertia or mass could be entirely accounted for in electromagnetic terms.[75] Larmor's form of theory was especially attractive to Lodge because it also proposed that electrons were structures in the ether, thus making his cherished medium antecedent to matter and of even greater cosmic significance. Although the experimental evidence for the electrical nature of an electron's mass was far stronger than evidence that all matter was built from electrons and was fundamentally electrical, Lodge felt that the shift toward explaining matter in terms of something immaterial yet physical constituted a clear tendency in physics away from the philosophical materialism of the previous century.[76] And the shift toward a broader and "really fundamental dynamics" based on electricity and ether further raised the possibility of physics embracing those topics that it had always sidestepped: life and mind.[77] In his opinion, physicists were now more likely to raise "matter and all existence to the level of life and mind" rather than reduce life and mind to ordinary matter in motion.[78]

One of Lodge's main preoccupations in the early decades of the twentieth century was speculating on how the potentially extraordinary, idealized or "glorified" mechanics of the ether might solve the problem of the interaction of mind and matter.[79] Common experience of life and mind suggested that some kind of mechanism or "physical vehicle" was involved in this interaction, but Lodge did not think this mechanism had to be limited to the domain of matter.[80] The results of psychical research made elucidating this immaterial mechanism more important to Lodge than ever before. His further séances with Piper and more notoriously, the communications he claimed to have received from his son Raymond Lodge, who was killed during the First World War, strengthened his personal conviction in the post-mortem existence and the need for a "mechanism of survival."[81]

Lodge had been aware since at least the 1890s that the results of psychical research also related closely to the question of free will because telekinesis and other effects dramatized the extent to which volition appeared to be independent of the body. Like many psychical researchers and spiritualists, however, Lodge believed that psychic phenomena were ultimately part of an intelligible, law-bound, and causal universe, the belief in which had been the cornerstone of scientific progress.[82] Lodge's hopes for the place of psychic phenomena in a causal cosmos informed his thinking on the general question of the conflict between free will and determinism. For Lodge the cosmos could be causal and mechanistic without being deterministic in the common, materialistic sense. In a short essay from 1903 he rejected the form of determinism that insisted that we were "wholly chained" by what we could apprehend of our environment, and the philosophy that insisted we were "*lawlessly* free, and able to initiate any action without motive or cause."[83] Both positions were blind to the existence of the cosmos as a whole. To understand how we could have the experience of freedom and yet be controlled, Lodge imagined being able to see the "totality of things," which would reveal that "everything was ordered and definite, linked up with everything else in a chain of causation, and that nothing was capricious and uncertain and uncontrolled" (4). What we in our "partitioned-off region" perceived as "unstimulated and unmotivated" freedom in mental and vital phenomena was in fact determined by a causal chain ultimately originating in a "transcendental world" whose obscurity had led it to be deemed "non-existent" (1–2, 4). Although Lodge admitted that humanity could never acquire the divine ability to apprehend the "totality of things," he maintained that it had the potential to discern vastly more about the hidden causes of human thoughts, emotions, actions, and overall evolutionary development (4). Lodge believed that his major contribution to this process was in exploring the possibility that the ether *qua* perfectly continuous universal medium and idealized mechanism guaranteed this cosmic chain of cause and effect and ultimately underpinned a cosmic intelligibility that would not be possible by focusing purely on its material aspects.

Lodge's voluminous writings of the 1920s and 1930s embodied numerous discussions of the ether's potential to give mechanical intelligibility to the mysterious ways in which life and mind interacted with matter. In *Beyond Physics: Or, the Idealisation of Mechanism* (1928) he upholds the ether as "something that feels like a key" to the "connexion between mind and matter" and to unify what he perceived as the mechanical tendencies in biology and idealistic tendencies in physics.[84] The ether constituted such a promising "key" because of the extraordinary physical properties that Lodge and many others of his generation now believed it had or suspected it to have. This was not the "old mechanical and engineering ether" of Lodge's youth and early scientific career that he now deemed "extinct," but a "new far less familiar but much more accommodating ether" (91). As something that appeared to be the seat of enormous energy, and which had such "perfect properties" as flawless continuity and no viscosity, it was closer to being an "ideal" rather than "material" mechanism, and vastly more adequate than matter as a "vehicle or physical instrument" of mind, whether those of individuals or of God (20, 46).

Lodge's less material and "more accommodating" ether seemed to fulfill the goals of *The Unseen Universe* better than the viscous type employed by Stewart and Tait. Although he probably read this book in the 1890s, he claimed that it took him several decades to agree with its argument that the "main realities" of the universe, including the memories and personalities of the dead, were somehow distributed in the ether.[85] It took long experiences of psychical research and a close friendship with Myers to persuade him that we have a larger and richer self beyond the material body, one that survives the death of this body; and it took years of experimenting on and brooding about the ether to conclude that it was a probably a perfectly frictionless continuum that never decayed.[86]

These ideas underpinned Lodge's most elaborate speculation on the ether's psychic functions. First hinted at in 1902, nurtured via conversations with the spirit of Raymond Lodge, and elaborated most fully in his last and most unrestrained book, *My Philosophy*, this was his scientific reinterpretation of an idea, widely circulated in spiritualist, theosophical, and occult circles, that we possessed an etherial as well as a material body.[87] Lodge argued that since the ether played a fundamental role in generating and cohering the material bodies of the cosmos, it was likely that all bodies, both inanimate and animate, had material and etherial counterparts. If the etherial body was made from materials infinitely more robust than ordinary matter, then it was a vastly more robust vehicle of our spiritual self: indeed, it survived the dissolution of our material bodies and helped us lead a fuller existence in the afterlife. The etherial body associated with human beings was Lodge's hypothetical "intervening mechanism" between mind or spirit and matter, and he hoped it would eventually give the otherwise unsatisfactory Christian ideas of the spiritual body "a definite and clear connotation" and help bring the "obscure communications and strange movements" of spiritualistic circles "in the orderly scheme of recognised science."[88]

Despite his hopes for the religious, philosophical, and scientific significances of the etherial body hypothesis, Lodge admitted that ether remained a physical entity with measurable mechanical properties, attributes that many of his critics believed rendered the ether too materialistic for psychical or spiritual duties.[89] Lodge had neither fully succeeded in finding ways of describing the ether in terms distinct from those of ordinary mechanics, nor in explaining how the supposedly extraordinary mechanics of the ether mediated psychical, spiritual, and nonphysical impulses. He fully accepted that he had only moved the explanation of the interaction between mind and matter a "step," but he felt it was one in the right direction.[90] One promising source of further steps was a relatively new development in physics that, in 1928, Lodge considered "the beginning of a comprehensive theory of the ether."[91] This was the development, by Louis De Broglie and Erwin Schrödinger, of wave mechanics, and this appealed to Lodge because it described the behavior of material particles in terms of waves, and thereby seemed to be another vindication from modern physics of the importance of a physical ether. But wave mechanics was most appealing because it hinted at possible "machinery" for resolving the problem he had faced since the 1890s of showing that immaterial guiding agencies did no mechanical work on physical systems.[92] Wave mechanics proposed that a particle could be represented as group waves that transmitted energy, but that the group waves were interlocked with and guided by constituent or "form" waves. Crucially, form waves carried no energy, were undetectable, and could travel significantly faster than group waves—faster, in principle, than the speed of light. Thus, Lodge identified in these etherial vibrations the most promising speculation to date for the "instrument" used by life and mind to interact with the physical universe and for an "idealistic interpretation of the universe, in which life and mind are supreme."[93]

* * *

Lodge's use of wave mechanics to promulgate an idealistic philosophy is significant because it appeared in *Beyond Physics*, which also contained criticism of one of the most prominent younger expositors of idealism in physics: Arthur Stanley Eddington.[94] Lodge had long regarded Eddington as one of the champions of relativity who, much to his chagrin, appeared to privilege mathematical abstraction over physically intelligible theory. However, he was delighted to find that Eddington, like Einstein, was not prepared to reject some kind of physical medium in space, and Lodge strongly sympathized with the younger scientist's goal of showing the fundamental importance of mind in the objective world, even if Eddington reached it via mathematics rather than through Lodge's route of psychical research.[95] Yet Lodge could not accept the way that Eddington had used the indeterminacy of quantum mechanics and other "curious developments of the latest speculations in physics" in arguments that free will was more plausible to, and religion now possible, for the "reasonable scientific man."[96] Lodge regarded this as a "preposterous" conclusion (73). The idea of a cosmos driven purely by chance rather than by discernible causes conflicted with his profound conviction in a cosmos that was ultimately intelligible to humanity and that the process of better understanding the causes, mechanisms, and purposes of the cosmos was the only way of bringing together scientific and religious outlooks (147). Lodge modestly accepted that none of his etherial speculations constituted a comprehensive physical explanation of free will and purpose, but to the end of his life he maintained that we were "only beginning" to understand the mechanism explaining the undeniable fact of that free will and purpose interacted with the material world (146).

A MARGINAL PHYSICIST?

In 1914 Lodge admitted to a close friend that his "occasional psychic utterances [did] harm to [his] scientific reputation."[97] There was no doubt that his cases for the reality of telepathy, telekinesis, survival, the immortality of the soul, and related questions were seen by many critics to be at best inconclusive and at worst deeply flawed because they were the result of sloppy investigative procedures and a delusive emotional commitment to the idea of the afterlives of his son Raymond and others.[98] For these critics Lodge's forays into psychical research were doubly problematic because they represented a subversion of increasingly rigid and fiercely defended disciplinary boundaries—in this case, they represented a physicist claiming to speak on questions of life and mind that were deemed the exclusive provinces of biology and psychology. Criticisms that Lodge had exceeded his authority also followed many of his religious utterances. Many Christian thinkers saw Lodge's attempts to update and liberalize Christian doctrine in light of the conclusions of modern physical, psychological, and biological sciences as weak, misguided, and hopelessly amateur.[99]

Lodge's ardent belief in the ether certainly affected the reputation he held among physicists. Edward Andrade spoke for many younger physicists when, in 1924, he suggested that Lodge's conception of the ether as a substance (albeit one that was confusingly described in terms of inertia, density, and other attributes of ordinary matter) had been abandoned by the "orthodox physicist," who, in the wake of relativity, denied that the ether had determinable points or any mechanical properties.[100] The ether's questionable existence and puzzling nature was seen by other critics to be a major weakness in his religious and philosophical arguments. As Lodge seemed to acknowledge, far from illuminating the problem of how mind and matter interacted, his conception of the ether seemed to compound or simply restate the problem.[101]

The foregoing discussion suggests that in some scientific, religious, and philosophical quarters Lodge occupied a marginal position. But Lodge's constituencies had always been broader than those from which his most incisive critics derived, and these audiences often took him to be the leading authority on "modern" physics and science per se.[102] Indeed, it was because of this reputation with popular audiences that many critics believed that his spiritualistic, psychical, and religious forays misled or even damaged the public mind.[103] The phenomenal success of Lodge's books, articles, lecture tours, sermons, and radio broadcasts suggests that despite his marginality in some quarters he was enormously effective in catering to the intellectual, philosophical, and emotional needs of other readers and listeners, many of whom would have demurred to the idea that relativity had killed off the ether. His conception of an etherial body was welcomed by spiritualists because it satisfied their desire for modern science to illuminate and lend credence to their long-held belief in some kind of invisible and subtle vehicle of the soul independent of the material body.[104] His preference for a physical ether and hostility toward abstract mathematical theories of space-time matched the outlook of many professional and amateur wireless enthusiasts who, perceiving no scientific consensus on the ether's reality or nature, followed one of their heroes of the invention of radio in assuming that some kind of physical medium was needed to make wireless transmission through space intelligible.[105]

One of the ingredients of Lodge's success as a communicator of physics, wireless telegraphy, psychical research, modernized Christianity, and other subjects was the lucidity of his expression and the breadth of his knowledge.[106] A key ingredient of this lucidity was his employment of the same pictures, analogies, and models that had long baffled and annoyed so many of his scientific peers. A contributor to a 1933 issue of the *Spectator* was probably right when, in a review of Lodge's last book, they suggested that as far as the lay reader was concerned, Lodge and other scientists of his generation had a "distinct advantage over the modern mathematical school of physicists" because their pictures and models were simply more effective than abstract hypothesis in communicating "how the physical processes of the cosmos work."[107] But it was also an approach that seems to have proved effective on emotional and intellectual levels.

One explanation of the astonishing success of *Raymond,* Lodge's account of the communications he had received from the spirit of his son, is that it reassured thousands of others who had lost loved ones in the First World War that there was scientific evidence of the post-mortem existence.[108] But another explanation is that in this book, as in so many of Lodge's that sold in the thousands during the 1920s and 1930s, he offered tentative but reassuring causal models of how the physical and psychical "processes" of the cosmos might "work." Death was "not a word to fear" because it probably involved a change from life in a semimaterial and semietherial state or body to life in a purely etherial state or body.[109] For some readers, the faint possibility that we do not die, because we already have an etherial existence connecting us seamlessly to the wider cosmos, gave Lodge's philosophy an intellectual and emotional value that the scientific cases against the ether could not easily diminish.[110]

NINE

THE CASE OF FLETCHER

SHELL SHOCK, SPIRITUALISM, AND OLIVER LODGE'S *RAYMOND*

CHRISTINE FERGUSON

IN A letter written just months after his arrival at the Belgian front in the bloody spring of 1915, Second Lieutenant Raymond Lodge reports his concern for friend and fellow officer Eric Fletcher, a casualty of a strange and seemingly new psychological phenomenon sweeping the ranks. Later collected for inclusion in his grieving father's spiritualist testimonial *Raymond: Or Life and Death* (1916), it observes: "[Fletcher] went off for a rest cure yesterday morning to a place about five miles from here. He is my greatest friend in the Battalion, so I miss him very much and hope he won't be long away. He will probably go back to England, however, as his nerves are all wrong. He is going the same way as Laws did and needs a complete rest."[1] Six months later, when the nervous condition described here was coming to be more widely known to the British public under the neologism "shell shock," the twenty-six-year-old Raymond Lodge was killed by a piece of flying shrapnel during the battle of Hooge Hill.[2] Fletcher would soon follow suit; within ten months, he too had joined the South Lancashire Regiment's list of fatalities.[3]

This letter, recording as it does the growing front-line military awareness of what has today become known as posttraumatic stress disorder, might initially seem like a trivial inclusion in a sensational spiritualist bestseller that presumably had much larger claims to stake.[4] *Raymond*, after all, was a survivalist manifesto that sought to prove nothing less than the continuing postlife existence, not just of its author's fatally wounded son but of all human beings, in an afterlife outlandishly occupied by libraries, scientific laboratories, and hospitals, and in which they could eat, smoke, and drink through the eons.[5] Yet shell shock functions in *Raymond* as far more than a ephemeral reference to a new pathology; in fact, it provides the foundational baseline for Oliver Lodge's modernized and counterintuitively disenchanted form of spiritualist belief during the war years, demonstrating with unique clarity the deep, if never wholly coherent, continuity between his scientific and spiritualist thought. Lodge's adoption of this emergent diagnostic category to explain the fragmented nature of séance communication also challenges both the meliorism and the insistence on the persistence of personal identity that remained at the core of the post-Victorian spiritualist movement.

While the resurgence of spiritualist belief in Britain during the war years has been characterized by some as a consolatory return to eclipsed nineteenth-century faith systems, Lodge was never so modern, nor arguably so vulnerable to the nihilism he aimed to reject, as when he attempted to fuse fresh insights from the field of war medicine into his model of the afterlife.[6] Shell shock's significance for our understanding of Lodge's spiritual-scientific outlook during the war years, and of the biomedical imagination of the period's spiritualist writing more broadly, remains immense and largely untapped. While other spiritualist believers sought consolation in a vision of the afterlife in which combat injuries were instantly or eventually cured, Lodge positioned battlefield trauma as a crucial, medically credible, and, most importantly, *enduring* condition through which communication between the living and the dead might be explained and catalyzed. In *Raymond*, life in the spirit spheres may ostensibly be more just and full of promise than that in the mundane world, but it relies on the lingering effects of the latter's violence to forge its communicative channels.

LODGE'S SPIRITUALISM: FROM SCIENCE TO SENTIMENT

By the time he published *Raymond*, Oliver Lodge was something of an old hand on the British spiritualist scene, having investigated psychical phenomena with Edmund Gurney and Frederic W. H. Myers in the early 1880s, sat with famous mediums Eusapia Palladino and Leonora Piper in the 1890s, and in the Edwardian years published for the popular market a series of three books on the scientific probability of postlife survival: *The Substance of Faith* (1907), *Man and the Universe* (1908), and *The Survival of Man* (1909).[7] These publications caused opprobrium in British rationalist and orthodox religious coteries alike; while the latter objected to Lodge's departure from Christian soteriology in his championship of a universal progressive afterlife open to all, scientific naturalists mocked his retention of an increasingly outmoded physical concept of the ether in which disembodied spirits could allegedly exist and communicate with the still incarnate.[8] Lodge himself denied any contradiction between the scientific and spiritualistic in these works, insisting in *Man and the Universe*, for example, that spiritual immortality was simply a logical extension of the law of "conservation of Value."[9] "Is it not legitimate to conjecture that while Matter and Energy neither increase nor decrease," he asks there, that "life too perhaps is constant in quantity, though alternating into and out of incarnation according as material organisms are put together and worn out"? (180). To reject the spiritualist hypothesis was thus to commit the supremely unscientific error of rejecting the first law of thermodynamics.

Raymond was immediately recognized as belonging to a different category than these earlier works, much more personal in its record of devastating family loss, far more willing to accommodate eccentric speculation and erratic evidentiary interpretation than Lodge's previous, cautiously argued approaches to spirit survival. This shift Lodge freely admitted, writing, "Hitherto, I have testified to occurrences and messages of which the motive is intellectual rather than emotional."[10] That said, he refused to abandon the pose of scientific objectivity, contending that for him "the hypothesis of continued existence" was "not a gratuitous one made for the sake of comfort and consolation" but rather a conviction just as stringently tested as "the foundation of atomic theory in Chemistry" (288). He also imported his increasingly scientifically outmoded theory of the ether into *Raymond*, although as evidence it bears far less importance than phenomenal corroborations such as the so-called Faunus message, in which Lodge retrospectively believed himself to have received a premonitory warning of Raymond's death through the mediumship of Piper, or the group photo of Raymond and his fellow officers that was described by the entranced medium Alfred Vout Peters before its existence was allegedly known.[11]

Such proofs seemed like no proof at all to *Raymond*'s many secular critics, who treated them with as much contempt as the spirit Raymond's fanciful descriptions of the newly dead drinking whisky sodas while regrowing lost arms and teeth.[12] London psychiatrist Charles Mercier was particularly damning in his assessment, writing of Lodge that "when I read his naïve and innocent account of his own simplicity, I wonder if Lady Lodge ever allows him to go out in the street without a nurse to see that he does not bring home a gross of sentry boxes, or chimney-pots, or left-hand gloves, or something equally profitable."[13] Mercier somewhat unfairly depicts as tenets of sincere belief the more outré mediumistic claims that Lodge himself recognized as "very non-evidential and perhaps ridiculous" and included only so as to present as complete, objective, and representative a record of the Raymond séances as possible.[14] A fixation on the material aspects of the afterlife had after all been a standard feature of Western afterlife reportage since at least Emanuel Swedenborg's *Heaven and Hell* (1758); the *Raymond* messages were by no means unique in this respect and are on the whole far less extravagant in their detail than competing spirit soldier biographies, such as J. S. M. Ward's *Gone West* (1917) and its sequel *A Subaltern in Spirit Land* (1920), in which half-human, half-animal monsters and ravenous antediluvian dinosaurs populate the astral plane and wait to attack the newly received spirits of the war dead. Few of Lodge's critics, however, were willing or able to distinguish the differing motives—fanciful and sensational, or simply dutifully inclusive—that inspired these accounts. *Raymond* and *Gone West* were thus routinely lumped together in contemporary accounts of the war's spiritualist revival, presented by nonbelievers as equivalent testimonies or reflections of the dangerous epidemic of delusion spawned by the conflict.[15] To promote such beliefs at a time when the nation was already in peril constituted, to some antispiritualist critics, an act of near sedition: "Sir Oliver Lodge has done a public disservice in issuing this book," claimed the *Saturday Review* in 1917, echoing Mercier's conviction that "the publication of *Raymond* at this psychological moment is much to be deplored."[16]

SHELL SHOCK AND ITS RECEPTION IN THE BRITISH SPIRITUALIST MILIEU

On the battlefields of Europe, another kind of nonsense in addition to the one lamented by Mercier was showing up with increasing regularity in field hospitals, manifesting in proportions too great to ignore.[17] Soldiers both on and beyond the front lines were exhibiting a strange set of often allied symptoms without apparent organic cause; they suffered unexplained amnesia; struggled to eat or sleep; experienced nightmares, headaches, and chronic panic; developed speech defects; or lost the ability to speak altogether. At first suspected of shamming or insanity, sufferers of this syndrome, known initially as "war neuroses" before its official medical designation as shell shock in 1915, began to be taken more seriously in the midwar period when their numbers reached into the tens of thousands.[18] An early pioneer of shell shock treatment, Sir Frederick Mott, in his 1916 Lettsomian Lectures, traced the condition to barely detectable but nonetheless material injuries to the central nervous system caused by proximity to high explosives or exposure to carbon monoxide, injuries that could be cured through rest and cheerfulness—ideally, in his opinion, provided without recourse to what he viewed as the more contemporarily dubious techniques of hypnosis or psychoanalysis.[19] By insisting, however ultimately incorrectly, on a somatic basis for shell shock, Mott sought to give the condition a legitimacy it might otherwise lack if it were designated as a purely mental phenomenon; he also rendered it a uniquely modern pathology engendered by the new combat technologies of submachine guns, trench construction, and gas shelling whose sustained effects on the human body had never been tested on so great a scale. Each of these contentions later gave way, without ever entirely ceding, to more complex understandings of the epidemiology and history of war-acquired posttraumatic stress disorder in soldiers and noncombatants alike. Despite the ultimate relegation of his etiology, Mott remains important here because, as we will see, Lodge consulted his lectures at the time of *Raymond*'s composition and found in their descriptions of consciousness a buttress for his own theory of disembodied communication.

Shell shock and spiritualism were aligned during the war years via their mutual positioning as mental phenomena induced or encouraged by the conflict, and ones that could, if left untreated, significantly undermine the Allied campaign. Beyond this analogical alliance as sources and symptoms of peril, the two were also linked in other ways, perhaps most strikingly by the earliness and frequency with which the condition featured in spiritualist writing. Fiona Reid has argued that shell shock became a significant trope in secular British literature only after the war, suggesting that, with the exception of Wilfred Owen's poetry, literary treatments of the topic were tacitly embargoed until approximately a decade after the conflict's end "in part . . . because many combatants did not know how to frame their responses."[20] Even if we accept Reid's thesis as true in regards to nonspiritualist literature—and works like Rebecca West's *The Return of the Soldier* (1918) act as significant exceptions—it certainly does not apply to spiritualist war writing, which addressed shell shock with varying degrees of explicitness both during and immediately after the war. *Raymond* is only one such example; others include *The Hidden Side of the War*, a compilation of automatically written scripts that features this account, produced during a March 1915 séance, of shell-shocked soldiers being treated by spectral medics on the other side. The communicating spirit writes:

> The most recent case which I was sent to help was that of a Welsh lad who was fighting in one of the Line Regiments, and who suffered terribly from the nerve shock consequent on the noise and general terror . . . he was in danger of becoming influenced by ignorant and undeveloped spirits unless protected. Therefore were some of us sent to help him, and this we did—first of all by surrounding him with his own guides, etc., so shutting him off from those spirits would have promoted his downfall. To strengthen and encourage him we endeavored to rouse the spirit within to fight manfully and resist fear and cowardice. Finally, to help his body physically, we influenced those in authority to protect him by removing him to a less exposed position, although they were unaware of our action in effecting this.[21]

Similarly, British Bahá'í spiritualist Wellesley Tudor-Pole's *Private Dowding* (1917), another automatically written account of posthumous solider existence, describes its subject's disorientation from shell shock in the runup to his death and eventual numinous cure.[22] Nonchanneled texts by spiritualists and psychical researchers were equally fascinated by the nascent diagnostic category and its potential to shed light on psychic life. In *Psychical Phenomena and the War* (1918), Anglo-American psychical researcher Hereward Carrington argued that shell shock's successful treatment by hypnosis pointed to the existence of extrasensory channels of influence.[23] Even Lodge, in his later spiritualist soldier biography *Christopher: A Study in Human Personality* (1918), cites it as one of the most horrific consequences of modern warfare: "The physiological collapse spoken of as shell-shock or war stress occurs among the bravest. War has become impossible and inhuman."[24]

Jay Winter has claimed that the spiritualism of the First World War represented a throwback to "well-established Victorian sentiments or conjectures concerning the nature of the spiritual world."[25] However, the pervasive presence of what was contemporarily recognized as a malady of modernity in the period's spirit soldier narratives complicates this assessment. Indeed, there are few places where we see more clearly the radical impact of modernity on the spiritualist imagination than in its channeled accounts of the effects of military technology on soldiers both living and dead. The willingness of spiritualist writers to engage with shell shock early on also distinguishes their work sharply from the competing popular supernatural war fictions produced by nonbelievers, most famously Arthur Machen's "The Bowmen." Published in the London *Evening News* on September 29, 1914, it tells of a solider within the badly outnumbered British ranks at the Battle of Mons who unintentionally summons the ghostly archers of Agincourt to repulse the German forces; he does so by spontaneously chanting the motto of Saint George—*Adsit Anglis Sanctus Georgius*—that he once saw inscribed on, of all places, a plate in a London vegetarian restaurant.[26] The patriotic story had an immediate and electrifyingly mythopoeic effect on the British public, one whose consequences have been well documented.[27] Readers within and beyond Britain's contemporary occultural milieu insisted that the incident had actually happened and that Machen, far from inventing it, "may have had, unconsciously to himself, some telepathic communication of the vision"; Machen, who despite his occultist leanings in the late 1890s had always hated what he viewed as the vulgarities of exoteric spiritualism, angrily rejected this slight to his creative powers.[28] Admittedly, Machen's patriotic tale was published before shell shock was recognized as an official medical diagnosis, but nevertheless, its deliberate suppression of any reference to the even then recognized psychological consequences of modern shell warfare is telling. In his First World War fiction, war held no mental dangers that the "stou[t] hearts" of the British troops could not handle, and supernatural entities were clearly on the side of the Allies.[29] Such a pleasingly pro-British and nontraumatic vision of the conflict was not always or necessarily offered in spiritualist writing, and arguably in *Raymond* least of all.

SHELL SHOCK IN *RAYMOND*

Lodge's *Raymond* is prefaced with a poignant declaration of devastating simplicity: "This book is named after my son who was killed in the war."[30] The book is composed in three parts: an opening "normal portion," which describes Raymond's predeath personality; a supernormal one, which records the afterlife messages he allegedly sent through London mediums Gladys Osborne Leonard and Alfred Vout Peters; and a philosophical final section in which Lodge argues for the scientific and moral value of spiritualism. Formally, it is a seething mass of composite paratexts; letters, séance transcriptions, photographs, memoirs, and personal testimonials from friends and family members of the fallen soldier.[31] Shell shock appears in all three of the book's sections, first, as we have seen, in the battlefield correspondence that Lodge includes as evidence of his son's character prior to his death; then implicitly in the spirit Raymond's discussion of his postlife verbal confusion; and finally in Lodge's Mott-derived speculations on the nature of disembodied communication.

In the letters written while he was still living, Raymond Lodge demonstrates an intense sympathy for sufferers of the war neuroses to which he presents himself as relatively immune. Thus in May 1915, just a month after his deployment to the front, he describes feeling "cheerful and well and happy as ever" (36), in sharp contrast to those around him, who he found "suffering from nerves and unwilling to talk about shells at all" (21). As the spring campaign lengthened, the young Lodge's concerns about the psychological effects of combat settled on his friend Eric Fletcher, who, having been in the field for longer than he, had suffered "some awful times in the winter campaign, and . . . the length of time one is exposed to the mental strain and worry make a difference" (42). Raymond Lodge hoped that a rest cure would provide Fletcher with the distraction and ease he needed, writing that "brooding is the very worst thing for him. He sees all the past horrors all over again; things which, at the time, he shut his mind to" (40).

This process of constant repetition without resolution is, of course, central to current definitions of trauma; it is also, fascinatingly, a plot device in many First World War spirit soldier narratives, which feature scenes of German and British soldiers repeatedly fighting and rekilling each other in the spirit world without realizing that they are no longer alive.[32] Raymond Lodge worried that this repetitive brooding was starting to infect Fletcher's fellow soldiers, including the company's senior subaltern Thomas, later killed during a June 1915 gas attack.[33] Clearly devastated by the loss, the young Lodge in his subsequent letters seems to defensively escalate the almost manic cheerfulness that pervades the "normal portion" of *Raymond*: "One can contrive to be light-hearted and happy through it all—unless one starts to get depressed and moody. And it is just what happened to Laws and Fletcher . . . none but the very thick can stand it" (51).

When Raymond's father came to compose the philosophical final part of the survivalist biography devoted to his son, this distinction between the thick- and the thin-skinned combatant would break down, and the two would emerge as fellow sufferers—and potential beneficiaries—of shell-shocked communication dysfunction. Lodge had clearly been reading extensively within the biomedical discourses of the war as he prepared the manuscript, respectfully referencing, for example, Peter Chalmers Mitchell's *Evolution and the War* (1915) as counterpoint to his own belief, contra to that of his scientific spiritualist predecessor Alfred Russel Wallace, that all forms of life, whether human or animal, were capable of progressive evolution in the etherial world (336).[34] More conducive to his own spiritualist ontology were Mott's recent lectures on shell shock, which seemed to offer a solution to what Lodge knew would likely be the most serious objection to the Raymond séance messages: not that his son spoke coherently about ridiculous or fanciful things (the saloons and lecture halls of the Summerland) but rather that he spoke incoherently or forgetfully about the things that should matter most, such as the names of his family members and fellow soldiers and the shared incidents of their mundane life together. Lodge was right to worry: the fragmentation and apparent triviality of the messages was indeed a chief target of their skeptical readers in the mainstream press.[35] "For what reason," demanded Bernard Sickert in the *English Review*, "are [the proofs] invariably introduced in cryptic form, as a kind of puzzle, or cryptogram, allusively, by initials; gradually by stages; incorrectly, by blunders? . . . All of the messages have this character, appearing to make the proceedings of the nature of a silly game, inappropriate to the occasion."[36] Even Lodge's mediumistic interlocuters recognized these problems, with Leonard's spirit guide Feda complaining in a November 1915 sitting with Raymond's brother Lionel that "[Raymond] has been trying to come to you at home, but there has been some horrible mix-up. Not really horrible, but a muddle . . . other conditions get through there, and mixes him up."[37]

For Lodge to win for these communications the scientific credibility he clearly sought, he needed to account for their errors and interruptions in a way that went beyond the familiar extenuating allusions to poor conditions or malign spirit interference favored by the Raymond mediums themselves.[38] He found such an improved and up-to-date rationale in Mott's etiology of shell shock. In the "Mind and Brain" chapter of *Raymond*, Lodge quotes Mott's contention in "The Effects of High Explosives upon the Central Nervous System" that "a continuous supply of oxygen is essential for consciousness" (329). From this principle, he extrapolates that a lack of oxygen only prevents consciousness from manifesting; it does not in his view suggest that consciousness did not exist. He then cites at length what was evidently Mott's most electrifying line of inquiry for the spiritualist hypothesis:

> Why should those men, whose silent thoughts are perfect, be unable to speak? They comprehend all that is said to them unless they are deaf; but it is quite clear that in these cases their internal language is unaffected, for they are able to express their thoughts and judgments perfectly well by writing, even if they are deaf. The mutism is therefore not due to intellectual defect, nor is it due to volitional inhibition of language in silent thought. Hearing, the primary incitation to vocalization and speech, is usually unaffected, yet they are unable to speak; they cannot even whisper, cough, whistle, or laugh aloud. Many who are unable to speak voluntarily yet call out in their dreams expressions they have used in trench warfare and battle. (330)

In Mott's shell shock victim, temporarily but not permanently impeded from communicating in a normal way, Lodge found an ideal and, for him, medically legitimate explanatory model for spirit subjectivity, one that the modern spiritualist movement had arguably been waiting almost seventy years to find. Both types of traumatized subject—the shell-shocked solider and the human who had recently died—might ultimately be restored to communicative normalcy, but only through the support of an adept practitioner who could provide the fleshy, material support for their unvoiceable experience. "It is through physical phenomena that normally we apprehend, here and now; and it is by aid of physical phenomena that we convey to others our wishes, our impression, our ideas, and our memories," he concludes. "Dislocate the physical from the psychical, and communication ceases. Restore the connection, in however imperfect a form, and once more incipient communication may become possible" (330). The spirit medium could here act as restorative prosthesis, reforging the connection between body and spirit required to both communicate and complete the cure of war fatalities in the spiritual spheres.

THE IMPLICATIONS OF LODGE'S SHELL-SHOCKED SPIRITUALISM

Lodge's appropriation of shell shock theory to explain the exigencies of postlife communication had significant implications for the spiritualist faith he so ardently championed, some positive, some considerably less so. First, by configuring the séance as a site not (just) for the consolation of the living by the dead but for the reparative cure of war casualties, he eroded the barrier between combatant and noncombatant, establishing survivors on the home front as equal participants in a military effort that only continued if not intensified in the afterlife. Such an imaginative expansion of the war's active constituency was indeed common in First World War spiritualist writing, epitomized in the insistence of dead judge David P. Hatch to his

American medium Elsa Barker during the conflict that "you have not been to the wars, either as a soldier or a nurse, *but you have been to the wars*."[39] By entering communion with the war dead, Barker's so-called "Living Dead Man" asserted, the living could not only manage their care but also, through an act of supernatural transport, visit the battlefield and observe the intermediating occult forces that spurred on the competing forces at firsthand. Yet as Lodge's interpretive model expanded the ranks of those who could consider themselves in active service, so too did it swell the population of shell shock sufferers to include not just the casualties of the recent European conflict but also all people who had ever lived on earth. After all, Lodge's description of death as the catalyst for communication impairment necessarily implicated not just soldiers but the entire constituency of the human dead who no longer had connection to or control over an individualized physical vehicle for speech or writing.

To view all spirits as potentially shell shocked was not to say that the armed forces did not constitute a special and particularly damaged category of the dead in *Raymond*: Lodge, unlike a number of his spiritualist peers, saw the death of young men in battle as a needless waste that no resulting cosmic glory or karmic recalibration could redeem. While other believers argued that the war's casualties were in fact members of a spiritual elite specially equipped for early death, and whose sacrifice would usher in a new, progressive millennium, Lodge took no such easy comfort.[40] He writes: "The spectacle of thousands of youths in full vigour and joy of life having their earthly future violently wrenched from them, amid scenes of grim horror and nerve-wracking noise and confusion is one which cannot and ought not to be regarded with equanimity. It is a bad and unnatural truncation of an important part of an individual career, a part which might have done much to develop faculties and enlarge experience."[41] Lodge would repeat this sentiment in *Christopher*, a tribute to the young spiritualist and war casualty Christopher Tennant that contains none of the séance evidence that had figured so prominently in *Raymond*. Here Lodge negatively contrasts the afterlife development potential of those killed violently in youth with that of those who pass peacefully in old age. "Old people pass away in the course of nature, prepared by experience and feebleness of the body for the transition," he writes. "But amid the whole destruction caused by inhuman war, strong, healthy, vigorous lives are exploded into apparent nonentity; and . . . the experience may occasionally be an unnerving shock, an experience incredible and prostrating."[42] Lodge's dead soldiers were the most impaired of a wider spirit constituency that was itself always already shocked and sensorily disordered, reliant on living mediums as prostheses.

In this positioning, *Raymond* is a world away from those triumphalist spiritualist accounts of the afterlife that pitch it as a place of immediate recuperation and hyper-ability. Little wonder then that some of the book's critics would be less concerned with what it implied about its writer's credulity than with what it suggested about the lamentable nature of postlife existence. Even accepting, if only temporarily, that the spiritualist hypothesis was true, wrote Alfred Martin, leader of Ethical Culture, in 1918, were these messages not a dismaying indictment of the future state? "To excuse the defects of obscurity, discontinuity, incoherence and incompleteness that mark of many of the alleged 'messages' on the grounds of 'amnesia,' the transition to 'the other side' causing forgetfulness—would involve a surrender of the sole remaining means for identifying the deceased. Now that bodily continuity has been destroyed, how, with loss of memory, shall identity be established?"[43] Lodge's scientospiritualistic appropriation of trauma, and of shell shock in particular, as explanatory device sat directly at odds with both the wider movement's optimistic cosmological schema and insistence on the retention of personal identity, echoing, in fact, the challenge to patriotic and triumphalist understandings of the Great War posed by living shell-shocked veterans in the secular public sphere.

The commitment to scientific inquiry and to the innovative new findings of war medicine that Lodge hoped would legitimize his spiritualism thus implicitly came to undermine the latter's progressive ethos, introducing into *Raymond* a dissonant nihilism at odds with its message of hope and consolation. This is nowhere more apparent than in the text's conflicted meditation on the lingering effects of battle injury on the newly dead, and specifically on their challenge to that most enduring of tropes in the First World War spiritualist soldier memoir; namely, the emphasis on persistent postlife service. In almost every specimen of the genre published in this period, the war dead are presented as continuing to fight for the Allies, sometimes simply by influencing the thoughts of the living, but in other instances through far more exacting duties. The dead wireless operator in Grace Boylan's anonymously published *Thy Son Liveth* (1918), for example, comforts his mother by explaining that he is now using the superior mental powers developed in his new life to study and intercept German code.[44] This trope of enduring service would have been well known to Lodge and his mediums through their participation in the spiritualist public sphere and is by no means wholly absent in *Raymond*; we can detect it, for example, in Feda's claim that the spirit Raymond's "work still lies at the war, in helping on poor chaps literally shot into the spirit world" and in Lodge's opening injunction to his readers: "Let us think of [Raymond], then, not as lying near Ypres with all his word ended, but rather, after due rest and refreshment, continuing his noble and useful career in more peaceful surroundings."[45] But in *Raymond*, at least, this convention of numinous labor is enclosed within a section of séance messages whose speculative descriptions of "the nature of things 'on the other side'" Lodge had described as "absurd" and "free-and-easy," worthy of inclusion only for the purposes of faithful transcription (172). Furthermore, the wistfully hortative mode in which Lodge himself describes his son's continued industry suggests caution rather than confidence: rather than simply stating that Raymond *is* furthering his career in the next life, Lodge instead proposes that we allow ourselves to think of him as such.

Raymond's discourse on war trauma provides plenty of reasons for this hesitancy. After all, if shell-shocked veterans were considered unfit for service in the mundane world—and remember that in this work, *all* of the dead could be considered shell shocked to some extent—were they any less so in the spiritual one? In opening this question, however implicitly, *Raymond* becomes an altogether more conflicted text than those competing spirit soldier memoirs that saw no limit to the recuperative power of death.[46] True, some of the book's séance messages proclaimed that lost or burned limbs would instantaneously be regenerated on the other side, but others not only pointed to but also claimed as their foundational condition the persistence of trauma.[47] The Raymond Lodge who appeared in the séance room was by no means always recognizable as himself or even seemingly cognitively normal, a fact emphasized in the explanations offered by Lodge's mediums for the frequent breakdowns in his communications. During an October 1915 session, Peters declared: "From his death, the work which he (I have to translate his ideas into words, I don't get them verbatum [*sic*])—the work which he volunteered to be able to succeed in,—no, that's not it. The work which he enlisted for, that is what he says . . . yet the very fact of his death will be the means of pushing it on. Now I have got it. By his passing away, many hundreds will be benefited" (103). Here we see the trope of continuing service broken down into stuttering missteps at the moment of its very utterance, the mode of expression not complementing but slyly subverting the mission it proclaims. If the work of the dead was to influence the living, it could not proceed successfully if the former were too shocked and disoriented to express themselves.

Further doubt about the fitness of the dead for postlife work surfaces in the more coherent Raymond messages produced during the séances with Leonard. In Oliver Lodge's first anonymous sitting with this medium, held just two weeks after Raymond's death, Leonard's spirit guide Feda reported, "He knows that as soon as he is a little more ready, he has got a great deal of work to do. 'I almost wonder,' he says, 'shall I be fit and able to do it. They tell me I shall'" (98). This moment of hesitation is given fuller context in the more detailed transcription of the sitting that Lodge included later in the volume, one that acknowledges the ongoing and, intriguingly, unknown to their victim, impact of Raymond Lodge's injuries on his new subtle body. "Feda feels like a string round her head; a tight feeling in the head, and also an empty sort of feeling in the chest, empty, as if sort of something gone . . . also a bursting sensation in the head. But he does not know he is giving this. He has not done it on purpose, they have tried to make him forget all that, but Feda gets it from him"[48]

(127). Even if Raymond's spirit guide F. W. H. Myers had succeeded in making his charge largely oblivious to the extent of his trauma—"I feel splendid," Raymond then cheerily reports, "... splendid!," his living communicants could feel that something was still broken, still absent (127). By choosing to include such messages in the volume, Lodge moots the possibility that once physical but now psychic wounds could linger on in the afterlife, disrupting spirit communications and threatening to overturn the comforting possibility of healing, improvement, or even simply metaphysical meaning, which made all such deaths bearable to surviving believers. Such communications also, in emphasizing the potentially enduring nature of war-induced impairment, undermine Victoria Stewart's reading of *Raymond* as an ultimately apolitical and sublimating text, one that, through its use of the trope of continuing service in the afterlife, finds "a means of by-passing, temporarily, the reasons why that service has become necessary."[49] On the contrary, behind the defiant cheerfulness of Raymond Lodge's war letters and alleged postlife communiqués, one finds constant reminders of the disorienting impact of modern Western militarism on the bodies, minds, and immortal spirits of the young soldiers it was now consuming.

The shell-shocked spiritualism of Lodge's *Raymond* is thus of a distinctively difficult and challenging nature, rejecting the easy comforts and indeed narrative coherence of those competing spirit soldier memoirs that proceeded without flagging their compiler's doubts and insisted that the dead had been metaphysically selected for their evolutionary fitness to direct human affairs from the afterlife, or that the slaughter would seed a new race, conquer materialism, karmically recalibrate the world, or catalyze the coming of a new millennium.[50] Ever the conscientious man of science even when he seemed most to have abandoned that role, Lodge insistently included data that countered his own desires for a progressive, or even meaningful, afterlife in which he might eventually reunite with his son. "I am aware that some of the records may appear absurd," he writes, noting that many contain "the kind of assertions which are not only unevidential but unverifiable, and which we usually either discourage or suppress."[51] In this collective "we," Lodge both identifies with and distances himself from his cobelievers, offering an unusually candid insight into the implicit editorial processes and decisions that guided the publication of other séance records. By rejecting such a tactic, even at the peril of his own proselytizing ambitions, Lodge saw himself not only as fulfilling the duties of accurate representation but also as preserving important if as yet uninterpretable data for future use when it might, "like traveller's tales . . . ultimately furnish proof more logically cogent than was possible from mere access to earth memories" (192). If the dead could be analogized to, or even directly diagnostically aligned with, the shell-shocked veterans of the Great War, then it should be no surprise if their speech occasionally descended into nonsense. The fragmented or incongruous utterances of the séance room and the field hospital alike had a value beyond their veridical function.

Lodge's engagement with shell shock in *Raymond* ultimately put him at odds not only with his more optimistic cobelievers but also with the hopeful fantasies of postlife recuperation for the conflict's casualties shared by heterodox and traditional Christian believers alike. Incorporating these into his 1915 short story "Fun!—It's Heaven" (1915), Ford Madox Ford suggests that the living have an ethical responsibility to imagine a pleasurable and active afterlife for the war dead, even if such a vision is a necessary delusion. "It would not be decent," remarks the story's world-weary doctor protagonist, "if Heaven wasn't like that for those poor young things. . . . They die for you and me. It's our business to invent a fit heaven for them."[52] Such consolation was not readily identifiable in the shell-shocked spirit who spoke in incoherent riddles and whose ability for continuing service lay in doubt. Yet in acknowledging this figure at all, however briefly, Lodge demonstrated his enduring commitment to a scientifically searching investigation of the afterlife and a more complex awareness of the relationship between personality and trauma than many of his spiritualist counterparts were willing to acknowledge. In making the case of the shell-shocked Fletcher the condition of all the dead, *Raymond* imagines a spiritual polity founded on universal trauma and united by communicative impairment.

TEN

BEYOND *RAYMOND*

THE THEOLOGY OF SPIRITUALISM AND THE CHANGING LANDSCAPE OF THE AFTERLIFE IN THE CHURCH OF ENGLAND

GEORGINA BYRNE

WHEN OLIVER Lodge published *Raymond,* the book of communications with his departed son, in 1916, he was engaging with readers already familiar with the language and paraphernalia of spiritualism. From its arrival in England in the autumn of 1852, spiritualism had captured the imagination of a diverse range of people: fashionable ladies in London salons—along with their servants—as well as secularists in industrial northern English towns, clergymen, journalists, politicians, and poets. The popularity of spiritualism waxed and waned over subsequent decades as newspapers enthusiastically reported the performances of the latest celebrity mediums and then charted their spectacular unmasking as frauds. The more serious-minded believers could hear lectures or attend services at which the philosophies and seven guiding principles that underpinned spiritualism were discussed, and, in true nineteenth-century fashion, societies sprang up in many towns and cities to bring together like-minded spiritualists, electing committees and holding regular meetings.[1] Whatever form or fashion spiritualism took, its central tenet remained the same: the spirits of the departed could and did communicate with the living. It presented an afterlife that was very different from the one offered by the established church. During the latter part of the nineteenth century, the Church of England was already reexamining its teaching about life beyond death, but had yet to find ways to frame it in something other than traditional language. It is my contention that *Raymond* was much more than a restatement of spiritualism; rather, it promoted newer metaphors for post-mortem survival, drawing on the language of evolution, domesticity, and human experience that were already in circulation, and, due to its notoriety, it brought to the attention of a wider audience some significant theological shifts.

Lodge, a man of science and loyal member of the Church of England, wrote *Raymond* as a grieving father and also as one convinced by the claims of spiritualism who wanted to offer reassurance to parents similarly afflicted.[2] *Raymond* was a best-seller. It was reprinted five times in 1916, twice in 1917, and again in 1918. New editions appeared in 1918 and 1922. At the 1919 Church Congress, in Leicester, one speaker, even as he denounced the book, gave an indication of its popularity when he argued, "If you have read, as many of you in this hall have read, *Raymond*, you will realise the pitiableness of such communications."[3] It cost eleven shillings for a cloth copy, which was not cheap, suggesting that those who bought it were not without means. Even those who could not afford it, though, clearly imbibed its message and its vivid descriptions. Thus, in 1947, a clergyman noted wryly to a Mass Observation survey that his congregation talked "glibly" about heaven, "like Oliver Lodge and Raymond . . . who was supposed to have said there was good whisky there. They hope there'll be good beer."[4]

Beyond this, the book provides an intriguing prism through which can be seen changing patterns of theology and wider discussions concerning the afterlife, not least because not only did it present the alleged communications of Raymond Lodge to his family but also in its final section Oliver Lodge took the opportunity to outline and expand the theological assumptions that underpinned his own spiritualist beliefs. During the Great War the Church of England, the established church, found itself ill prepared for mass bereavement and unable to offer adequate spiritual comfort through its public prayers and pastoral offices. The fact that *Raymond* was quoted by churchmen such as those at the Church Congress as well as spiritualists suggests that it met a need not provided by the church to a population already sixty years familiar with the cadences of spiritualism. The same sixty-year period had seen a great liberalizing of the Church of England's theology of heaven and hell, but this had taken place mostly within academic circles. When the Great War brought this theology into the public sphere, it was beginning to be clothed, in part, with language that would not have sounded out of place in spiritualist assemblies. *Raymond* played a significant part in sharing that more liberal vision of the afterlife with the masses, even as it attracted attention with its promise of evidence for spirit communication.

This chapter examines how *Raymond* fit into the changing patterns of theology within the Church of England. Firstly, early twentieth-century spiritualism, when stripped of its more flamboyant Victorian expressions, offered a coherent and attractive view of the afterlife that contrasted with that still being preached by traditional Anglicanism. Lodge's writing on spiritualism, and *Raymond* in particular, exemplified this. Secondly, although spiritualism was condemned or criticized for a variety of reasons by churchmen, there were clergy—even senior clergy—who were drawn to it, finding its teaching conducive to their own. At the same time, Anglican theologians were rethinking the nature of the afterlife, even though traditional expressions concerning heaven and hell still held sway in the language of the church's liturgy and teaching. Thirdly, given that there had been a real and substantial shift in academic thinking, by the time of the war some theologians began to explore alternative ways of framing their theology by reference to biology, anthropology, and psychology. *Raymond* captured something of these shifts and explorations, even as it spoke to those experiencing bereavement in a very particular time and situation.

THE "SIMPLE" TEACHING OF SPIRITUALISM

The Great War revived spiritualism. In the second half of the nineteenth century it had been, by turns, a parlor game, a fraud's trick, a music hall performance, a scientific experiment, or a new esoteric philosophy, but by the end of the century its popularity had waned. With the arrival of war and the widespread experience of bereavement, spiritualists were moved to a greater seriousness and began to offer their services. Thus Gladys Osborne Leonard, one of the mediums visited by Oliver Lodge upon the death of his son, wrote of how, in 1914, she turned to professional mediumship because she had been "deluged" by spirit messages telling her that "something big and terrible" was about to take place.[5] Arthur Conan Doyle, who had long been convinced by the claims of spiritualism, set off around the country at the end of the war giving lectures to packed halls about the teachings of spiritualism. He wrote of his exertions:

> A great body of information which has come to us is information which purports to have come from the dead. When I formed the opinion that this was true I saw at once the enormous importance of it. I thought nothing that I could do in connection with it would be too much trouble. Here were many people in this land needing consolation so badly; so many mothers who had lost sons and wives who had lost their husbands, Rachels mourning and without comfort. If only they could keep in touch with their dear ones what a comfort it would be.[6]

His choice of phrase "Rachels mourning" is significant, being a reference to the slaughter of the innocents in Matthew's gospel and the sound of mothers grieving their lost sons. As a bereaved father and husband, Conan Doyle had an authentic voice. That, and his considerable fame as a writer, meant that his lectures were well attended.

Conan Doyle noted that the phenomena that had been associated with nineteenth-century spiritualism had been useful for drawing attention to the existence of the spiritual realm, but, for him, such excitements were "really nothing at all." "They were sent as an alarm to the human race, something to call their attention, but nothing more; to shake them out of their mental ruts and make them realise that there was something beyond all this" (3). Lodge, like Leonard and Conan Doyle, sought to divert attention from the flamboyant excesses for which spiritualism had been known and toward the teaching it espoused. This, along with the communications themselves, could offer comfort in a time of grief.

The teachings of spiritualism were (and, indeed, still are) set out in the communication known as "The Seven Principles of Spiritualism" received by Emma Hardinge Britten in 1871, allegedly from the spirit of Robert Owen.[7] They were adopted by the National Association of Spiritualists, founded in 1873, and were:

1. The Fatherhood of God
2. The Brotherhood of Man
3. The communion of the spirits and the ministry of angels
4. The continuous existence of the soul and its personal characteristics
5. Personal responsibility
6. Compensation and retribution hereafter for good or evil deeds done here
7. The path of eternal progress is open to every human soul.

Beyond these bare principles there were clear emphases within the testimony of convinced spiritualists and the communications of the alleged spirits.[8]

In the first place, death was of supreme importance in spiritualism because it brought an individual, seamlessly and painlessly, to the point of entry into the afterlife. Accounts were given by the departed spirits of how, on their deathbeds, they had been surrounded not only by living family members but also the spirits of departed loved ones, who offered encouragement as transition approached.[9]

Once the departed spirit reached its destination it began to acclimatize to the new surroundings, in a place known variously as "Summerland," "the Spirit World," or "Spirit Land." Some spirits presented this acclimatization as similar to waking after a good and refreshing sleep. Other spirits explained that they were initially confused by their surroundings and did not realize that they had died because they awoke in a place that appeared similar to the earth they had left. Oliver Lodge described how the spirits were "puzzled" at first by this, but then recognized that this assisted their passage between life and death.[10] Gradually the newly arrived spirits were "weaned" from their earthly tastes and took in the new surroundings.[11] Many of the communicating spirits described the sheer beauty of what they saw as they developed in their spiritual life. What they saw was "earth made perfect."[12]

The spirit was expected not merely to wonder at its surroundings but to grow in spiritual understanding. Whereas the Church of England offered two locations, heaven and hell, in which a soul dwelled for eternity, in spiritualist teaching the afterlife was stratified into "spheres" or "realms" and a spirit was considered to have made "progress" when it rose upward through these realms. The higher the realm, the more refined and perfect in spiritual knowledge the spirit had become. There were also, according to *Raymond*, "halls of learning" where spirits were taught and encouraged in the ways of the higher spheres—to help them rise.[13]

The precise number of spheres varied according to different accounts, but the afterlife was more broadly divided into three sections. The communicating spirits tended to be in the middle section and had little knowledge of life in the highest realms, or in the lowest, darkest realms, where lurked the obstinate ones. "More like a reformatory," is how *Raymond* describes it, where even those full of vice were "given a chance" (230). Even from here there was the possibility of spiritual growth and forgiveness.

The traditional Christian doctrine of eternal punishment was "almost always explicitly denied," and many spirits communicated that hell did not exist.[14] Nor was there any divine judgement. Instead, when a person died their spirit simply entered a sphere that was appropriate to their earthly spiritual development. This bore little relation to a person's professed faith, but rather to their habits of life and thought on earth. "It isn't what you professed, it's what you've done," is how *Raymond* puts it.[15]

God, when mentioned, was an all-pervading force, the supreme spirit, or father of all spirits. He was believed to be loving and merciful, rather than judgmental, but the communicating spirits did not feel it necessary to mention him much. The spirit of Raymond told his parents that he had been taken into the higher spheres and felt himself "faint" with delight at the wonderful experience. He was "charged" with "something—some wonderful power" (232–33). Christ, he said, didn't "mingle" but the departed spirits were "always conscious of his presence" (260). Spiritualists and the communicating spirits were unable to agree about the nature of Christ's divinity.

This teaching, remarkably consistent in spiritualism from the middle of the nineteenth century onward, stood in contrast to what some spiritualists characterized as the church's teaching about the afterlife, and, more particularly, the place of hell and the possibility of eternal damnation. The spiritualists' more extreme caricatures lacked accuracy as well as subtlety, but it was certainly possible to find clergy, in the second half of the nineteenth century, who upheld the doctrine of everlasting punishment for the wicked. In 1864, for example, eleven thousand clergy signed a petition claiming that the everlasting punishment of the wicked was part of Christ's teaching as made plain in the Bible. This came as a response to the judgment against Henry Wilson in 1862, whose contribution in *Essays and Reviews* (1860) stood counter to the final "damnatory clauses" of the Athanasian Creed in the Book of Common Prayer. By the late 1920s everlasting punishment could no longer command such a level of support.[16] Spiritualists had denied the doctrine of everlasting punishment from the outset, preferring an afterlife in which every person had opportunity to progress to perfection and joy.

Raymond was by no means radical spiritualist propaganda; indeed, in an addendum to later editions Lodge advised people to avoid mediums. Lodge wrote as a bereaved father in search of comfort and as one who believed that the claims of spiritualism illuminated and supported the Christian faith rather than superseding or negating it. It is, in fact, a rather stilted and at times confusing book, one that places a great deal of importance on what, to the reader, might seem trifling matters and discussing them at length with alleged spirits, through the mediums and their "controls."[17]

The success of the book most likely lay in the combination of the personality of Raymond Lodge and the overwhelming conviction that ran through the communications from him and from others: that beyond death the newly departed were much as they were on earth. Those who had "passed over" were described in such a way that they were recognized: the color of their hair or shape of their face, for example. A young girl is described as "humming" (187). The mortal Raymond was playful, kind, and brave; beyond the grave he continued to be a vital and loving character, concerned for his family and hopeful that his communications would set their minds at rest. He spoke of meeting old friends and departed family members, and of other soldiers, newly arrived, learning to adjust to their new situation. The language was domestic and simple, redolent with jokes and filial concern. Raymond instructed his father, in one passage, to "tell mother she has her son with her all day on Christmas Day. There will be thousands of us back in the homes on that day . . . please keep a place for me" (207).

Although not written explicitly as theology, *Raymond* brought Lodge's hitherto private theological opinion out into the open. In the third section of the book, he displayed his credentials as a man of science, who had spent years investigating psychic phenomena, but also as one who had engaged with spiritualism and acknowledged its importance. This was what he most wanted to communicate to his audience: that the faith he had encountered through spirit communications was a simple and hopeful one. There was, he argued, nothing in spiritualism that was contrary to a "simple" understanding of Christ's teaching, and he was of the opinion that "no matter how complex and transcendentally vast the Reality must be, the Christian conception of God is humanly simple. It appeals to the unlettered and ignorant; it appeals to 'babes'" (392). Lodge was saying nothing new: it had long been a convention among Christian spiritualists that the teaching of Jesus was "simple" and that the church had overcomplicated it.

Lodge's promotion of spiritualism's simplicity prompted criticism from the scientific community as well as the church. Writing toward the end of his life, Lodge noted that his "sane and permanent belief, which has stood the test of some forty years," had occasioned "ridicule and dislike."[18] Opposition from scientific quarters was to be expected, though, as pioneers of knowledge always found themselves tested and criticized. He had been drawn to it, in the first place, as a scientist struck by William Crookes's experiments, although he concurred that such matters were not "the kind of things for which a technical language has been invented" (346). As the nature of the communications, or attendant phenomena, tended to be ordinary and domestic in character, he argued that the language used to describe them was, necessarily, "commonplace." For Lodge this was entirely understandable, not least because the responsibility for reporting psychic investigations had been left for the most part to people who were untrained in scientific research. It is clear, however, that he found it rather regrettable that other men of science could not get beyond the clumsy domestic language and failed to grasp that the manner in which such a thing was described bore no relation to its veracity. His critics remained "aloof" and "contemptuous" whereas he sought only to learn from the facts (347).

Churchmen who spoke against spiritualism dissected the domestic, familiar, and commonplace language in the second section of *Raymond.* Raymond's claims concerning the availability of meat, cigars, and whisky sodas in the heavenly realms "cheapened" the church's own conception of the future life.[19] On February 14, 1917, Viscount Halifax, a leading Anglo-Catholic layman and president of the English Church Union, gave a lecture at Saint Martin in the Fields titled "Raymond: Some Criticisms." In it he described such passages as "foolish" and a "miserable substitute" for the church's doctrine of the Communion of Saints, advising his hearers to read John Henry Newman's *Dream of Gerontius* if they wished to know of eternal life.[20] At the Church Congress of the same year, William Ralph Inge, the dean of Saint Paul's, described spiritualism as "trifling" and having created a "child's-picture-book" heaven.[21] Lodge countered that any inference that the next world was similar to this was greeted with skepticism—and had been since the time of Swedenborg. The departed spirits who were describing their surroundings were, he argued, describing them in a natural way because they still retained their ordinary powers of observation.[22]

Lodge's theology was not "simple" to nontheologians; indeed, the final section of *Raymond* is neither a short nor an especially easy read. It was his critics, such as those above, who drew on the controversial or domesticated imagery to describe it as such, thereby casting doubt over the author's understanding of theology. Lodge's explicit desire was to communicate what he had learned from spiritualism to the widest readership, in the most attractive and compelling way that he could. By writing about post-mortem survival in language that was removed from the traditional cadences of theology, he stepped into a broader theological debate that had been taking place in the church for some years. Theologians were beginning to explore new ways of articulating that debate, as shall be seen below, but by drawing attention to the "simple" faith offered in spiritualism, Lodge also implicitly cast his critics as making it overly complicated.

THE CHURCH OF ENGLAND AND SPIRITUALISM

Unlike the Roman Catholic Church, which was unequivocal in its condemnation of spiritualism, the Church of England's response had been decidedly mixed even from the 1850s.[23] Some clergy expressed distaste not only for the language of the alleged messages but for the flamboyant séance phenomena, one describing them as "debasing and puerile."[24] George Longridge, of the Community of the Resurrection, described "the whole atmosphere" as "repugnant," noting that no profound spiritual ideas had ever emerged from spiritualism.[25] Others marked the biblical injunction against necromancy or were anxious that séances might conjure up demons.[26] In the period when Lodge was writing *Raymond,* the strongest criticism was that fraudulent or predatory mediums were preying on the weak and vulnerable in the time of their bereavement. The *Church Times,* reviewing *Raymond,* claimed that most mediums were frauds, delighting in the pain and grief of others.[27]

There was no "official" Church of England response to spiritualism until 1920, when the Lambeth Conference discussed it briefly and concluded that there were "grave dangers" in "the tendency to make a religion of spiritualism."[28] Despite this, curious clergy attended séances and meetings discussing spiritualism, and their presence was noted by spiritualist newspapers.[29] Others, among them the bishop of Carlisle, Harvey Goodwin, and the bishop of Ripon, William Boyd Carpenter, joined the Society for Psychical Research (SPR), of which Lodge was member and president.

Some were content to be known as interested supporters. Percy Dearmer, editor of the *English Hymnal* (1906) and a canon of Westminster Abbey from 1931, became involved with spiritualism in part through membership of the SPR and in part from his wife's interest in automatic writing. In 1920 Nancy Dearmer published a book called *The Fellowship of the Picture,* delivered to her hand via the spirit of a departed friend, killed in France in 1918. Percy Dearmer put his name to the publication and wrote the introduction for the book, saying that it was written "as from one who was urgent to give a message to the world, who was the friend we had known, and whose identity was familiar and unmistakable."[30] Walter Matthews, dean of Exeter and then Saint Paul's, was another senior clergyman who declared a quiet interest in spiritualism. Having been fascinated by it as a subject since childhood, he joined the SPR as soon as he could afford the subscription, sat with mediums, and attended séances. Matthews saw spiritualism as offering proof of life beyond death, but, more than this, saw it as presenting a call to the church to reconsider a more explicit teaching of the Communion of the Saints.[31] Worthy of note is the Anglican layman and prime minister W. E. Gladstone, who, as well as attending séances and joining the SPR, read spiritualist literature. That he read such texts on Sundays, the days that he reserved for religious reading, suggests that he was interested in the theology of spiritualism more than the phenomena.[32]

Parish clergy who took to advocating spiritualism were not deprived (that is, dismissed from their positions) for their beliefs, although some of them met with episcopal disapproval. George Vale Owen, vicar of Orford in the diocese of Liverpool and author of several books about spiritualism, displeased his bishop (Francis Chavasse) but left his parish in order to concentrate on his lecturing and writing—not, as Conan Doyle intimated, because he was hounded from his office.[33] Graeme Maurice Elliott, rector of Saint Peter's Cricklewood, cofounder of the Confraternity of Clergy and Spiritualists and author of *Angels Seen Today* (1919), made a nuisance of himself for the bishop of Guildford by holding spiritualist meetings within the diocese and encouraging other clergy to attend, but was given nothing more than a warning to desist.[34] On the whole, such clergy were less interested in the phenomena associated with séances and far more concerned with engaging with the sort of ideas and imagery that Lodge dealt with in the final section of *Raymond*.[35]

They were not alone in their desire to see the Church of England reframe its theology of life after death. In fact, Lodge and the clergy spiritualists of the early twentieth century were entering late into a debate that had been fermenting in academic circles since the middle of the previous century. In 1853 Frederick Denison Maurice, professor of theology at King's College London, published *Theological Essays*, in which he challenged the Church of England's "traditional" understanding of the afterlife. The publicly accepted (Calvinist) view was that beyond death a soul was judged but then waited in a "static" state, aware of its ultimate and eternal destiny beyond the final resurrection but unable to alter this. The righteous would spend eternity in bliss, but the wicked would be condemned to an eternity of punishment and torment. There was no room for a middle state, since the doctrine of Purgatory had been explicitly denied by the Reformers and proscribed by Article XXII of the Thirty-Nine Articles of Religion. God, who was unchangeable, would not change his mind once judgment had been passed, so the eternal destiny of each person was fixed.

By the seventeenth century this idea of a static state was being questioned, largely, although not entirely, from outside the Church of England. Cambridge Platonists, such as Peter Sterry, began to explore the possibility of post-mortem repentance, as did the nonconformist Jeremiah White and Unitarians David Hartley and Thomas Belsham.[36] These men suggested that post-mortem punishment, although necessary, might be reformatory rather than retributive, thus allowing for the possibility of repentance and even salvation. Comments voiced privately were made public by Maurice in his essays. His argument was rooted in a discussion of the Greek word for "eternal"; he confronted the prevailing notion of "eternal" as meaning "everlasting future duration," arguing that this was a product of Enlightenment teaching rather than Christianity. "Eternal," he claimed, was a characteristic of God, not a period of future time, and therefore it was dynamic rather than static. Most controversially, he suggested that if eternity were dynamic then, beyond death, it was possible for an individual to progress in faith. Death removed the obstacles of human life and offered "perpetual growth in the knowledge of God and in the power of serving him."[37] It meant, ultimately and importantly, that the wicked might not be condemned to everlasting punishment after all, but able to repent.

Maurice's work was widely published and the newspapers, according to his son and biographer, teemed with articles in response. Maurice was criticized for his essays and subsequently dismissed from his post at King's.[38] The college council considered that his words regarding the future punishment of the wicked would unsettle the minds of students, conveying a dangerous possibility of salvation for all (191). His dismissal only served to make this more of a cause célèbre in academic circles, and others soon took up the baton. The Broad Churchman Henry Wilson, in his contribution to the liberal *Essays and Reviews* (1860), suggested that beyond death a soul might develop and be saved. In 1877 Canon Frederick Farrar of Westminster Abbey decided similarly that there might be the opportunity for post-mortem repentance. Evangelicals Andrew Jukes and Thomas Rawson Birks seemed to agree, and by 1889 the publication of another collection of essays, *Lux Mundi,* suggested that a hope of post-mortem repentance was becoming a commonplace in academic circles.[39]

In 1915, Hastings Rashdall gave the Bampton Lectures at Oxford, choosing the doctrine of the Atonement as his subject. Rashdall argued that, for the present age, the church needed to rethink its traditional theology about the afterlife and, in particular, the eternal punishment of the wicked. God, he argued, "could not have designed everlasting, meaningless, useless torments as the sole destiny in store for the great bulk of his creatures. That doctrine is dead, though much of the language which really implies it is still repeated in the church, the school, and the theological classroom."[40] Instead, and in the sort of language not out of place in spiritualist circles, Rashdall argued that it was right intentions rather than right beliefs that were commended by Jesus as the way of entering the Kingdom of Heaven. Indeed, the Kingdom of Heaven was open to those who were beyond the reach of Protestant Christianity. This was not a new and radical idea: drawing on the subjective and experiential theology of the twelfth-century thinker Peter Abelard, he claimed that it lay at the very heart of the teaching and the self-sacrificial death of Christ. The character of God was love and his desire was to save, meaning that Jesus's teaching did carry with it a "latent Universalism" (19). Rashdall went further, imagining salvation as progressive and dynamic—and, significantly, beginning in this world. Training and education in love and faith continued, he thought, beyond the grave. Growth and development in this life was a gradual process, so there were grounds for thinking that, after death, it might be completed gradually and by "degrees" (461). A clear sign of how far theological debate had moved since the publication of Maurice's *Theological Essays* is that Rashdall was able to use such provocative language without fear of losing his job.

New theological descriptions took the place of the static post-mortem state and unchanging judgmental God, but a new language was needed to express them. The language of punishment, dread, and torment did not fit well with dynamism and growth. Theologians turned away from and even distanced themselves from "traditional" language. So Canon Burnett Streeter wrote in 1917 of the need to find new ways of presenting the Christian hope. Traditional images of heaven and hell were "morally revolting."[41] He was not alone in this assessment. The liberal theologian Leslie Weatherhead said in 1923 that "in this age, fortunately, we have got away from the old doctrine which taught that all man's destiny had been arranged for him, willy-nilly, by some inscrutable and immutable power."[42] Writing in 1932, the archdeacon of Westminster, Vernon Storr, said that when it came to the everlasting torment of the wicked, "our whole outlook has changed" from that of the sixteenth-century reformers and nineteenth-century Evangelicals.[43] The church's theology had developed considerably, but theologians had yet to find a way of framing it in non-traditional language, meaning that, however altered it was, it still sounded less comforting, inspiring, or hopeful than spiritualism.

THE DEVELOPMENT OF NON-TRADITIONAL TROPES IN THE THEOLOGY OF LIFE AFTER DEATH

If the traditional language of heaven and hell was unhelpful in conveying a new liberal theology, then an alternative framework was needed. Beyond the church, developments in biology, anthropology, and psychology offered possibilities. Theories of evolution and of human instinct and experience chimed, in the minds of some, with a deeper theological truth; domestic and familial aspirations were gain-ing respect. Lodge himself was not content simply to repeat the spiritualist language of the previous century; he drew on the same tropes as some contemporary theologians to press his claims for spiritualism.

Evolution demonstrated that change was a natural part of existence and "growth is the law of all life."[44] It showed that God's design was process and progress—and therefore human existence must inevitably be a movement toward something qualitatively better than this life. Human beings progressed to what Storr called a "higher significance."[45] This could not take place within the span of a mortal life. The full development of a person could surely happen only if eternity were taken into account. As Storr noted, "When we consider the incompleteness of man's earthly life, and how he might travel much further along the road of perfecting his personality if further opportunity were given him, we feel that death cannot be the end. The purpose of God in creating man defeats itself if man's possibilities of progress are limited to the threescore years and ten of earthly life" (95). Instead, it was clear that a person would spend a dynamic life beyond the grave developing his or her character, making use of the raw materials that such a life had given. The higher ideals present within mortal existence were but "gleams of a better world of which we are heirs" (66).

There were limits to the application of evolution, as far as some were concerned. Thus Inge turned away from the "proofs" offered by science. Instead, he looked to what he deemed the permanent attributes of goodness, truth, and beauty, which, though demonstrated only imperfectly in the world, were signs of a real and "eternal" world that was "around us and within us."[46] John Baillie, professor of theology in New York and Edinburgh, argued that theology had become too entranced by evolution and that theologians had decided that "evolutionary progress is the law of all being" and interpreted everything in the light of growth. It was, he cautioned, "parochialism and anthropomorphism" to suggest that human beings could progress beyond death in such a way that would bring them ever closer to divine perfection and "infinity by approximation."[47] Baillie did not exclude the possibility of post-mortem growth for those who had not found fellowship with God in this life, but he was wary of a simplistic connection between evolutionary theory and Christian theology.

Lodge, on the other hand, comfortably accommodated the language of evolution in his theology. Though he was a physicist, whenever he wrote of the significance of spiritualism more broadly his language encompassed the progress and development that he observed in the natural world. He wrote about this extensively in the third section of *Raymond*, "Life and Death," to demonstrate how the body's natural pattern of growth and decay suggested that beyond death there would be personal continuation. He had already applied the theory of evolution more widely, as in *Man and the Universe* (1908) when he wrote concerning the doctrine of Atonement that "we are beginning to realise that we are part of nature, parts of a developing whole, enfolded in the embracing love of God."[48] Indeed, it was when he looked to evolution that he apprehended the reality of God's plan:

> For consider what is involved in the astounding idea of Evolution and Progress as applied to the whole universe. Either it is a fact or it is a dream. If it be a fact, what an illuminating fact it is! God is one; the universe is an aspect and a revelation of God. The universe is struggling upward to a perfection not yet attained. I see in the mighty process of evolution an eternal struggle towards more and more self-perception, and fuller and more all-embracing Existence—not only on the part of what is customarily spoken of as Creation—but, in so far as Nature is an aspect and revelation of God, and in so far as Time has any ultimate meaning or significance, we must dare to extend the thought of growth and progress and development even up to the height of all that we can realise of the Supernal Being. (314)

Elsewhere he wrote of life beyond death, "Let us realise that love continues, and that our friends are still rejoicing in our successes, lamenting our failures, appreciating our efforts, regretting our lamentations and hoping that we may be happy, even as they are, in doing our work and 'carrying on' in a worthwhile spirit till the time comes for joyful reunion."[49]

If evolution did not sufficiently demonstrate personal survival, then "experience" supported it, according to some. "Experience," or, more particularly, the instinctive human apprehension of the divine, was offered as a counter to "tradition" or "authority" in theology more generally. Thus Walter Matthews argued that "apologetics of the old sort are of little value." Instead, Christian teaching needed to be written in light of modern knowledge and experience.[50] Experience did not replace thought but was its stimulus. More than this, religious experience was universal and therefore not confined to Christianity. He argued that "religion, in spite of its wide diversity of forms, possesses a recognizable character throughout. It appears to constitute one movement of the human spirit, and though it has proved difficult to define, it is not hard to recognize. Indeed, the student of comparative religions is often startled by resemblances between its highest and lowest forms" (3). James Welldon, the dean of Durham, drew on the work of anthropologists like James Frazer to show that the desire for God and the longing for immortality were a natural part of the human instinct.[51] Human ideals, such as truth, justice, mercy, and virtue, were, he argued, "essentially immortal"—indicating that the soul did not die (27). Storr said that the complex ancient burial sites of primitive religions demonstrated that human beings had always seen themselves in the light of an eternal existence beyond death.[52] The experience of human nature itself demonstrated that we were being trained for something beyond mortal existence.

Weatherhead described the instinct for post-mortem survival as "reasonable."[53] The bishop of London, Arthur Foley Winnington-Ingram, similarly thought that immortality was a "persistent instinct" of human beings from all ages.[54] In fact, he argued that it would be "unreasonable" for God to work so hard on the universe to produce "so little," if this life were all that existed for individuals (x). Storr deemed it "unthinkable" that if man's highest achievement was to know God, that same God would cut off further progress in knowledge at the point of death.[55]

Lodge, in *Raymond*, drew on the human instinct for survival and the reasonableness of post-mortem spiritual development to support the claims of spiritualism. He took the example of the human experience of the present, which was always suffused with memories of the past and anticipations of the future. The future dictated the actions of the present, even among the higher animals. "Without any idea of the future our existence would be purely mechanical and meaningless: with too little eye to the future—a mere living from hand to mouth—it becomes monotonous and dull."[56] Human beings, transcending the animal kingdom, reasonably regarded the greater questions of "whence" and "whither" with interest. The future life, or concern for it, affected the present state of life, and it was entirely appropriate that people gave it serious thought.

Alongside the arguments from evolution (drawn from science) and religious instinct or experience (drawn from anthropology) came an argument for post-mortem growth and progress laden with domestic and familial language that was born out of the experiences of the Great War itself and becoming increasingly acceptable, for all its sentimentality. For many Christian preachers it was simply inconceivable that young men who had not had the opportunity to fully consider the claims of the Christian faith might find themselves judged severely by God at their death. Storr argued that such men were "spiritually immature" and God, surely, would not condemn them. Indeed, "God may have other training grounds for immature souls."[57] Winnington-Ingram, preaching to those bereaved by war, claimed that the son, husband, or brother who was manifestly growing in character in this life would be "carrying on his education" and growing in "the sunny land of Paradise."[58] Elsewhere he argued that those who had been loved would be recognized in heaven: "What is clearly taught in our religion is that the future life is some sort of spiritual body which will be as well adapted to the Spirit world as our present body is to this world, that we shall recognize one another, as we do here, and that we are bound to those who have gone before by the same intimate tie of love and mutual prayer which binds us here." The departed spirits would be "near" to us, whispering "sweet thoughts" into our minds. For this reason, he argued, "it is perfectly impossible that it is wrong (as used to be supposed) for the mother to pray for her son who has gone before, or for that son to cease praying for his mother in Paradise" (xi–xiii).

Theologians as well as preachers took, as point of reference, the imperfect departed soul, known to every bereaved person. Cyril Emmet, New Testament scholar, wrote that "even in the worst we know, we ourselves can always find some spark of goodness, some traits of love and unselfishness. We dare not abandon the hope of progress and forgiveness after death for such a soul. . . . The Good Shepherd will not rest until he has saved the goats."[59]

Lodge argued that while young men killed in war had suffered an "unnatural truncation" of their development into maturity, they continued to grow in spirit beyond death. Bodily wounds were temporary marks, but habits of character, for good or ill, marked their souls.[60] In his communications, the alleged spirit of Raymond suggested that learning and education were indeed significant aspects of the spiritual realm. He spoke of spirits reading books that had not yet been published on earth, even though not all spirits were allowed to read them (209). More important, though, was growth in righteousness. After a short time in Summerland, Raymond communicated: "I can feel my ideas altering. I feel more naturally in tune with conditions very far removed from the earth plane" (234).

Theological argument dressed in the language of evolution, experience, and the personal tragedies of bereavement became more common in presentations of the afterlife. There was still no "official" doctrine from the Church of England, beyond what was laid out in the Book of Common Prayer. When the archbishops set up a commission in 1922 to "survey the whole field of theology and produce a systematic treatise," they included, among many other controversial subjects, the idea of judgment and the future life.[61] When the commission finally made its report in 1938 it affirmed the prevailing arguments for spiritual progress and the possibility of postmortem growth and repentance. However, it upheld what it regarded as the more traditional teaching as well. "There is great peril in the easy-going sentimentality of some modern Christianity, which supposes all who have departed this life to be forthwith in 'joy and everlasting felicity'. . . . Such a view is inconsistent with the solemn warnings of scripture and especially the Gospels themselves, and converts the hope of immortality from moral stimulus to moral narcotic" (217). The commission noted that there were passages in scripture that, taken by themselves, might be "universalist in tendency" (although not which ones they were), and also that it would be difficult to find a basis in the New Testament for abandoning all hope of opportunity beyond death. However, even though God's nature was to love and forgive, judgement and the possibility of eternal death were so significantly present that a clear and definitive move away from traditional teaching was not advisable.

It is, in conclusion, easy to read *Raymond* in isolation because of its notoriety, or see it simply as part of the revival of spiritualism that occurred at the time of the Great War. In fact, the well-known vivid passages offering Raymond's observations on life beyond death represent only a part of a fuller, worked-out system of belief that had been common to the majority of spiritualists since the end of the nineteenth century. More than this, *Raymond* appeared at a time when the Church of England was already reshaping its language better to reflect the shift in the theology of the afterlife that had been taking place since the middle of the previous century. Although the domestic and commonplace language of *Raymond* was criticized and derided by some in the church, significant theologians and preachers were testing out commonly understood contemporary tropes, such as evolution, progress, and experience, to describe the spiritual possibilities beyond the grave. Some were even prepared to speak and write of the afterlife in such warm and domesticated terms that would not have been out of place in spiritualist circles—and would certainly have been familiar to readers of *Raymond.* In many respects, *Raymond* remains a work of its own very particular time, the story of one family coming to terms with loss during the Great War. The beliefs that underpin it, however, fit into a much larger perspective, shaped not only by widespread experiences of tragic bereavement but by a longer struggle to reframe an emerging theology of life beyond death in a new and attractive way.

ELEVEN

OLIVER LODGE'S ETHER AND THE BIRTH OF BRITISH BROADCASTING

David Hendy

> Driving alone at night, in the darkened car, reassured by the nightlight of the dashboard, or lying in bed tuned to a disembodied voice or music evokes a spiritual, almost telepathic contact across space and time, a reassurance that we aren't alone in the void.
>
> Susan Douglas, *Listening In: Radio and the American Imagination*

IN HER book on the place of radio in the American imagination, Susan Douglas conjures a scene many of us will recognize from a lifetime of listening. The aesthetic parallels between radio broadcasting and notions of spiritualism are striking. Radio involves tuning in to signals that are otherwise invisible and inaudible. It allows a voice to become untethered from the human body and travel freely across the open air to be heard—at the very same moment—hundreds, perhaps thousands of miles away. "As if by magic," one is tempted to add. Because, while communication across the electromagnetic spectrum is no longer mysterious in the strictly scientific sense, when radio first impinged on public consciousness in the opening years of the twentieth century, it felt to many like the electrical equivalent of telepathy—a kind of ghost trick. As Gillian Beer points out, a world suddenly full of disembodied sounds becomes "the quintessence of the magical."[1]

Those who witnessed the earliest wireless voices certainly deployed psychic metaphors. Take the following account from a young man working as a wireless operator on a small ship off the coast of Ireland just after the First World War, listening out for the dots and dashes of Morse code. Bored and distracted one night, he suddenly heard someone actually saying, "Hullo." "The voice sounded muffled, and as I had the 'phones on my head I took them off to hear better. There was nobody in the cabin, however, and I resumed my watch. In a few moments somebody said 'Hullo' again. I picked up a heavy ebony ruler, and looked round again. I thought first of a mad stowaway, then Conan Doyle and his 'spirits' and lastly of my past sins. The voice spoke again, and I awoke to the fact that it was Radiotelephony."[2]

It was precisely this uncanny quality that stirred myriad imaginations in the closing years of the nineteenth century and the opening years of the twentieth. Writers such as Rudyard Kipling, Mark Twain, and Arthur Conan Doyle all played creatively with the phenomenon of disembodied entities floating, unbidden through the air. The classic example of this burgeoning genre is Kipling's 1902 tale "Wireless," written not long after the author's young daughter, Josephine, had died, and when he still imagined sensing her as a radiance lurking in his garden.[3] Set in a pharmacist's shop on the south coast of England at night, "Wireless" tells the story of a young radio enthusiast trying desperately—though ultimately failing—to pick up a message from across the bay. This original message goes astray and our hero witnesses a different, entirely unbidden message, arriving from *nowhere*: a consumptive apothecary drugged up on a chloroform cocktail starts burbling the words of a poem by Keats, scribbling down sentences like a human Morse code receiver.[4] It was as though wireless somehow lifted the veil between the material world and another, more mysterious one filled with the wandering souls of the dead. Scholars of media history like this story because it points so neatly to the broader relevance of radio's arrival as a novel—though potentially unsettling—form of communication. Jeffrey Sconce, for instance, writes eloquently of how radio's sound "without material form" effortlessly evoked the idea of it being a gateway to "electronic otherworlds," a notion that neatly conjoined a technology of communication riding the electromagnetic spectrum with vaguer and older themes around the transfer of consciousness.[5] John Durham Peters explores how, among psychic investigators across the centuries, "'communication' was a concept that straddled the line between physical transmissions . . . and spiritual ones."[6] And in his recent book on the links between early electronic media and literary modernism, David Trotter suggests that the appearance of "connective" media such as early wireless energized writers into probing the possibility of new social relationships being forged, especially for those previously separated in time or space.[7]

In all these instances, the bedrock idea—what links science, spiritualism, and radio so deliciously—has been that invisible connective tissue, that all-pervading medium, the ether. Of course, scientific discoveries—beginning with the Michelson-

Morley experiments of the 1880s and ending with Albert Einstein's two overarching theories of relativity published in 1905 and 1915—had brutally dispensed with the need for a thing such as the ether to exist at all by the time radio itself was established as a widely used medium. But well into the twentieth century it kept on being talked about *as if* it still existed: it was simply too useful to be dispensed with as an *idea*. For a start, it was a neat shorthand for the electromagnetic spectrum, something that *did* exist but nevertheless represented a new and (for laypeople) difficult concept. More crucially, the ether was a vague enough entity to become a repository for all sorts of imaginings and ideas, not the least of which was that notion of it being an electronic otherworld where spirits might reside.

These magical and metaphorical properties are familiar enough tropes. But I want to focus in this chapter on something a little less well-attended, what I will describe as the ether's *ethical* dimension, and its implications for broadcasting in the 1920s. As we know, it was Oliver Lodge who most consistently advocated the existence of the ether in the second half of the nineteenth century and indeed persisted in doing so well after Einstein's theories had gained currency. As other essays in this collection show, the reasons for this determined advocacy ranged from the strictly scientific—his determination to maintain a nonatomic approach to physics—through to the deeply personal—his desire to understand how his son, killed in the First World War, might somehow have survived death. These seemingly disparate concerns had in common the notion of the ether as an all-enveloping space capable of connecting different realms of existence. In writing of an ethical dimension to the ether, I want to take inspiration from Sconce's groundbreaking analyses of what he calls "the symbolic import and cultural resonance" of the ether.[8] But I want to nudge my own discussion toward a more specific focus on knowledge and its publicness.

My starting point is this: that tangled up in Lodge's ruminations on telepathy, post-mortem existence, and the ether was a new model of consciousness. Sconce, focusing on the United States, detects an "anxious, pessimistic, and melancholy" dimension to this model—popular and creative feelings unsettled by the sheer eeriness of it all.[9] We find this in Britain, too—as Kipling's ghost story testifies. But I suggest that, perhaps chiefly because of the trauma and dislocation of the First World War, we also find Lodge's ideas about consciousness speaking to a more optimistic—even utopian—debate about the possibilities for sharing knowledge as a means of responding to the trauma of war.

First, then, I intend to tease out from his own published writings on the ether Lodge's model of etherial *consciousness*, and, more precisely, the implications this had in his own time for ideas about the nature of knowledge. Second, through examining some of the popular literature about wireless published between approximately 1913 and 1922, I want to suggest that these ideas about etheric consciousness and knowledge were also to be found in a broader range of popular writing, and that, through being embedded in discussions of wireless as an evolving phenomenon, they sustained what might otherwise have been regarded as redundant Victorian concepts right through until 1922, when wireless was transformed into institutionalized broadcasting in the form of the British Broadcasting Company. Third, and as a result of this, I wish to argue that the BBC's sense of purpose in the 1920s, and in particular the public service ethos espoused by its founder, John Reith, was articulated in ways that directly echoed (indeed, were arguably informed by) the same etheric notions established by Lodge many years before. In short, I want to suggest that we cannot fully understand the founding ethos of the BBC as a public service broadcaster unless we attend to the place of the ether in the discourse of early twentieth-century Britain. Conversely, we might also conclude that Lodge's etheric thinking was to have its lasting ethical legacy in the form of a public broadcasting system that survives to this day.

LODGE, THE ETHER, AND THE NATURE OF KNOWLEDGE

Lodge had questioned the nature of knowledge from an early age. His own childhood experience of grammar school—what he called "the fag-end of an old tradition"—had been one of bullying, beatings, and dull, dispiriting rote learning.[10] For him, insight came from experiments he conducted in his bedroom or from his regular Friday evening visits as a teenager to lectures at the Royal Institution by the celebrated physicist John Tyndall. Lodge was attracted to the Royal Institution's inclusive approach, speaking accessibly to those with, as yet, no formal scientific training. But Tyndall's lectures also stimulated his interest in the *invisible* forces shaping the physical world. After one such occasion, Lodge recalled walking back through the streets of London "as if on air," with a freshly kindled "sense of unreality in everything around."[11] He later claimed it was then that he experienced "an opening up of deep things in the universe, which put all ordinary objects of sense into the shade": "the square and its railings, the houses, the carts, and the people, seemed like shadowy unrealities, phantasmal appearances, partly screening, but partly permeated by, the mental and the spiritual reality behind" (78). "I decided," Lodge said, "that my main business was with the imponderables"; indeed, "The Imponderables" was the title he gave to his notebook from this time (343–44).

Although much of Lodge's subsequent work on the ether had a specific scientific purpose—namely, to explain the movement of energy such as light through space—it clearly also had a broader motivation: to pull together a range of thinking about how true meaning resided in a dimension of existence beyond that which our "ordinary" senses could apprehend. Indeed, in his 1933 book, *My Philosophy: Representing My Views on the Many Functions of the Ether of Space*, Lodge makes clear that he had "upheld the reality of the ether of space," *despite* the damning implications of Einstein's theories precisely because he found it "necessary to a *philosophic* contemplation of the sensory universe." "When in my old age I came to write this book," he added, "I found that the Ether pervaded *all* my ideas, both of this world and the next."[12]

In his famous address to the British Association in Birmingham twenty years earlier, Lodge had quoted the theoretical physicist and polymath Henri Poincaré: "Whether the ether exists or not matters little . . . what is essential for us is that everything happens as if it existed."[13] For Lodge, what mattered was the omnipresence of the ether. He went on to speak of it as "the universal connecting medium which binds the universe together and makes it a coherent whole, instead of a chaotic collection of independent isolated fragments" (33). In the narrow scientific sense, Lodge used his 1913 speech to challenge what he called the "modern tendency" to "emphasize the discontinuous and atomic character of everything" and to assert in its place a "principle of continuity" (22). But in thinking of the various elements of the universe as being seamlessly joined in this way, he was also positing a view of the material world of humanity's existence as something always and everywhere embedded in and acted upon by the nonmaterial, invisible ether. "Matter," he suggested, only served "as an index or pointer, demonstrating the unseen activity all around."[14]

The nature of this "unseen activity" was elaborated in Lodge's most widely read book, which told the eponymous story of his dead son, Raymond, communicating with him from beyond the grave. In it, his account of spiritualist séances is followed by a more ruminative section arguing that what he had just described was based on a "coherent system of thought."[15] His chief concern was to explain scientifically the idea of life after death, a central notion being that the material body is by no means the "complete organism": after death, a nonmaterial part persists even when the material portion we recognize with our everyday senses ceases to be "animated" (289–91). One key implication was that consciousness existed separately from the body; indeed, since it was unbounded by the body it extended physically *beyond* it. This, for Lodge, was what explained phenomena such as telepathy. But, crucially, it also posited the ether as a kind of sea of consciousness: a space without hierarchy, without a center or a boundary, in which personalities and their thoughts endured and moved about freely. Just as the singular ether joined together separate realms of *being*, it also joined together separate realms of *thought*. An "atomic" approach to knowledge kept ideas isolated. But Lodge advocated something akin to what we would now call interdisciplinarity: "No human being can always be satisfied with any one department of knowledge; there are times when he must seek a more comprehensive view."[16] For Lodge, it was of course the ether that nurtured and sustained this comprehensive view: its continuous and omnipresent nature was what created the possibility of intellectual *coherence* in the face of fragmentation.

The notion of coherence—and specifically, the unity of knowledge—lies at the heart of Lodge's "philosophical" ether, then. And an interest in the public benefits of unifying otherwise distinct fields of thought consequently colored his whole intellectual outlook. There was, for instance, Lodge's membership of the aptly named Synthetic Society, a London-based discussion group founded in 1896 as a kind of successor to the old Metaphysical Society, and designed, in Lodge's words, "to get together as many sides of the intellectual world as could be managed, and to give them free utterance."[17] And then there was his own assessment of the significance of wireless, and specifically, its ability to send signals over the horizon. This potential for communicating human thought and intelligence around the globe, he reckoned, was something providential "to those who were working for international conversation and co-operation" (233–34). In both instances, Lodge extolled a vision of human improvability built largely on the idea of connecting the otherwise disparate: "water-tight compartments" represented an approach to intellectual life that had long "retarded mutual understanding."[18]

THE RADIO ETHER

How does this conception of the ether, fully fledged by the outbreak of the First World War, help us to understand the origins of the BBC nearly a decade later? The chronological gap we have to account for is in fact greater than one single decade, for as David Wilson has suggested—and as Richard Noakes explores elsewhere in this collection—Lodge's thought, though articulated in the twentieth century, bore "the imprint of late nineteenth-century ideas."[19]

One way to begin unpacking how an essentially Victorian conception entered the DNA of a "modern" 1920s institution is to disclose some of the subterranean debates around the purposes of wireless in those nine years between Lodge's 1913 address and the establishment of the BBC in October 1922. During this period, we see wireless technology develop from an experimental form of point-to-point communication into the prototype of a widely scattered, public medium of entertainment and information, the forerunner of what we today recognize as "broadcasting." This was a transformation that took place against the backdrop of a still-contested notion of what the ether itself represented; not so much scientifically, but rather culturally and politically. To be specific, Lodge's vision, with its profound commitment to ideas of coherence and universality, had to compete with a rather narrower conception advocated by his great rival in developing wireless technology, Guglielmo Marconi. The contest between these two competing etherial conceptions was not only very much alive between 1913 and 1922; radio's future pathway greatly depended on which version of would prevail.

For Marconi—someone racing to develop radio as an entirely private channel of communication between sender and receiver—the ether was more an obstacle to be overcome than something to be embraced as universal, accessible, and binding. In the very same year that Lodge addressed the British Association on his principle of continuity, we find, for instance, the Marconi Company's house magazine *Wireless World* explaining to its readers how the aerial masts of Marconi House in London represented a "new power . . . over the elements and forces of Nature," how wireless was something capable of *annihilating* space: governments could communicate with the outposts of empire, generals could direct their armies from afar, and ordinary paying subscribers could speak confidentially to friends or relatives on the other side of the globe.[20] For Marconi, envisaging radio as a form of narrowcasting for subscribers, the ether's ability to disperse radio signals freely in all directions was problematic. Unsurprisingly, therefore, *Wireless World* cultivated a notion of the ether that was far from the fully shareable realm others imagined. In 1913 it reported the complaint of an amateur wireless operator, "CQ," one of several thousand then scattered around Britain excitedly listening in on homemade receivers to the messages flying freely around them. The Marconi company had set up "wireless schools" to train operators, mostly for all the ships then hurriedly equipping themselves with radio in the wake of the *Titanic* disaster. Trainees would learn the basics of Morse code, how to tap out messages at speed, and how to decode any messages that were incoming. Naturally, this involved lots of trial transmissions. The London-based schools, CQ claimed, were "a serious nuisance," sending out signals of such strength that it was all but impossible for amateurs like him to hear anything else. The editorial response of *Wireless World* to this complaint was blistering. Those being trained at Marconi's schools, it explained, were "taking up wireless seriously as a profession": "In a short time, perhaps, they will be in charge of the wireless plant of some big liner, and on their efficiency and nerve and self-forgetfulness may depend the lives of hundreds. Have not they, then, a thousand times the prior claim on the ether? Is it 'good form'—to say the least of it—for an amateur who is often merely amusing himself in his spare time to try to 'put his spoke in their wheel?'"[21]

Wireless World sought constantly to impose order and hierarchy on the otherwise organic, grassroots phenomenon of amateur experimenters. Etiquette apparently demanded that if one person's signals interfered unduly with those of others, he (it *was* usually a he) would be sent a stream of dots and dashes calling on him to cease forthwith. On this basis, the magazine suggested: "No one should be allowed to work a transmitting station who is not able to read Morse rapidly and accurately enough to learn at once if he is in any way interfering with traffic; and, of course, anyone who learns this and fails to attend to this information is unworthy to be allowed the 'freedom of the ether.'"[22] As for big business interests, a "vigorous campaign of litigation" was pursued to protect any of the Marconi Company's patents against "even the hint of competition."[23] Licenses were enforced on various business partners and subsidiaries to ensure exclusive use of Marconi equipment, and the company refused to interconnect with any competing systems—wireless or cable.[24] The ether itself was not always invoked explicitly here, but the implicit conception was of something to be harvested, carved up, fought over, seized, protected.

Set against this monopolistic approach, Lodge's own etheric thinking stands out even more starkly for its ethical dimension. Not that Lodge was entirely consistent. The syndicate he formed with the London-based telegraph manufacturer Alexander Muirhead between 1887 and 1911 regularly asserted its own exclusive rights over technology; Lodge himself had "begun planning to patent his own system of syntonic wireless telegraphy well before Marconi's patents were finally sealed and publicly issued."[25] Yet this did not prevent his more philosophical outpourings articulating the ether as something *beyond* individual ownership or control. Moreover, the "ethical ether," as we might call it, was now being shaped conceptually by a range of voices broader than just Lodge's own—indeed, by myriad imaginings and rhetorical flourishes. What unites them all is the fundamental assumption that the ether is unavoidably public and open: that, as a result, it is a space capable of encompassing a multitude of messages and thought.

We get a whiff of such thinking—despite the magazine's usually relentless pro-Marconi tone—in *Wireless World* itself. In 1917, for instance, it reported on the activities of German wireless stations, "radiating" what it called "vast floods" of false news reports. These "ceaselessly streaming" communiqués, it pointed out, were being received by American stations and then instantly recirculated "throughout the New World."[26] In this global space of communication, disinformation was spreading like the "poison gas" first unleashed in the trenches to devastating effect back in 1915.[27] It was suggested that Britain and the USA should contemplate inventing "a wireless device" of their own to "ensure that nothing but truth reaches the receiving station: this would involve "super-saturating the ether with truth ions, capable of performing the necessary filtration of the wireless waves as they pass through it.[28] Fanciful stuff, but in deploying metaphors of contagion—in acknowledging that wireless could spread information in an *un*controlled way—it implicitly positions the ether as a medium in which rival viewpoints are inevitably forced to coexist and compete.

This more capacious conception is equally implicit in the stream of bizarre phenomena reported by early wireless experimenters. In October 1913, one British amateur described hearing transmissions from Paris via a metal bedstead, a piano, even a nest of cake tins.[29] Several years later, *Popular Science* magazine revealed a farm building in Ohio seemingly alive with mysterious sounds after the opening of a nearby transmitter: a waterspout in one corner humming the strains of an orchestra, a tin roof in another making political speeches.[30] There were even reports from Sweden of a "singing" coal shovel.[31] At a time when so few members of the public understood the workings of the electromagnetic spectrum, it was easy to be mesmerized by the idea of the air groaning with messages passing through walls and people—lurking, perhaps for ages, invisibly and unheard—until alighting on anything that might act as a kind of impromptu receiver and suddenly, dramatically, making their presence felt. It was but a fictional echo of the more "scientific" rationale erected by Lodge himself during the war in trying to explain the messages he apparently received from Raymond. Telepathy and radio were both understood as dependent on the same underlying medium of the ether. And in the case of radio, we sense a familiar space of random communication, in which, thanks to the vagaries of the ionosphere and the recurrence of "stray" signals, nothing was ever fully bounded. In this sketchy world, the ether was like a vast "ocean" of air, with no center and no periphery, filled with all manner of sounds and spirits, thoughts and messages, a sea of possibility *to* which all might feasibly contribute and *from* which all might feasibly draw a rich and eclectic stream of information.

THE ETHER AND THE BBC

In the aftermath of the First World War, this "ethical ether" was to have more traction than the "gated" model implicit in Marconi's earlier vision and in some of Lodge's own business enterprises. To some extent, this was because, in pursuing its own sectional interests so nakedly, the Marconi Company had alerted the government to the danger in one company achieving a commercial monopoly. The four years of conflict had also demonstrated the effectiveness of public control of utilities such as health, coal, or food.[32] More profoundly, there was a general horror at the slide into barbarism that seemed to characterize the war—a horror that now prompted all manner of strategic—if overheated—thinking about how to avoid future conflict. Richard Overy calls the interwar period in Britain "The Morbid Age" because of the palpable mood of impending disaster. "Dismay," he writes, "was a mainstream concern."[33] Biologists worried about genetic degeneration; economists and politicians worried about whether the human race had run out of ideas for how to organize society; psychologists worried about the chaos of instincts and urges that were now understood to lurk just beneath the surface of the rational mind; and Oswald Spengler's widely discussed book *The Decline of the West* predicted unavoidable and all-embracing collapse in civilized life.

There were others grasping eagerly at anything that might reinforce the essential unity of human existence. In his 1869 extended essay, *Culture and Anarchy*, Matthew

Arnold had argued that in a divided society where self-indulgence ruled, and that was never more than a hairsbreadth away from open revolt, the road to salvation would come through culture, or, as Arnold termed it, "sweetness and light."[34] It was in pursuing this sweetness and light that we would transcend our base instincts to find our "best self" (6, 80–81). Importantly, it was no good *individuals* improving themselves through culture, he thought, for it was precisely such individualism that had led people to indulge their baser instincts in the first place. Instead, Arnold believed in a spirit of generosity toward others, "the noble aspiration to leave the world better and happier than we found it." "The expansion of our humanity," he argued, must be "a general expansion" (12). Sweetness and light did not just need to *exist*—it needed to *prevail.*

If *Culture and Anarchy* was quintessentially Victorian in its scope and optimism, the "Morbid Age" of the 1920s would nevertheless see its call for a "general expansion" having renewed relevance. Indeed, given that the postwar era in Britain had ushered in a dramatic expansion in voting rights, the capacity of the public at large to resist the siren voices of demagoguery had become a very real cultural problem. The pressing issue was discerning the *mechanism* for spreading goodness to all and sundry. The universities were too exclusive, the church a declining influence, newspapers too class-bound and, anyway, tarnished by their sensationalist reporting of supposed enemy atrocities during the war.[35] Radio, however, could reach into every home, indifferent to class or reading ability: it was its dispersed, freely radiating and accessible nature—in other words, its *etheric* nature—that made it a household utility capable of rising to the occasion and delivering cultural and political enlightenment to all.

The geographically expansive nature of radio communication, as a public rather than purely private medium, had already been alluded to in the *London Standard* newspaper in 1912. It pointed out then that hundreds of amateur listeners in and around the capital were having "no difficulty whatever in receiving the daily time signals and weather reports from the Eiffel Tower and in picking up odds and ends flying backwards and forwards between the Admiralty and the fleets, merchantmen and Lloyd's, and the several coastal stations under the control of the Post Office." Receiving equipment could even be bought that allowed one to hear "all the great signal stations of Northern Europe." Anyone listening, the article suggested, was fully "entitled to any pleasure he may obtain by eavesdropping."[36]

After the Armistice, the idea of the ether as a publicly accessible resource for pleasure was soon accompanied by equally clear notions of it as a resource for political harmony. The author of that 1912 article was Arthur Burrows, a young journalist who, during the war, worked on and off for the Marconi Company, providing nightly news bulletins for the navy. He was later seconded to Geneva, where he helped run a radio service covering the first sessions of the League of Nations, and in 1922 he joined the BBC; indeed, as director of programs, he was one of its very earliest employees. But in 1925 he was drawn back to Geneva, becoming the secretary-general of the International Broadcasting Union. His brief experience of military life—he had been at the front in Courtrai in 1918 before being hospitalized with influenza—had, according to an unpublished account of his life, made him "a pacifist for the rest of his life."[37] And now, back in the home city of the League of Nations, he had "high hopes of being able to promote broadcasting in the service of international understanding."[38] For Burrows, it was the intrinsic qualities of radio as an etherial means of communication that made for a natural alliance between broadcasting and global peace. In a talk delivered in 1923, when he was still at the BBC, he articulated a vision that stood in profound contrast with Marconi's earlier call for radio to be deployed as a "weapon of war." Burrows talked instead of radio as a force for "disarmament"—something that would "assist the progress of civilization."[39] He explicitly accused newspapers of being highly partisan, failing "to circulate in the homes of all classes of Society and people of all tastes and temperaments." Radio, by contrast, could be "received in any house or institution." As such, it was inclined to being more open and eclectic in its content, embracing a potpourri of voices and opinions. In "radiating amusement and instruction," it would thus help "in spreading throughout the world a doctrine of common sense."[40]

We can detect, side by side with such high-flown rhetoric, a steady shift in the way the ether is discussed in popular culture. And again, the postwar tone reflects an unmistakable continuity with Lodge's earlier thinking. Whereas *Wireless World* had presented the ether as a troubling domain to be battled over, a new publication appearing from 1922, the *Broadcaster*, was infused with a more utopian perspective. The very first article in the very first issue was titled "Making the World Smaller." A few pages later, there is another item called "When Wireless Dreams Come True." "When broadcasting becomes universal," its author writes, radio will bring not "a hardening influence" but "a humanizing note when scattered people are brought into closer contact"; it will banish boredom, and, once universal and ubiquitous, will "make for a happier world" as it forges "the closer unity of mankind."[41]

One interesting aspect of this growing acceptance of the ether as an intrinsically connective space accessible *by* everyone—and of potential benefit *to* everyone—is the way in which privatizing approaches now start to get marginalized. Back in 1913, as I noted earlier, *Wireless World* had castigated a reader for daring to suggest that Marconi Company trainees should dial down the power of their transmissions so that everyone else might be heard. In 1922, we find the *Broadcaster* criticizing someone for daring to suggest that "inexperienced" amateurs should be forced to give way to elders and betters like him. "The complainant has the right of the enjoyment of the aether," the *Broadcaster* opines, "but not the uninterrupted use of it."[42] Also running through several issues of the new magazine was a lively debate as to whether the term "listeners-in" was any longer appropriate, since it implied that a listener was eavesdropping on communications intended for someone else rather than actively participating in communications intended for everyone. One reader wrote in suggesting the term "Etherites": the word, he suggested, "would apply to all those who own wireless apparatus, whether their set be an expensive receiver and transmitter or a simple crystal set, for they all make use of the same invisible wonder—ether."[43]

As with Lodge, the ether is understood here as something not just connective but *equalizing*. Again and again, it is precisely this quality—made manifest through its ubiquity and accessibility—that allowed the pioneers of broadcasting to think of radio, too, as an intrinsically democratizing force. One of these pioneering figures was Cecil Lewis, a wartime Royal Flying Corps pilot, who joined the BBC in 1922 as Burrows's deputy. Wartime had left Lewis, too, with distinct pacifist tendencies—he described looking down at the trenches from his aerial vantage point "alarmed and disgusted" by what he called "the fixity with which men pursued immediate trivialities," and, ever since had been looking for anything that "might promise even a slim chance of a better way of life."[44] In the spirit of Matthew Arnold, Lewis reckoned that it was art—"a link between the impossible and the possible"—that provided the best hope for healing humanity's woes. And it was broadcasting that provided the best hope of disseminating it (116). Before broadcasting, he asked, "who went to concerts, who went to the opera? Only the people who could afford it." Meanwhile, "the poor, the ordinary people in the street never heard an opera in their lives, never heard a symphony concert. . . . And then suddenly there it was, in their ears . . . they hadn't known: now they began to know."[45] Another early BBC figure, the producer Lance Sieveking, even channeled Lodge's telepathic thinking by writing of radio achieving "sudden mental contact" between the producer in the studio and the listener at home. Broadcasting, he claimed, was all about transmitting one person's ideas to another by means of "thought waves." "How far is harnessed telepathy possible?" he wonders, "How far, if at all, [does] it occur in every radio transmission?"[46]

Perhaps the single most striking echo of Lodge's etheric thinking to be heard in the rhetorical construction of broadcasting in the 1920s comes directly from the BBC's most senior figure: its first managing director and founding father, John Reith. In 1924, Reith published an extended personal manifesto, *Broadcast over Britain*. The book contains what is perhaps Reith's most famous phrase, succinctly articulating the direct application to his own organization of the Arnoldian ideal. The BBC's responsibility he declared, was not just to broadcast "everything that is best in every department of human knowledge, endeavor and achievement"; it was to "carry" all this "into the greatest possible number of homes."[47] The echoes of Lodge, though, are just as strong as the echoes of Arnold. Take the chapter titles: "Uncharted Seas," "The Voice from the Silence," "The Great Multitude," "Beyond the Horizon." The final one is called "In Touch with the Infinite," and as we turn its closing pages we sense a man—uncannily like Lodge—seeking to square his embrace of new technology with a deeply held Christian faith. Lodge's own vision of radio's ultimate purpose—"the harnessing of electrons and making them help in the work of the world"[48]—is here rearticulated by Reith in the most grandiloquent of terms. Broadcasting, he suggests, would forge "a new and better relationship between man and man" precisely because it represented "a reversal of the natural law, that the more one takes, the less there is left for others." A broadcast, he went on, "is as universal as the air":

> It does not matter how many thousands there may be listening; there is always enough for others, when they too wish to join in. It is the perquisite of no particular class or faction. Most of the good things of this world are badly distributed and most people have to go without them. Wireless is a good thing, but it may be shared by all alike, for the same outlay, and to the same extent. The same music rings as sweetly in mansion as in cottage. It is no respecter of persons. The genius and the fool, the wealthy and the poor listen simultaneously, and to the same event, and the satisfaction of the one may be as great as that of the other. . . . Broadcasting may help to show that mankind is a unity and that the mighty heritage, material, moral and spiritual, if meant for the good of any, is meant for the good of all. . . . So our desire is that we may send broadcast through the ether, which is universal, the universality of all that is good . . . so all may receive without let or hindrance."[49]

Not only could "all receive without let or hindrance"; all could receive a near-infinite *variety* of signals, voices, and messages. For Matthew Arnold culture involved cultivating "all sides of our humanity."[50] Now, echoing both Burrows and Lodge as well as Arnold, Reith suggests that a variety of stimulation was precisely what allowed us to grow into our better and less conflictual selves. "In entertainment or edification, in enlightenment or education, in all the manifold influences of its activity," he wrote, "there is a *consolidating* influence at work."[51] In Reith's expansive vision, broadcasting—provided it genuinely offered a rich mix of programs—made *tangible* Lodge's concept of an etherial space of intellectual coherence and continuity.

Finally, on the very last page of his book, in a passage that stands, in effect, as homage to Lodge himself, Reith conjures a supernatural claim of his own. "Wireless," he writes, "is in particular league with ether": "[We] speak of sounds or colours . . . [but] we are missing infinitely more than we are receiving, and we shall continue to function defectively until, with limitations overcome and with the necessity for interpreting senses removed, we shall be introduced to fresh and amazing realms of activity." Thus, he continues, the "music of an orchestra is unreal, symbolic and transitory. Thought is probably permanent, and a means may be found to ally thought with ether direct and to broadcast and communicate thought without the intervention of the senses or any mechanical device, in the same manner as a receiving set is today tuned to the wavelength of a transmitter so that there may be a free passage between them" (223–24). Nowhere does Reith explicitly acknowledge the influence of Lodge. But the parallels are hard to ignore. There, in Reith's generous conception, are almost all the elements of Lodge's own thinking on the ether: its ubiquity and universality, its openness, its extrasensory wonder, its status as a space without center or hierarchy, its role as a repository of consciousness, as a manifestation of continuity in thought. Rather aptly, we might say, it is as though the ethical vision of one man had been transferred telepathically into the mind of the other.

LODGE'S ETHER, BROADCASTING, AND THE PUBLIC SPHERE

The most obvious parallel between twentieth-century broadcasting and nineteenth-century etherial thinking—one that struck so many contemporaries that it rapidly became a commonplace—was straightforwardly physical. Radio had the ability, as Reith put it, to "cast a girdle round the earth."[52] But it is in the ethical—rather than the physical—realm that the parallels are not just striking but enduring. The ether provided a structure for rethinking ideas about knowledge in the early twentieth century: how knowledge was to be dispersed; how it might be formed into something coherent. Lodge's notion of the ether, let us recall, was not just that it was a shareable space with no obvious center or hierarchy, or a medium for achieving the transference of thought; it was also, implicitly, a space through which otherwise discontinuous bodies of *knowledge* could be conjoined. And if that concept felt somewhat metaphorical in Lodge's own articulation, it took more tangible form through broadcasting. Indeed, we might say that broadcasting incarnated the ether; in doing so, it also put into practice some of the utopian ideals associated with it.

Or at least, that was broadcasting's promise. As *Vox* magazine pointed out in 1929, the vast canvas created by radio, its profusion of voices and ideas, could just as easily result in chaos as in order. So, it was not quite broadcasting as such that embodied the ethical ether: it was broadcasting of a particular type. It was "by impartially balancing the admission to microphone publicity of one point of view, one type of belief, against another," *Vox* explained, that broadcasting "does away with sectarian blindness and bitterness and tends increasingly to create a *synthetic* view of differing attitudes to all great questions."[53] Here, *Vox* is not just describing an abstract ideal: it is summarizing the BBC's own policy. Reith presided over a schedule that deliberately mixed entertainment, information, and education—that offered a carefully calibrated juxtaposition of uplift and relaxation, of virtue and pleasure. For Reith, broadcasting was "part of a systematic endeavor to re-create, to build up knowledge, experience and character," something only achieved when we are gently but persistently "brought into contact" with music, ideas, voices, or opinions with which we are otherwise unfamiliar.[54] That Lodge himself saw the value in this mission is surely indicated by his decision to join the BBC's Central Council for Broadcast Adult Education. He was also a regular and enthusiastic broadcaster at the BBC microphone. For him, it seems, it was broadcasting that now offered the best hope for democratizing knowledge.

One final thought follows from this. If we look at Lodge's work—backward, as it were—through the prism of the 1920s BBC, we might be prompted to recognize in his version of the ether a prototype of the new "public sphere" that broadcasting later embodied. When Jürgen Habermas first outlined his concept of the public sphere in 1962, the emphasis was on an essentially "bourgeois" realm of social intercourse as it had evolved in the late seventeenth and eighteenth centuries, manifested most palpably in the coffee houses and public squares of Western Europe and America. For Habermas, if there was a role for the modern mass media, it was a largely destructive one, effectively degrading the public sphere as a space for rational debate.[55] With Lodge in mind, one reasonable objection to this theory is that it neglects the possibility of a meaningful role for spaces created through *irrational* thinking, what we might categorize as "magical," uncanny, even spiritual longings.[56]

Moreover, through broadcasting a new idea of "the public" emerged. The BBC aspired to address not a segment of the British public but the nation at large—and it did so through the vehicle of fleeting, but pervasive sound: the public it forged was therefore "more intermixed, promiscuous and democratic" than ever.[57] The effect was that "material for thought that had earlier pointed towards a particular audience with particular expertise now had to be re-thought as 'general knowledge'" (151). Stefan Collini has noted one longer-term implication of this: that public intellectuals became increasingly engaged in "a discourse of general ideas."[58] But, arguably, such intellectual restructuring worked at all levels of British society simply because of broadcasting's ability to create—and then curate—a public *space* for the widest possible sharing of knowledge.

As I hope to have shown in this chapter, the outlines of this intellectual shift were actually in evidence before broadcasting was fully established. The ether of Oliver Lodge had already evolved from a simple "connecting medium" explaining the transmission of light and energy into something far looser but also far grander—an ocean of dispersed consciousness able to contain and even bind together disparate realms of thought and existence, and, by its very nature, accessible to all. In short, Lodge's ether was the modern, democratic public sphere in gestation.

TWELVE

"BODY SEPARATES: SPIRIT UNITES"

OLIVER LODGE AND THE MEDIATING BODY

JAMES MUSSELL

IN 1918 Oliver Lodge wrote another book dedicated to the memory of a young man killed in the war. *Christopher*, like *Raymond* (1916), was an expression of faith in human survival, the book serving as both memorial to the life lived and an apparatus through which to comprehend ongoing life beyond the grave. Christopher Tennant was the nephew of Frederic Myers, psychical researcher and Lodge's friend, and the book's subtitle, "A Study in Human Personality," deliberately evoked Myers's *Human Personality and Its Survival of Bodily Death* (1903). For Lodge, death was not separation but a different form of connection. "For we have not really lost them," he writes. "They feel themselves to be nearer to us than before; death is not estrangement, it has been felt by many as a kind of reunion. Body separates: spirit unites."[1] Inscribed within the body of the book, the personality of the dead soldier is given integrity enough to survive in the world beyond.

Embodiment requires edges, differences that dictate where one body ends and another begins. Lodge, as both physicist and psychical researcher, dedicated his career to making the immaterial material, to giving the intangible bodily form. In his autobiography, *Past Years*, he claimed that he had always been interested in the "imponderables," "the things that worked secretly and have to be apprehended mentally."[2] Chief among these was the ether, the elusive medium thought to account for all electromagnetic phenomena including light. To think the ether, Lodge gave it edges, a provisional materiality so that it could act like a medium and produce difference. In Lodge's tactile imagination, spirit had a body of its own.

Matter individuates, but spirit threatens to dissolve difference in radical oneness.[3] In his work, Lodge kept returning to embodied forms in the ether in order to make it comprehensible. In what follows I explore what such thinking has to offer us as we puzzle out embodiment in a digital age. Born in 1851, Lodge was a Victorian, at home among steam, print, and rail; yet his work on electromagnetism meant that he was at the forefront of the media technologies that would characterize twentieth-century modernity. The wires of the telegraph had introduced a mode of signaling that was potentially instantaneous; with the introduction of wireless telegraphy this medium was universalized, putting everything, potentially, in touch with everything else. More than any other technology, wireless brought the ether into the world, but the ether, as a perfect medium, not only dissolved the boundaries between things but also suggested that they were not boundaries at all. Lodge was a pioneer of these teletechnologies and recognized the philosophical consequences of the new regimes of connection they introduced. As he grappled with the ineffable, always just beyond the bodies pressed into service, Lodge speculated on the relations between matter and spirit, medium and message, form and content, body and soul.

This chapter details Lodge's response to an ether that would decompose the world. Lodge's scientific work gave the ether shape, delineating its forms while allowing the pure, unmodified ether to reconstitute itself once more out of reach. Yet Lodge also looked to the ether to sustain the soul and turned to material media to ensure the soul's coherence in a medium that otherwise denied difference. For Lodge, consciousness was underpinned by an etheric essence that, because it was etheric, was the true seat of the self; however, whereas the ether, as opposed to matter, had the potential to overcome difference, in the spirit realm such properties constituted a threat to the persistence of the individual soul. While he investigated a wide range of technologies to make contact with the spirits of the departed, from séance sessions with trance mediums to table tapping and automatic writing, it was to the book that Lodge turned to embody the dead.

Body separates; spirit unites. In Lodge's conception of the universe, ether complemented matter, the other term in a binary pair that created difference. However, as he worked to make the ether sensible, it could not but become embodied, leaving a purer, rarefied ether to be pursued anew. Today, the ether has not really gone away. Early fantasies of cyberspace were based on an etheric world offered as an alternative to the "meatspace" that we habitually occupy. Equally, the rhetoric of the cloud insists that not only is data of the ether but it can also be stored there, temporarily given form in whatever device is at our fingertips. Lodge has much to teach us about such thinking, about its pleasures and its repressions. He was fixated on the transcendent, whether the soul overcoming the bounds of the body or the message its media. Yet no matter how much he insisted on continuity, of an ether that effaced difference, he still longed for a world in which discrete parts had some sort of integrity, a world in which personality, in particular, could persist. Standing on the brink of the electrical age, Lodge utilized new technologies to maintain the ether and so shore up the universe against those who insisted on its radical discontinuity. While this reliance upon the ether was increasingly anachronistic—in 1913 Lodge himself called it "the sheet anchor of nineteenth-century physics"—it should not simply be dismissed as a reactionary effort to keep modernity at bay.[4] As he used the ether to surpass the bounds of the material world, Lodge came to realize that a body of some sort was the precondition of difference. The magic of media—the way that mediating objects somehow transcend the here and now—depends upon material forms.

THE MATTER WITH ETHER

In 1913, Lodge gave the presidential address at the British Association for the Advancement of Science on the subject of continuity.[5] Issued later that year as a short book, *Continuity*, Lodge's address was a plea for the integrity and connectedness of the physical universe against those who would break it up into parts. For Lodge, this was an important moment. Founded in 1831, the British Association was intended to provide a public audience for the latest in scientific progress. Lodge himself had been a keen attender of its meetings since 1873 and now, forty years later, he held the presidency. Of all years, this was the one in which he should not have occupied the chair. Lodge had been principal of the University of Birmingham since its foundation in 1900, and it was a longstanding convention that the president not host the association in his own city. However, the sudden death of the sitting president, Sir William White, meant the chair fell to Lodge when plans were already in place.[6] Speaking on behalf of the association in which he believed and in the city he had made his home, Lodge maintained that the ether was the connecting medium that bound the universe together. "I am myself an upholder of *ultimate* Continuity," Lodge argued, "and a fervent believer in the Ether of Space" (32). Whereas those invested in the new physics, in relativity and quanta, could only divide things, describing but not explaining, the ether provided the means to ensure that the universe and everything in it were one.

The problem was that the ether—frictionless, elastic, and everywhere—proved to be stubbornly elusive. In the early 1890s Lodge had made an attempt to detect it: his ether whirling machine, built into the bedrock beneath University College Liverpool, tested to see whether a moving mass could drag the ether with it through space and so explain the famous null result of the Michelson-Morley experiment.[7] Since then, Lodge had reconciled himself to the ether's intractable tendency to remain out of reach, even as he made it ever more central to his conception of the universe. In his address he argued that the ether's elusiveness was an inevitable consequence of its properties. "We must have difference to appeal to our senses," Lodge told his audience, "they are not constructed for uniformity." He added, "It is the extreme omnipresence and uniformity and universal agency of the ether of space that makes it so difficult to observe. To observe anything you must have differences."[8] The ether represented disembodiment, a perfect medium that was defined by its uniformity, yet this meant that it was impossible to think, let alone detect. Without differences of its own the ether remained imponderable, its only existence as the shadowy other to matter.

Lodge's presidential address was not just a defense of the ether; it was also authorized by the ether's conceptual slipperiness. While his scientific reputation was founded on his work in electromagnetism in the 1880s and 1890s—research that, for Lodge, made the ether tangible—it was for his more speculative uses of the ether that he was better known among the wider public. Since he had become principal at Birmingham, Lodge had developed a distinctive philosophy, articulated in a string of popular books, that united science, spiritualism, and Christianity, with the ether providing the necessary common ground.[9] Since becoming interested in psychical research in 1883, Lodge had been attuned to the boundary between orthodox science and spiritualism and carefully avoided the kind of work that might get him labeled a crank. Nevertheless, as his public reputation grew as a man of science, he was increasingly comfortable making the case for broadening the scope of scientific investigation to include psychical phenomena. In 1891, when he was president of Section A at the British Association meeting in Cardiff, Lodge used anticipated opposition to the National Physics Laboratory as a pretext to defend "free and open inquiry, for the right of conducting investigation untrammelled by prejudice and foregone conclusions."[10] In that spirit, he took the opportunity to argue that phenomena such as thought transference—and so what Lodge described as "the whole borderland between physics and psychology"—be subjected to scientific scrutiny.[11] Twenty years later, as president of the British Association and with a string of successful books behind him, Lodge put the ether's capacity to sustain his synthetic philosophy to the ultimate test.

* * *

In the final stages of his 1913 address, Lodge insisted that the president of the British Association "should not be completely bound by the shackles of orthodoxy, nor limited to beliefs fashionable at the time" and then went on to set out his position: "In justice to myself and my co-workers I must risk annoying my present hearers, not only by leaving on record our conviction that occurrences now regarded as occult can be examined and reduced to order by the methods of science carefully and persistently applied, but by going further and saying, with the utmost brevity, that already the facts so examined have convinced me that memory and affection are not limited to that association with matter by which alone they can manifest themselves here and now, and that personality persists beyond bodily death."[12] For Lodge, the ether warranted these audacious speculations. For those who read his books, such remarks would have been utterly predictable, but even his "present hearers" would not have been that shocked: after all, Lodge's connection with the Society for Psychical Research was well known and he had long appealed for broadening the scope of legitimate scientific enquiry. But Lodge did not just want to state, publicly, his belief in survival; he also wanted to make the case for an etheric realm resistant to scientific analysis but there nonetheless. The "occurrences" Lodge described were "occult" not just in the sense that something currently hidden would eventually be explained through conventional scientific methods but in that they could only be known indirectly, through their effects on matter. For Lodge, the evidence "goes to prove that discarnate intelligence, under certain conditions, may interact with us on the material side, thus indirectly coming within our scientific ken; and that gradually we may hope to attain some understanding of the nature of a larger, perhaps etherial, existence, and of the conditions regulating intercourse across the chasm."[13] At stake here was the possibility of a materialist practice, science, not only to deal with things like "memory and affection" but also the basis on which such things were made present "here and now." Only open-minded scientific endeavor would be able to register the influence of this etherial existence upon matter, tracing a hinterland in negative that could complete the scheme of the universe and so offer a bulwark against a crude materialism that would deny the ether its very existence.[14]

Posited in opposition to matter, the ether provided the common ground from which to oppose materialism while nonetheless making sense of the material universe. In a later essay, "Ether, Matter and the Soul" (1918), Lodge maintained that a "duality runs through the scheme of physics—matter and ether"; however, the way he formulated the ether meant that it kept undoing the very difference it was supposed to sustain.[15] Nowhere was this more clear than in his understanding of electricity. In the first edition of *Modern Views of Electricity* (1889) Lodge, true to the Maxwellian tradition, maintained that all electromagnetic phenomena derived from conditions in the ether: all that was needed, he argued, was to put it into motion. The question "What is ether?" ("*the* question of the physical world at the present time") might not be easy to answer, according to Lodge, but it was not unanswerable: "If a continuous incompressible fluid filling all space can be imagined in such a state of motion that it will do all that ether is known to do; if simply by reason of its state of motion, it can be proved capable of conveying light and of manifesting all electric and magnetic phenomena which do not depend on the presence of matter; and if the state of motion so imagined can be proved stable and such as can readily exist, the theory of free ether is complete."[16] There were four classes of electricity for Lodge—static (at rest); current (in locomotion); magnetism (in rotation); and light (in vibration)—and all that was needed was a model of the ether that could account for them simply by the way it moved. Throughout the book Lodge provided a series of models, some of which he built, to which the behavior of the ether might be compared.[17] These became most fully elaborated in chapters 10 and 11, when Lodge offered mechanical models for the magnetic field and current induction.[18] Here the ether became an interlocked set of cogwheels and the various electromagnetic phenomena represented by changing the gearing or direction of motion. Lodge situated his model in a tradition that went back to Descartes, but it had more immediate antecedents in the work of Maxwell and Kelvin in the 1860s.[19] The clarity of his cogwheel models allowed Lodge to demonstrate that electromagnetism could be an effect of motion, but they did so by making the ether mechanical.

Lodge never built a full cogwheel ether, but it was so clearly realized on paper that he did not need to. This slippage between the ether and the cogwheels, between the elusive medium and its representation, obscured the distinction between free ether and matter with which Lodge began. The distinction was further undermined by Lodge's speculations as to the constitution of matter itself. Lodge was particularly taken with Kelvin's vortex model of atomic structure and, in a lecture reprinted in *Modern Views of Electricity*, described it for his readers: "The atoms of matter, according to it, are not so much foreign particles imbedded in the all-pervading ether, as portions of it differentiated off from the rest by reason of their vortex motion, thus becoming virtually solid particles, yet with no transition of substance; atoms indestructible and not able to be manufactured, not mere hard rigid specks, but each composed of whirling ether; elastic, capable of definite vibration, of free movement, of collision."[20] On the one hand the etheric strains called electromagnetic force were the effect of rotating matter; on the other, that matter itself was produced by rotating ether. In Lodge's mechanical ether difference was performative, the result of motion in the universal substrate rather than any essential distinction of being.

In *Modern Views of Electricity* Lodge understood electricity as an etherial phenomenon caused by the shearing of the ether into positive and negative parts. The emergence of the electron, however, gave electricity mass, a body, and so, when it came to the third edition of *Modern Views* (1907) he could not account for it through motion alone. In the preface, Lodge insisted that the doctrines expounded in the book—"the electrical nature of light, a recent theory of matter, and an ethereous view of electricity"—were unchanged, and, sure enough, electricity was still produced when the ether was sheared.[21] However, whereas in 1889 the sheared ether simply produced difference, positively and negatively charged portions, now it could produce matter, able to move about independently. "The negative electricity," Lodge writes, "when separated, is freely mobile and easily isolated: it is what we experience as an electron."[22] A few years previously, in his Romanes Lecture of 1903, Lodge had noted that the electron "displaced the so-called atom of matter from its fundamental place of indivisibility."[23] No longer the place where the ether acts as if other than itself, the atom had now been opened out. "An atom is not a large thing," Lodge argued, "but if it is composed of electrons, the spaces between them are enormous compared with their size—as great relatively are the spaces between the planets in the solar system."[24] At the end of the final chapter of the new edition of *Modern Views of Electricity*, Lodge summarized his position:

> Throughout the greater part of space we find simple unmodified ether, elastic and massive, squirming and quivering with energy, but stationary as a whole. Here and there, however, we find *specks of electrified ether*, isolated yet connected together by fields of force, and in a state of violent locomotion.
>
> These "specks" are what, in the form of prodigious aggregates, we know as "matter"; and the greater number of sensible phenomena, such as viscosity, heat, sound, electric conduction, absorption and emission of light, belong to these differentiated or individualized and dissociated or electrified specks, which are either flying alone or are revolving with orbital motion in groups.[25]

With mass and charge, it was electrons, electrified specks swirling in the ether, that now marked the point at which matter became differentiated, creating aggregated clumps that themselves caused etheric effects.

From things that could be touched, to the atoms that accounted for molecular behavior, to the electrons that marked the atomic interior: what appeared to be discrete entities were revealed to be connected at a deeper level. Yet while matter decomposed when thought, there was always the unmodified ether, from which the mechanism was composed, to stop it from disappearing completely. Given Lodge's preoccupations, it is unsurprising that there was something theological about the way the ether functioned as a kind of ontological safety net. Whenever some sort of action was attributed to the ether, it necessarily took on shape, becoming other to the undifferentiated ether from which it was composed. Even Lodge's mechanical models betray this tendency. In the 1889 edition of *Modern Views of Electricity* Lodge noted, in passing, that "since magnetic induction can spread through a vacuum quite easily, the wheel-work has to be largely independent of material atoms."[26] This was never simply a mimetic model of atomic structure, in other words, but a generalizable model of mediation that could apply at molecular, atomic, subatomic, or etheric levels. In the 1907 edition the chapters containing the models remain unchanged and, even though the existence of the electron meant the end of the atom as vortex ring, the lecture in which he endorsed Kelvin's model was also reprinted. In a nod to progress, Lodge also inserted a more recent lecture in which he noted that nobody "believes now that an atom is simply a vortex ring of ether, and that the rest of the ether is a stagnant fluid in which the vortex rings sail about."[27] Yet Lodge did not give up on vorticity entirely, regretting at that year's British Association meeting that Kelvin had "of late shown a tendency to destroy his own children."[28] An adaptable technology for producing difference, the vortex was compelling as it produced matter from motion even while it left what moved undefined.

"Matter," as Lodge put it in "Ether, Matter and the Soul," "appeals to our senses, but the unmodified ether makes no such appeal; it is so inaccessible that its existence even has been denied."[29] In the first edition of *Modern Views of Electricity* he laments that some were inclined to imagine the ether was "still a hypothetical medium whose existence is a matter of opinion," a lament repeated in the preface to the 1907 edition.[30] While the book demonstrates the effects of the ether on matter, it could not go further and give it substance. In his presidential address to the British Association in 1913, Lodge claimed that "matter it is not, but material it is; it belongs to the material universe and is to be investigated by ordinary methods."[31] This was a methodological approach necessitated by the ether's properties: as matter was, according to Lodge, "what appeals to our senses here and now," then materialism was "appropriate to the material world; not as a philosophy but as a working creed, as a proximate and immediate formula for guiding research" (73). With only material resources to hand, the ether could only exist in provisional, embodied forms behind which pure, unmodified ether remained tantalizingly out of reach.

Lodge could not detect the unmodified ether as it did not constitute an object. Instead, it was always just beyond whatever form lay at hand, its existence posited through its otherness. In this way, the unmodified ether, that imponderable uniform medium, behaved like Derrida's supplement: the center that establishes the form of a structure while nonetheless being located elsewhere.[32] The unmodified ether was not even virtual—a virtual form, like one of Lodge's mechanical models, had integrity and edges—yet, when held in abeyance, could serve as the source of being, of presence, and so fill the various forms of the universe at the expense of itself Lodge expressed a similar idea in his book *Atoms and Rays* (1924), noting that "the apparent discontinuity which the Atomic Theory suggests, the discontinuous nature first of matter then of electricity, is supplemented or replaced by the absolute continuity of the connecting Ether."[33] What is important here is the process, the way in which in striving to reach the ineffable Lodge's imagination produced a universe made up of cascading binary pairs, in which continuity gave way to discontinuity and so on. In his 1913 address he had described it like this: "On the surface of nature at first we see discontinuity; objects detached and countable. Then we realise the air and other media, and so emphasise continuity and flowing quantities. Then we detect atoms and numerical properties, and discontinuity once more makes its appearance. Then we invent the ether and are impressed with continuity again. But this is not likely to be the end; and what the ultimate end will be, or whether there is an ultimate end, is a question difficult to answer."[34]

Lodge's recognition that the ether, too, might give way points to the primacy of material, the way this universal medium was produced rather than given in advance. In *Atoms and Rays,* Lodge speculates that it "will probably turn out that there is some kind of structure even in Ether, but that structure has not yet be ascertained; and when it is ascertained, it is quite unlikely that it will be of a discontinuous character."[35] Each time Lodge gave shape to the ether, bringing it into the material world, he also produced the unmodified ether, that guarantee of presence kept tantalizingly at arm's length. Lodge thought that the ether was "the great engine of continuity," but, in the way that it functioned, it was more like a difference engine, producing useful mediating form on one side while reestablishing elusive, uniform, and transcendent formlessness on the other.[36] In its perpetual absence, the unmodified ether anchored presence for Lodge, standing for the possibility of Being, even for God himself, but, no matter how hard he pursued it, could never serve as content. The unmodified ether could only be realized in opposition to some sort of form, produced by difference and so intelligible to the human mind.[37]

RAYMOND/*RAYMOND*

Lodge's ether was necessarily supplementary. It took on form against the undifferentiated uniformity out of reach, while at the same time it established that uniformity *as* out of reach: in abeyance but thinkable through difference. Lodge drew upon this supplementariness to make sense of human relations, too. Etheric connection undid the boundaries that made the world meaningful and its oneness filled the gaps between people just as it did between particles. Lodge's science, for instance, allowed him to mix with eminent figures on the floor of the Royal Society, but it also connected him to electrical engineers such as his collaborator Alexander Muirhead. Similarly, through his psychical research Lodge was introduced to Cambridge philosophers such as Frederic W. H. Myers and Henry and Eleanor Sidgwick. Through them he became friends with an aristocratic set that included Arthur Balfour (prime minister from 1902 to 1905, and Eleanor Sidgwick's brother) and Percy Wyndham. Lodge spent his Easters at Wyndham's house, Clouds, with the group known as the Souls.[38] In Birmingham, Lodge mixed with the local Liberal elite (the Chamberlains, most obviously), but also important Quaker families such as the Cadburys.[39] Lodge claimed that his politics were of no party and instead professed a kind of casual Fabianism, informed by the political economy of John Ruskin.[40] He knew Sidney and Beatrice Webb, and they published his lecture to the Ancient Order of Foresters, "Public Wealth and Corporate Expenditure," as Fabian Tract 121.[41] The ether, which was everywhere, was radically democratic. In a world where everyone was connected, all actions contributed to the whole and so were everybody's business.

Lodge understood society in etheric terms. In his Lockyer Lecture, "The Link between Matter and Matter" (1925), he attacked materialism for its orientation to exteriority: "There are those who think that these material bodies represent ourselves,—our personality, our memory, and our character. If they can work out everything on that hypothesis, by all means let them do so. But let them also take the whole of the facts into account. The atoms are not isolated, and we are not isolated. We are members of one another. There is a link between the atoms. Human beings are connected, and in societies and families, and for purposes of mutual co-operation,—or, if lunacy seizes them, of extermination."[42] By cautioning against an emphasis on outsides, on edges, Lodge also cautioned against a crude notion of autonomy. Venerated in Victorian liberalism, the individual as self-governing, rational subject had come under sustained pressure in the latter part of the nineteenth century as a number of factors—the emergence of mass culture, the rise of socialism, the flu pandemics—suggested that people were connected in mysterious ways. And, as Lodge's evocation of violence makes clear, such connections were not always benign.

In the same period, Lodge's friend and mentor in psychological and spiritual matters, Myers, had put both connection and multiplicity at the heart of subjecthood. For Myers, the conscious self was simply that best adapted to present circumstances and lurking beneath was a much broader subliminal consciousness that he understood as a spectrum. It was "indefinitely extended at both ends": in one direction were the rudimentary and archaic forms no longer needed in day-to-day life; in the other, toward "the superior or psychical end," it was capable of telepathic and clairvoyant impressions.[43] The subliminal consciousness meant that people were constantly in touch with one another and in ways not accessible to the conscious self: in *Human Personality and Its Survival of Bodily Death* Myers would extend this thesis to show how such communications continued once the body was no more.

Lodge's understanding of selfhood was grounded in Myers's thought, but, because Lodge associated spirit with ether, departed from it in significant ways. Initially skeptical about survival, Lodge had come to believe in its reality after a trance sitting with the medium Leonora Piper in 1889.[44] In the sitting, Lodge's aunt Anne, the woman who had introduced him to science as a boy, spoke to him in her own voice through Piper. The fact that she spoke in her own voice was particularly important, as it was what differentiated her from any other disembodied phantoms and, of course, the embodied Piper. The problem was that such vestiges of embodiment had no place in the ether from which she apparently spoke.

The coherence of the self was maintained by its difference, imagined or otherwise, from others, yet the possibility of perfect, etheric connection threatened such distinctions. Spirit might unite, but Lodge insisted that spirits retained their individuality despite dwelling in a perfect medium in which everything was connected and nothing ever ceased. He proposed an explanation in "Ether, Matter and the Soul." Written the same year that *Christopher* was published, the essay describes an "etherial body," a counterpart to the physical body that could provide individuated matter in the spirit world. After explaining how energy passes from kinetic to static forms, with kinetic energy characteristic of matter and static energy of ether, Lodge suggested "that every sensible object has both a material and an etherial counterpart. One side only are we sensibly aware of, the other we have to infer."[45] As always, the ether, out of reach, was constituted by its otherness; however, because there were things that had an etherial existence but not a material existence, the etherial, no matter how elusive, came first. "I foresee a time," writes Lodge, "when the term soul will be intelligible, and I think it will be found that soul is related to the ether as body is related to matter. I suggest that it will turn out to be a sort of etherial body, as opposed or supplemental to our obvious material body" (258). Preceding the embodied self and living on after it had perished, the etherial body was the unconscious spiritualized. Like the Freudian unconscious, its existence could only be inferred indirectly by its effects; however, because Lodge needed the dead to resemble the living, whatever personality resided in the ether had to resemble its mortal counterpart. Once granted a body, in other words, the unconscious was no longer distressingly other.

Granting the soul a body was a way of ensuring continuity between the living and the dead, but it also ensured that the dead kept their integrity in the spirit realm. For personality to survive in this environment, to escape radical oneness, it had to become material. The etherial body allowed Lodge's spiritualism to mediate between his science and his faith, yet the shift from soul or spirit to etheric body is telling. By technologizing the soul, granting it a body that could differentiate itself from the unmodified ether, Lodge promised humanity immortality by making spiritual existence material. As there was nothing in the ether that suggested wear and tear, the etherial body, the true seat of selfhood, could live on after the physical body had perished. Not only live on, but be rejuvenated. "Freed from the disabilities and imperfections of matter," Lodge writes, "it can lead a less abstracted and livelier existence" (258). The cost of such liberation, though, was loss of contact with the physical universe. Whereas the etherial bodies of the living were indirectly disclosed through the actions of their material bodies, the dead could only become known by pressing other people's bodies into service.

For Lodge the etherial body came first, the material body merely a temporary dwelling place for a more mysterious life elsewhere. In practice, however, the etherial body was only ever known as an effect of matter. This meant that the contours of the etherial bodies of the dead could be traced only by disentangling them from material bodies of various kinds. Lodge favored table tapping as the most unmediated way to contact the dead. Acknowledging that it was old-fashioned, hearkening back to when spiritualism was little more than a parlor game, Lodge preferred it as it cut out both medium and spirit guide.[46] But it was an awkward mode of communication: like telegraphy without a telegraph code, messages were spelled out letter by letter, only making sense once the tapping had stopped. The sender that resulted from such encounters, while tantalizingly close, was unbearably inarticulate and hardly took shape at all. The cost of proximity to the other was diminished contact: for a richer encounter with the dead they had to be mediated and to understand how mediation shaped presence, Lodge turned to paperwork.

For Lodge the most convincing evidence for survival came in the form of cross-correspondences. These were messages that "referred to some theme in an obscure way through different mediums, and by different sentences, at about the same time, so that no one medium should understand the meaning of what was being transmitted; but yet they appended some distinct sign or mark which later might be interpreted as indicating that there was a connexion between the fragments; so that, when all the messages were sent up to a central office and compared, the connexion between the different portions should be apparent, and the meaning of the whole reference become ultimately clear."[47] Such occurrences were much prized, as they apparently triangulated a specific personality as a point of origin, insisting on its integrity in the otherwise undifferentiated ether. As they only made sense when collated, they relied upon writing things down, posting messages to a central location, and collating them when they came in. All this sorting was a prelude to the work of interpretation: the content of the messages was often cryptic, a kind of riddle that could only be solved by those in the know in order to establish both the identity of the sender and the authenticity of the message. To convince others, both stages—the story of the messages and their interpretation—had to be narrated, put in order so that the sender could cohere.

Accounts of cross-correspondences were published in the journal and proceedings of the Society for Psychical Research, and Lodge contributed to both publications. When he set out to offer his fullest and most personal account of survival, however, Lodge turned to the form of the book. *Raymond; Or Life and Death* (1916) records the life, death, and continuing survival of Lodge's youngest son, killed at

Ypres on September 14, 1915. Published on November 2, 1916, it went through six editions before the end of the year and six more before 1919.[48] The book was an attempt to allow Raymond to cohere, to give him a body, while eliding the work done by Lodge, the mediums, and their respective spirit controls. It is organized into three sections: the first, the "Normal Portion" consists mainly of Raymond's letters from the front. The second, the "Supernormal Portion," consists of a series of accounts of sittings where Raymond established contact with Lodge or one of his family. The final section of the book is titled "Life and Death," and contains philosophical and speculative essays on various subjects. This composite structure, made up of disparate material, makes clear that Raymond's etheric body not only depended on doing things with material media—transcribing sittings, letters, reprinting photographs—but also transforming them, binding them into the linear form of the codex.

What separated the first section from the second was Raymond's body. The letters in the "Normal Portion" are arranged chronologically and so establish a narrative that leads, inexorably, to Raymond's death. Their purpose, according to Lodge, was to "engender a friendly feeling towards the writer of the letters, so that whatever more has to be said in the sequel may not have the inevitable dullness of details concerning a complete stranger."[49] However, as letters, they also signal Raymond's writing body, something that is absent in the "Supernormal Portion" that follows. Raymond's original letters, handwritten with a signature on the bottom, located his body in time and space through the tangible marks offered to the recipient. Without a material body to put into service, however, Raymond's etherial body had to rely on spirit controls, mediums, and the apparatus of the séance to establish its presence. To ease the transition between these two portions and so compensate for the absent material body, Lodge utilized the form of the printed book. Firstly, as we approach the crisis of Raymond's death at the conclusion of the "Normal Portion," the linear form of the codex drives us on into the "Supernormal Portion" beyond. Secondly, as we have read Raymond's letters only in print, the appearance of the page remains unchanged from one portion to the next. *Raymond* makes an argument for Raymond's continuity by encompassing both portions within its covers and making them parts of a whole.

While Lodge's editorial hand is present in the "Normal Portion," arranging letters and adding the odd comment, the absence of Raymond's body means that Lodge has to do more editorial work in the "Supernormal Portion" that follows. At stake in this section is the origin of messages and so Lodge provides a paratextual commentary on the trustworthiness of the various mediating links while suppressing his own agency as much as possible so as to let Raymond speak for himself. The bulk of contact recorded in *Raymond* was provided through two mediums, Gladys

Osborne Leonard and Alfred Vout Peters, and their respective spirit controls: Leonard called upon a young girl called Feda; while Peters relied on a man quixotically named Moonstone. At such sittings there were, then, at least two intermediary links, one alive and one dead, between the sitter and the spirit with whom he or she wished to speak. Lodge describes the difficulties of such a mediating chain with a characteristic reference to communications technology:

> The confusion is no greater than might be expected from a pair of operators, connected by a telephone of rather delicate and uncertain quality, who were engaged in transmitting messages between two stranger communicators, one of whom was anxious to get messages transmitted, though perhaps not very skilled in wording them, while the other was nearly silent and anxious not to give any information or assistance at all; being, indeed, more or less suspicious that the whole appearance of things was deceptive, and that his friend, the ostensible communicator, was not really there. Under such circumstances the effort of the distant communicator would be chiefly directed to sending such natural and appropriate messages as should gradually break down the inevitable scepticism of his friend. (87)

In an essay for the *Fortnightly Review* in 1893, Lodge imagined the telephone, along with the electroscope and galvanometer as new sense organs, technological prostheses with which to detect the ether.[50] In *Raymond*, while Lodge is concerned with the content of the communications (after all, they might be messages from his son), his concern is sublimated in the interest of establishing the channel between this world and the next. Despite his desperation for Raymond to cohere, Lodge privileges medium over message. Raymond is repressed in the hope that he might take shape.

This is represented typographically on the page. Figure 12.1 shows a typical example of a transcript. The sitter's questions are attributed to "A.M.L." (i.e., Lodge's son, Alec), but the responses are unattributed.[51] Unanchored by a name, these complex utterances—Leonard reporting Feda who in turn reports Raymond—are left to stand for themselves. Lodge's interventions are in square brackets, marking them as commentary on the exchange rather than placing him in the scene as participant. There are times, though, when this textual order is disturbed. In chapter 20 of the "Supernormal Portion," Lodge gives details of a sitting between his wife, Mary, and Leonard, in which Raymond, via Feda, reports meeting Jesus. Lodge is a little squeamish about this, and strikes out the descriptions of both Jesus and the Highest Sphere in which he dwells (figure 12.2): "Until the case for survival is considered established, it is thought improper and unwise to relate an experience of a kind which may be imagined, in a book dealing for the most part with evidential matter. So I have omitted the description here, and the brief reported utterance which followed. I think it fair, however, to quote the record so far as it refers to the youth's own

210 PART II—CHAPTER XVIII

There are hundreds of things he will think of after he is gone.

He has brought Lily, and William—the young one——

(Feda, *sotto voce*.—I don't know whether it is right, but he appears to have two brothers.)

[Two brothers as well as a sister died in extreme infancy. He would hardly know that, normally.—O. J. L.]

A. M. L.—Feda, will you ask Raymond if he would like me to ask some questions?

Yes, with pleasure, he says.

A. M. L.—A little time ago, Raymond said he was with mother. Mother would like to know if he can say what she was doing when he came? Ask Raymond to think it over, and see if he can remember?

Yes, yes. She'd got some wool and scissors. She had a square piece of stuff—he is showing me this—she was working on the square piece of stuff. He shows me that she was cutting the wool with the scissors.

Another time, she was in bed.

She was in a big chair—dark covered——

This refers to the time mentioned first. [Note B.]

A. M. L.—Ask Raymond if he can remember which room she was in?

(Pause.)

He can't remember. He can't always see more than a corner of the room—it appears vapourish and shadowy.

He often comes when you're in bed.

He tried to call out loudly: he shouted, 'Alec, Alec!' but he didn't get any answer. That is what puzzles him. He thinks he has shouted, but apparently he has not even manufactured a whisper.

A. M. L.—Feda, will you ask Raymond if he can remember trivial things that happened, as these things often make the best tests?

He says he can now and again.

A. M. L.—The questions that father asked about 'Evinrude,' 'Dartmoor,' and 'Argonauts,' are all trivial,

FIRST LEONARD SITTING OF ALEC 211

but make good tests, as father knows nothing about them.

Yes, Raymond quite understands. He is just as keen as you are to give those tests.

A. M. L.—Ask Raymond if the word 'Evinrude' in connexion with a holiday trip reminds him of anything?

Yes. (Definitely.)

A. M. L.—And 'Argonauts'?

Yes. (Definitely.)

A. M. L.—And 'Dartmoor'?

Yes. (Definitely.)

A. M. L.—Well, don't answer the questions now, but if father asks them again, see if you can remember anything.

(While Alec was speaking, Feda was getting a message simultaneously:—)

He says something burst.

[This is excellent for Dartmoor, but I knew it.—A. M. L.] [Note C.]

A. M. L.—Tell Raymond I am quite sure he gets things through occasionally, but that I think often the meaning comes through altered, and very often appears to be affected by the sitter. It appears to me that they usually get what they expect.

Raymond says, "I only wish they did!" But in a way you are right. He is never able to give all he wishes. Sometimes only a word, which often must appear quite disconnected. Often the word does not come from his mind; he has no trace of it. Raymond says, for this reason it is a good thing to try, more, to come and give something definite at home. When you sit at the table, he feels sure that what he wants to say is influenced by some one at the table. Some one is helping him, some one at the table is guessing at the words. He often starts a word, but somebody finishes it.

He asked father to let you come and not say who you were; he says it would have been a bit of fun.

A. M. L.—Ask Raymond if he can remember any characteristic things we used to talk about among ourselves?

FIGURE 12.1. From Oliver Lodge, *Raymond: Or, Life and Death* (London: Methuen, 1916), 210–11.

feelings, because otherwise the picture would be incomplete and one-sided, and he might appear occupied only with comparatively frivolous concerns."[52] The inserted passage is unbearably dispassionate about Raymond (he is merely "the youth') while also deeply concerned to convey his character. But what is most striking is the way the page has been set, reproducing both the space of the description and its absence. Where the composite voice of Leonard, Feda, and Raymond should be is Lodge's own editorial intervention.

By offering his editorial presence instead of Raymond's, the disembodied author who cannot write for himself, Lodge short-circuits the complicated chain that links Raymond—via Feda, Leonard, Mary Lodge, and Oliver Lodge—to the reader. For Raymond to cohere Lodge must remain in the margins; here, however, Lodge appears where Raymond should be, no longer the book's editor but its author instead. This change of subject effects another, as the book is no longer about Raymond but what is repressed throughout: Lodge's own unarticulated grief. In his anxiety over content, Lodge fills the empty form of this piece of paperwork with writing that points back to himself.

230 PART II—CHAPTER XX

street, you don't bother yourself to say ' how-do-you-do.'

M. F. A. L.—There are streets, then ?

Yes. He was pleased to see streets and houses.

At one time, I thought it might be created by one's own thoughts. You gravitate to a place you are fitted for. Mother, there's no judge and jury, you just gravitate, like to like.

I've seen some boys pass on who had nasty ideas and vices. They go to a place I'm very glad I didn't have to go to, but it's not hell exactly. More like a reformatory—it's a place where you're given a chance, and when you want to look for something better, you're given a chance to have it. They gravitate together, but get so bored. Learn to help yourself, and immediately you'll be helped. Very like your world; only no unfairness, no injustice—a common law operating for each and every one.

M. F. A. L.—Are all of the same rank and grade ?

Rank doesn't count as a virtue. High rank comes by being virtuous. Those who have been virtuous have to pass through lower rank to understand things. All go on to the astral first, just for a little.

He doesn't remember being on the astral himself. He thinks where he is now, he's about third. Summerland—Homeland, some call it. It is a very happy medium. The very highest can come to visit you. It is just sufficiently near the earth plane to be able to get to those on earth. He thinks you have the best of it there, so far as he can see.

Mother, I went to a gorgeous place the other day.

M. F. A. L.—Where was it ?

Goodness knows !

I was permitted, so that I might see what was going on in the Highest Sphere. Generally the High Spirits come to us.

I wonder if I can tell you what it looked like !

UNVERIFIABLE MATTER 231

[Until the case for survival is considered established, it is thought improper and unwise to relate an experience of a kind which may be imagined, in a book dealing for the most part with evidential matter. So I have omitted the description here, and the brief reported utterance which followed. I think it fair, however, to quote the record so far as it refers to the youth's own feelings, because otherwise the picture would be incomplete and one-sided, and he might appear occupied only with comparatively frivolous concerns.]

I felt exalted, purified, lifted up. I was kneeling. I couldn't stand up, I *wanted* to kneel.

Mother, I thrilled from head to foot. He didn't come near me, and I didn't feel I wanted to go near him. Didn't feel I ought. The Voice was like a bell. I can't tell you what he was dressed or robed in. All seemed a mixture of shining colours.

FIGURE 12.2. From Oliver Lodge, *Raymond: Or, Life and Death* (London: Methuen, 1916), 230–1.

THE ETHER AS OTHER

The book remains a fetishized object, perhaps more so now it is laced with nostalgia. In 2012 scientists at the University of Leicester printed the human genome in 130 volumes.[53] With its own system of sameness and difference, the genome constitutes a new media for producing the human. Just as *Raymond* was intended to establish the continuity of one particular etherial body, Raymond Lodge, so that it could establish a new understanding of humanity, so these volumes also witness an etherial body of sorts, a life already written that precedes the individual and will continue to define the life after they have gone. However, while the printed book, for Lodge, was the only object capable of encompassing a life, the bound volumes of the genome point to the book's insufficiency. In a comment on the trance of alphabetization, the way that knowledge is held to exist unmediated by printed letterforms, the volumes of the printed genome are legible but make little sense; furthermore, their sheer number exposes the limitations of the book-sized quantities into which knowledge is usually consumed. The fact that the genome was printed at all suggests that the book still promises enclosure and remains a monument to wholeness, yet these volumes can only present the book ironically.

In 1934 Lodge gave his last broadcast for the BBC. At age eighty-two, he suspected it would be his last and so wished his listeners goodbye.[54] A writer and broadcaster as well as an inventor of telegraphic systems, loudspeakers, and microphones, Lodge was at the cutting edge of the media technologies that would define the twentieth century. Yet looking back he seems peculiarly old-fashioned. Lodge's spirits are not even the ghosts of early twentieth-century modernity but seem to belong to the world of Victorian table tappers and Mr. Sludge the Medium. Yet there is, nonetheless, something modern about Lodge's concern for individuation and embodiment. Our digital imaginary prefers immaterial bits flowing freely around the world over the wires, hardware, and terms of access that sustain them; equally, the democratic potential of the web is hitched to the neoliberalism of the wisdom of the crowd. There is something compelling about Lodge's etheric imagination, not because it insists on the existence of higher realms but because it cannot but figure those higher realms on the basis of provisional material forms. To pass on messages and nourish the dead, the ether could not be left opposed to matter, on one side of a binary pair, but had to constitute a mediating technology in its own right. And whenever it did, it became necessary for Lodge to reimagine the unmodified ether as still there but out of reach. Lodge wanted the ether to come first, an elementary something out of which matter could take shape, but the universe didn't play along and the ether, which promised to make spirit matter, proved shadowy, fugitive, and occult. No matter how much Lodge wanted the ether to serve as the basis for matter, his work reveals that it was material that took precedence. Reading Lodge reminds us not only that the apparently immaterial is always in some way embodied but also that it is by doing things with embodied forms that we transcend them.

NOTES

INTRODUCTION: OLIVER LODGE

Epigraph: Oliver Lodge, *Past Years: An Autobiography* (London: Hodder and Stoughton, 1931), 340. Insertion Lodge's own.

1. John Ruskin, extract from *Fors Clavigera*, in *Theory of the Glaciers of Savoy*, by M. Le Chanoine Rendu, ed. George Forbes, trans. Alfred Wills (London: Macmillan, 1874), 206–7. The extract is from letter 34, "La Douce Dame" (1873); see *Fors Clavigera: Letters to the Workmen and Labourers of Great Britain*, vol. 1 (London: George Allen, 1907), 624–43, reprinted in *Library Edition of John Ruskin*, ed. E. T. Cook and Alexander Wedderburn, vol. 27 (London: George Allen, 1907).

2. For the original dispute, see Nanna Katrine Lüders Kaalund, "A Frosty Disagreement: John Tyndall, James David Forbes, and the Early Formation of the X-Club," *Annals of Science* 74, no. 4 (2017): 282–98.

3. Lodge, *Past Years*, 5.

4. Lodge, *Past Years*, 345.

5. Oliver Lodge, *Continuity: The Presidential Address to the British Association Birmingham 1913* (London: J. M. Dent and Sons, 1913).

6. Richard Noakes, "Thoughts and Spirits by Wireless: Imagining and Building Psychic Telegraphs in America and Britain, circa 1900–1930," *History and Technology* 32, no. 2 (2016): 137–58.

7. See Stathis Arapostathis and Graeme Gooday, *Patently Contestable: Electrical Technologies and Inventor Identities on Trial in Britain* (Cambridge, MA: MIT Press, 2013), 141–74.

8. J. D. Root, "Science, Religion, and Psychical Research: The Monistic Thought of Sir Oliver Lodge," *Harvard Theological Review* 71, no. 3/4 (1978): 245–63.

9. Graeme Gooday and Stathis Arapostathis, "Electrical Technoscience and Physics in Transition, 1880–1920," *Studies in History and Philosophy of Science A* 44, no. 2 (2013): 202–11.

10. Discussion of Myers runs throughout the spiritualist discussions of *Past Years* with a dedicated discussion on p. 279. Lodge describes the effect of hearing Tyndall on pp. 55–56 and 78. Muirhead does not even appear in the index.

11. *Past Years*, especially chapter 17, "Beginnings of Wireless," 225–36.

12. W. P. Jolly, *Sir Oliver Lodge* (London: Constable, 1974), 222–23.

13. Lodge, *Past Years*, 181. See Oliver Lodge, *Lightning Conductors and Lightning Guards: A Treatise on the Protection of Buildings, of Telegraph Instruments and Submarine Cables and of*

Electrical Installations Generally from Damage by Atmospheric Discharges (London: Whittaker, 1892). For the Maxwellian context, see Bruce J. Hunt, *The Maxwellians* (Ithaca, NY: Cornell University Press, 1991), 146–51.

14. See Arapostathis and Gooday, *Patently Contestable*, 148–50.

15. Oliver Lodge, *Signalling across Space without Wires: Being a Description of the Work of Hertz & His Successors* (London: The Electrician Printing and Publishing Company, 1898).

16. Jolly, *Sir Oliver Lodge*, 224.

17. Lodge acknowledged that Mary contributed an illustration of hydraulic bellows to his only formal textbook, *Elementary Mechanics: Including Hydrostatics and Pneumatics* (London: W. and R. Chambers, 1879). See *Past Years*, 74. She is also often mentioned in the various accounts of séance sessions in his spiritualist writing.

18. Lodge Plugs Limited, as it became, was sold to Smith Industries in the 1960s; Lodge Fume Deposit Company became Lodge Cottrell in 1921 and still exists today. K. R. Parker, *The History of Lodge Cottrell Limited: Development of Electrostatic Precipitation in the United Kingdom*, unpublished typescript, n.d. [1989?], personal communication with the author.

19. Lodge, *Past Years*, 270.

20. For instance, see Walter Cook, *Reflections on "Raymond": An Appreciation and Analysis* (London: Grant Richards, 1917); Paul Hookham, *"Raymond': A Rejoinder Questioning the Validity of Certain Evidence and of Sir Oliver Lodge's Conclusions Regarding It* (Oxford: B. H. Blackwell, 1917); Viscount Halifax [Charles Lindley Wood], *"Raymond": Some Criticisms* (London: A .R. Mowbray, 1917).

21. For details of these talks, see Peter Rowlands, *Oliver Lodge and the Liverpool Physical Society* (Liverpool: Liverpool University Press, 1990), 117–20.

22. Oliver Lodge, "Radium and Its Lessons," *Nineteenth Century and After* 54 (1903): 78–85; Oliver Lodge, "The Lessons of Radium," in *Modern Views of Electricity*, 3rd ed. (London: Macmillan, 1907), 463–77.

23. Oliver Lodge, *Public Service versus Private Expenditure* (London: Fabian Society, 1905), repr. in *Socialism and Individualism*, by Sidney Webb, Bernard Shaw, Sidney Ball, and Oliver Lodge (London: A.C. Fitfield, 1908).

24. Oliver Lodge, *Life and Matter: A Criticism of Professor Haeckel's "Riddle of the Universe"* (London: Williams and Norgate, 1905); Oliver Lodge, *The Substance of Faith* (London: Methuen, 1907); Oliver Lodge, *Man and the Universe: A Study of the Influence of the Advance in Scientific Knowledge upon Our Understanding of Christianity* (London: Methuen, 1908); Oliver Lodge, *The Survival of Man: A Study in Unrecognised Human Faculty* (London: Methuen, 1909). See also Jolly, *Sir Oliver Lodge*, 175–78.

25. Lodge's article on Tyndall for the tenth edition of the *Encyclopaedia Britannica* (1902–1903) prompted threats of litigation from Tyndall's widow, Louisa, and the article was replaced by an anonymous writer for the eleventh. See O.J.L. [Oliver Lodge], "John Tyndall," in *Encyclopaedia Britannica*, 9th/10th ed. (Edinburgh: A. and C. Black, 1902–3), 33:517–21; and Noakes, this volume.

26. For Lodge, Balfour, and The Souls see *Past Years*, 220–24; and Jolly, *Sir Oliver Lodge*, 115–18.

27. See the clippings in the Lodge scrapbooks, book XIV, 46, Cadbury Research Library, University of Birmingham.

28. Lodge scrapbooks, book XVI, 71.

29. Ruskin quoted in Lodge, *Past Years*, 340.

30. Lodge, *Past Years*, 232, 233.

31. Lodge, *Past Years*, 234–36. J. Clerk Maxwell, "A Dynamical Theory of the Electromagnetic Field," *Philosophical Transactions of the Royal Society* 155 (1865): 459–512.

32. For an account of Lodge and Marconi's rival patents, see Arapostathis and Gooday, *Patently Contestable*, 141–74.

ONE. COMMUNICATION, (DIS)CONTINUITIES AND CULTURAL CONTESTATION IN SIR OLIVER LODGE'S *PAST YEARS*

Epigraph: Oliver Lodge, *Past Years: An Autobiography* (London: Hodder and Stoughton, 1931), 11.

1. For discussion of Charles Darwin's autobiography see David Amigoni, "Between Medicine and Evolutionary Theory: Sympathy and Other Emotional Investments in Life Writings by and about Charles Darwin," in *After Darwin: Animals, Emotions and the Mind*, ed. Angelique Richardson (Amsterdam: Rodopi, 2013), 172–92.

2. See Charles Taylor, *Sources of the Self: The Making of the Modern Identity* (Cambridge: Cambridge University Press, 1992), in particular part III, chapter 13.

3. Galton's first work on "hereditary genius" was focused on the Fellowship of the Royal Society. The Galton Laboratory under Karl Pearson researched the genealogies of the leading scientific families of the early twentieth century. For example, see the career of Edgar Schuster, the first Galton Research Fellow: W. D. M. Paton and C. G. Phillips, "E. H. J. Schuster (1879–1969)," *Notes and Records of the Royal Society of London* 28 (1973): 111–17.

4. William Wordsworth, "Intimations of Immortality from Recollections of Early Childhood," ll.137–48, in *Wordsworth: Poetical Works*, ed. Tomas Hutchinson, rev. Ernest de Selincourt (Oxford: Oxford University Press, 1969), 460–62. De Selincourt was Lodge's senior professorial colleague at the University of Birmingham.

5. See Taylor, *Sources of the Self*, chapters 20–24.

6. See Wordsworth, "Miscellaneous Sonnets," "XXXIV. A Volant Tribe of Bards on Earth Are Found," 5–6, in *Poetical Works*, 206. See also Bernard Lightman, *Victorian Popularizers of Science* (Chicago: University of Chicago Press, 2007), 327–29. John Tyndall, *Address Delivered Before the British Association Assembled at Belfast* (London: Longman, Green, 1874).

7. Lodge, *Past Years*, 91.

8. David Vincent, *Bread, Knowledge and Freedom: A Study of Nineteenth-Century Working-Class Autobiography* (London: Methuen, 1981); Donna Loftus, "The Self in Society: Middle-Class Men and Autobiography," in *Life Writing and Victorian Culture*, ed. David Amigoni (Aldershot: Ashgate, 2006), 67–85; Linda H. Peterson, *Victorian Autobiography: The Tradition of Self-Interpretation* (New Haven, CT: Yale University Press, 1986).

9. Regenia Gagnier, *Subjectivities: A History of Self-Representation in Britain, 1832–1920* (Oxford: Oxford University Press, 1991).

10. Lodge, *Past Years*, 11.

11. High academic achievement and leadership was shared among Lodge's siblings: Alfred Lodge (1854–1937), mathematician; Francis Heawood Lodge (1857–1912); Sir Richard Lodge (1855–1936), historian; and Eleanor Constance Lodge (1869–1936), historian and principal of Westfield College, London. For Alfred Lodge, see Stanley, this volume.

12. Alfred Russel Wallace, *My Life: A Record of Events and Opinions* (London: Chapman and Hall, 1908), vii–viii.

13. Charles Darwin, *Autobiography*, ed. Nora Barlow (London: Collins, 1958), 120.

14. Wallace, *My Life*, 124.

15. Lodge, *Past Years*, 20.

16. Lodge, *Past Years,* 19; T. H. Huxley, "Autobiography," in *Autobiography and Selected Essays,* ed. Ada Snell (Boston: Houghton Mifflin, 1909), 5.

17. Lodge, *Past Years,* 28.

18. For an account of Bastian, Wallace's support, and the X Club rejection of Bastian's research, see Martin Fichman, *An Elusive Victorian: The Evolution of Alfred Russel Wallace* (Chicago: University of Chicago Press, 2004), 146–47.

19. Lodge, *Past Years,* 64–65. Thomas Day's work was one of the most influential examples of children's literature throughout the nineteenth century, in which Tommy Merton, a child of wealthy slave owners, is transformed from spoiled child into virtuous young man. Wallace recalls it as one of his influential early reading experiences. Wallace, *My Life,* 13.

20. Lodge, *Past Years,* 11–12. For Moorhouse's life, see E. A. Know and F. B. Smith, "Moorhouse, James (1826–1915)," *Oxford Dictionary of National Biography,* last revised September 23, 2004, https://doi.org/10.1093/ref:odnb/35093.

21. Oliver Lodge, *Pioneers of Science* (London: Macmillan, 1893), vii–viii.

22. See Lodge, *Past Years,* 150.

23. Stathis Arapostathis and Graeme Gooday, *Patently Contestable: Electrical Technologies and Inventor Identities on Trial in Britain* (Cambridge, MA: MIT Press, 2013), 147–49.

24. Lodge, *Past Years,* 114.

25. Lodge, *Pioneers of Science,* 161.

26. Lodge, *Pioneers of Science,* 274–95. For Clerke, see H. P. Hollis and M. T. Brück, "Clerke, Agnes Mary (1842–1907), *Oxford Dictionary of National Biography,* last revised September 23, 2004, https://doi.org/10.1093/ref:odnb/32444.

27. Lodge, *Pioneers of Science,* 276.

28. Lodge, *Past Years,* 83.

29. See the archive entry for Tyndall: L. C. Tyndall, "Tyndall, John (1820–1893)," *Oxford Dictionary of National Biography* archive, published in print 1898, https://doi.org/10.1093/odnb/9780192683120.013.27948.

30. Lodge, *Past Years,* 75.

31. John Tyndall, *Faraday as a Discoverer* (London: Longman, Green, 1868), 3.

32. Tyndall, *Faraday as a Discoverer,* 170.

33. Lodge, *Past Years,* 76.

34. Frank M. Turner, *Contesting Cultural Authority: Essays in Victorian Intellectual Life* (Cambridge: Cambridge University Press, 1993), 136.

35. Lodge, *Past Years,* 139.

36. Lodge, *Past Years,* 185, 228–29. For more on these lectures see Rowlands and Hunt, both in this volume.

37. Oliver Lodge, "Experiments on the Discharge of Leyden Jars," *Philosophical Transactions of the Royal Society of London* 50 (1892): 28. Emphasis mine.

38. Lodge, *Past Years,* 338–39.

39. Madeline Mason, "A Noble Life," *Saturday Review of Literature,* April 30, 1932, 696.

40. Lodge, *Past Years,* 234–36.

41. Arapostathis and Gooday, *Patently Contestable,* 145, 148–52.

42. Lodge, *Past Years,* 233.

43. Arapostathis and Gooday, *Patently Contestable,* 161.

44. Lodge, *Past Years,* 230.

45. Wallace, *My Life,* 336.

46. Lodge, *Past Years,* 316.

47. Walter R. Brooks, "New Books," *Outlook,* March 1932, 193. The review refers to Oliver

Lodge's reputation for "illogicality," though notably exonerates the autobiography from this taint, referring positively to the way in which the narrative accounts for the "steps" that took him to a form of spiritualism.

48. Peterson, *Victorian Autobiography*, 93–119.

49. Lodge, *Past Years*, 270–71.

50. Lodge, *Past Years*, 273–74. The article is Oliver Lodge, "An Experiment in Thought-Transference," *Nature* 30, no. 763 (June 12, 1884): 145. See also Rowlands, this volume.

51. Lodge, *Past Years*, 289.

52. See, for example, the way in which it figured in the title of Frederick Myers's classic work of spiritualism, *Human Personality and its Survival of Bodily Death* (London: Longmans, Green, 1903).

53. Karl Pearson, *The Life, Letters and Labours of Francis Galton*, vol. 2, *Researches of Middle Life* (Cambridge: Cambridge University Press, 1924), 359.

54. Lodge, *Past Years*, 118–19.

55. Francis Galton, "The History of Twins, as a Criterion of the Relative Powers of Nature and Nurture," *Fraser's Magazine*, November 1875, 566–76.

56. Lodge, *Past Years*, 247.

57. See Oliver Lodge, *The Ether of Space* (London: Harper and Brothers, 1909).

58. Lodge, *Past Years*, 342.

59. Brooks, "New Books," 192. There were other, indirect ways in which Lodge's reputation and identity were drawn into the cultural debate about the social goods associated with eugenics. Lodge's (unidentified) association of eugenic methods with "the stud farm" was an indication of his ambivalence, prominently cited in G. K. Chesterton's polemical *Eugenics and Other Evils* (London: Cassell, 1922), 13–14 (and resented by Galtonians).

TWO. BECOMING SIR OLIVER LODGE

1. Quoted in Oliver Lodge, *Past Years* (London: Hodder and Stoughton, 1931), 212.

2. John Belchem, *Merseypride: Essays in Liverpool Exceptionalism* (Liverpool: Liverpool University Press, 2006); Thomas Baines, *History of the Commerce and Town of Liverpool, and of the Rise of Manufacturing Industry in the Adjoining Counties* (London: Longman, 1852), 840; J. Langton, "Liverpool and its Hinterland in the Late Eighteenth Century," in *Commerce, Industry and Transport: Studies in Economic Change on Merseyside*, ed. B. L. Anderson and P. J. M. Stoney (Liverpool: Liverpool University Press, 1983), 1–25; Brian Hatton, "Shifted Tideways: Liverpool's Changing Fortunes," *Architectural Review*, January 2008, 39–50.

3. "The Commercial Ports of England," *Bankers' Magazine and Journal of the Money Market*, December 1851, 783. See also W. O. Henderson, "The Liverpool Office in London," *Economica* 1 (1933): 479.

4. Cutting from *Liverpool Daily Post*, June 11, 1935, Press Cuttings 1934–1936, S2520, 301, Special Collections and Archives, University of Liverpool.

5. Oliver Lodge, *Advancing Science: Being Personal Reminiscences of the British Association in the Nineteenth Century* (London: Ernest Benn, 1931), 16–22.

6. Lodge, *Past Years*, 34.

7. *Introductory Address Delivered at the Opening of the Liverpool Royal Infirmary School of Medicine as the Medical Faculty of University College, Liverpool, on Monday, 3rd October, 1881* (London: Harrison and Sons, 1881).

8. J. Ambrose Fleming, "Obituary Sir Oliver Lodge, F.R.S.," *Nature* 146, no. 3697 (September 7, 1940): 328.

9. "Address of the President Sir William Bragg, O.M., at the Anniversary Meeting, 30

November 1940," *Proceedings of the Royal Society A* 177 (December 31, 1940): 14; and *Proceedings of the Royal Society B* 129 (December 31, 1940): 426.

10. J. Ambrose Fleming, "Lodge and the Physical Society," *Proceedings of the Physical Society* 53 (1941): 58; R. A. Gregory and Allan Ferguson, "Oliver Joseph Lodge (1851–1940)," *Obituary Notices of Fellows of the Royal Society* 3 (1941): 559.

11. Gregory and Ferguson, "Oliver Joseph Lodge (1851–1940)," 559.

12. Fleming, "Lodge and the Physical Society," 58.

13. R. G. Roberts, "The Training of an Industrial Physicist. Oliver Lodge and Benjamin Davies. 1882–1940" (PhD diss., University of Manchester Institute of Science and Technology, 1984), 2–18.

14. Cutting from *Liverpool Daily Post*, September 9, 1940, Press Cuttings 1938–1941, S2522, 261, Special Collections and Archives, University of Liverpool.

15. Personal communication from D. N. Edwards (ca. 1990), based on an examination of materials in the University of Liverpool Special Collections and Archives.

16. Lodge, *Past Years*, 158.

17. Lodge, *Past Years*, 105; W. P. Jolly, *Sir Oliver Lodge* (London: Constable, 1974), 92.

18. Lodge, *Past Years*, 252.

19. Jolly, *Sir Oliver Lodge*, 90.

20. David Edwards, "A Victorian Polymath," in *Oliver Lodge and the Invention of Radio*, ed. Peter Rowlands and J. Patrick Wilson (Liverpool: PD Publications, 1994), 22.

21. Oliver Lodge, "The Ether and its Functions," *Nature* 27, no. 691 (January 25, 1883): 304–6; and "The Ether and its Functions," *Nature* 27, no. 692 (February 1, 1883): 328–30.

22. Oliver Lodge, "Electrical Accumulators or Secondary Batteries," *Engineer* 53 (May 19, 1882): 365; "Electrical Accumulators or Secondary Batteries," *Engineer* 53 (May 26, 1882): 373; "Electrical Accumulators or Secondary Batteries," *Engineer* 53 (June 16, 1882): 439; "Electrical Accumulators or Secondary Batteries," *Engineer* 53 (June 26, 1882): 457; "Electrical Accumulators or Secondary Batteries," *Engineer* 54 (July 7, 1882): 11; "Electrical Accumulators or Secondary Batteries," *Engineer* 54 (July 14, 1882): 30–31; "Electrical Accumulators or Secondary Batteries," *Engineer* 54 (September 29, 1882): 230; "Electrical Accumulators or Secondary Batteries," *Engineer* 54 (October 6, 1882): 249–50; "Electrical Accumulators or Secondary Batteries," *Engineer* 54 (December 8, 1882): 436.

23. Oliver Lodge, "Dust," *Nature* 31, no. 795 (January 22, 1885): 265–69.

24. See Oliver Lodge, "An Experiment in Thought-Transference," *Nature* 30, no. 763 (June 12, 1884): 145; Oliver Lodge, "An Account of Experiments in Thought-Transference," *Proceedings of the Society for Psychical Research* 2 (1884): 189–200; Edmund Gurney, "Thought-Transference and the Laws of Probability," *Proceedings of the Society for Psychical Research* 2 (1884): 257–62. See also Amigoni, this volume.

25. Lodge, *Past Years*, 182. See also Hunt, this volume.

26. Oliver Lodge, "Dr. Mann Lectures. Protection of Buildings from Lightning. Lecture I—Delivered March 10th 1888," *Journal of the Society of Arts* 36 (June 15, 1888): 867–74; and "Dr. Mann Lectures. Protection of Buildings from Lightning. Lecture II—Delivered March 17th 1888," *Journal of the Society of Arts* 36 (June 22, 1888): 880–93.

27. Oliver Lodge, "On the Theory of Lightning-Conductors," *Philosophical Magazine*, 5th series, 26 (1888): 217–30. Theodore Besterman, *A Bibliography of Sir Oliver Lodge F.R.S.* (Oxford: Oxford University Press, 1935).

28. See Hunt, this volume. For FitzGerald's announcement, see G. F. FitzGerald, "Address," *Report of the Fifty-Eighth Meeting of the British Association for the Advancement of Science* (London: John Murray, 1889), 557–62.

29. Lodge, *Past Years*, 185.

30. Lodge, *Advancing Science*, 88.

31. See Peter Rowlands, *Oliver Lodge and the Liverpool Physical Society* (Liverpool: Liverpool University Press, 1990), 3.

32. Minute Books of the Liverpool Physical Society, 2, p. 44, Victoria Gallery and Museum, University of Liverpool (for numbers); Benjamin Davies to Lodge, February 18, 1895, Lodge Collection, Archives and Special Collections, University of Liverpool (for illness). See Rowlands, *Oliver Lodge and the Liverpool Physical Society*, 129.

33. Minute Books of the Liverpool Physical Society, 2, pp. 69–71, Victoria Gallery and Museum, University of Liverpool; Rowlands, *Oliver Lodge and the Liverpool Physical Society*, 139. Lodge, *Past Years*, 150.

34. FitzGerald, "Address," 557–62; G. F. FitzGerald, "Address to the Mathematical and Physical Section of the British Association," in *The Scientific Writings of George Francis Fitzgerald*, ed. Jospeh Larmor (London: Longmans, Green, 1902), 240.

35. For more on this address, see Noakes, this volume; and Mussell, this volume.

36. Oliver Lodge, "Address," in *Report of the Sixty-First Meeting of the British Association for the Advancement of Science* (London: John Murray, 1892), 547–57.

37. Oliver Lodge, "Aberration Problems," *Philosophical Transactions of the Royal Society A* 184 (1893): 727–804.

38. John Hay, "The Centenary of the Liverpool School of Medicine: The Address of Commemoration 11 May 1934," hand-corrected typescript, SPEC.LUP.934.HAY(PC), Special Collections and Archives, University of Liverpool.

39. Oliver Lodge, "The Work of Hertz," *Nature* 50, no. 1284 (June 7, 1894): 133–39; and "The Work of Hertz," *Nature* 50, no. 1285 (June 14, 1894): 160–61.

40. For Righi and Marconi, see letter from Righi to Lodge, June 18, 1897, University College, London, quoted in Peter Rowlands, "Radio Begins in 1894," in *Oliver Lodge and the Invention of Radio*, ed. Peter Rowlands and J. Patrick Wilson, 94. For Popov, see J. Ambrose Fleming, *Principles of Electric Wave Telegraphy* (London: Longmans, Green, 1906), 425, quoted in Rowland F. Pocock, *The Early British Radio Industry* (Manchester: Manchester University Press, 1988), 93–94n1. For Bose, see Jagadish Chandra Bose, "On the Determination of the Indices of Refraction of Various Substances for the Electric Ray. I. Index of Refraction of Sulphur," *Proceedings of the Royal Society* 59 (December 12, 1895): 60–67. For Jackson, see W. P. Jolly, *Marconi* (London: Constable, 1972), 90n3. See also Rowlands, "Radio Begins in 1894," 84–86, 94.

41. For more on the early history of wireless, see Stathis Arapostathis and Graeme Gooday, *Patently Contestable: Electrical Technologies and Inventor Identities on Trial in Britain* (Cambridge, MA: MIT Press, 2013), chapter 6.

42. Oliver Lodge, "On the Question of Absolute Velocity, and on the Mechanical Functions of an Æther, with some Remarks on the Pressure of Radiation," *Philosophical Magazine*, 5th series, 46 (1898): 414–26.

43. Reported in the student magazine the *Sphinx* 9 (1901), 104, Special Collections and Archives, University of Liverpool, quoted in Rowlands, *Oliver Lodge and the Liverpool Physical Society*, 217–18.

44. Cutting from *Liverpool Daily Post*, December 1, 1935, Press Cuttings 1934–1936, S2520, 243, Special Collections and Archives, University of Liverpool.

45. Cutting from *Liverpool Daily Post*, August 23, 1940, Press Cuttings 1938–1941, S2522, 258, Special Collections and Archives, University of Liverpool.

THREE. LODGE IN BIRMINGHAM

1. "Mr Chamberlain on a University of Birmingham," *Times* (London), November 19, 1898, 10.

2. Quoted in Peter Marsh, *Joseph Chamberlain: Entrepreneur in Politics* (New Haven, CT: Yale University Press, 1994), 669.

3. Joseph Chamberlain Papers, JC4/5/96, Cadbury Research Library, University of Birmingham, quoted in Marsh, *Joseph Chamberlain*, 443.

4. See T. W. Heyck, *The Transformation of Intellectual Life in Victorian England* (London: Croom Helm, 1982), 53–57; Sheldon Rothblatt, *The Revolution of the Dons: Cambridge and Society in Victorian England* (London: Faber and Faber, 1968); Robert Anderson, *British Universities: Past and Present* (London: Bloomsbury, 2006).

5. T. H. Huxley, "Science and Culture," in *Science and Culture and Other Essays* (London: Macmillan, 1882), 1–23.

6. D. R. Jones, *The Origins of Civic Universities: Manchester, Leeds, and Liverpool* (London: Routledge, 1988).

7. Stathis Arapostathis and Graeme Gooday, "Electrical Technoscience and Physics in Transition, 1880–1920," *Studies in History and Philosophy of Science A* 44, no. 2 (2013): 202.

8. Arapostathis and Gooday, "Electrical Technoscience," 203.

9. See Ives, Drummond, and Schwarz, *First Civic University*, 149–50; *Annual Reports: Principal*, UC7/iii (1900–1901), 10, Cadbury Research Library, University of Birmingham.

10. See Ives, Drummond, and Schwarz, *First Civic University*, 149.

11. *Annual Reports: Principal*, UC 7/iii (1902–1903), 12, Cadbury Research Library.

12. *Annual Reports: Principal*, UC 7/iii (1900–1901), 10, Cadbury Research Library. See Ives, Drummond, and Schwarz, *First Civic University*, 149.

13. Ives, Drummond and Schwarz, *First Civic University*, 143.

14. Cadman to Beale, March 17, 1910, UC 14/ii, Principal's Letter Book OL37, Cadbury Research Library.

15. B. M. D. Smith, *Business Education in the University of Birmingham, 1899–1965* (Birmingham: Birmingham University Business School, 1990), 18. Quoted by Ives, Drummond, and Schwarz, *First Civic University*, 148.

16. Oliver Lodge, *University Development, and Survey of the Sciences: Appendix to University Development: Consideration of the Relation of the University of Birmingham to its Central and Suburban Sites by the Principal* (privately printed pamphlet, 1902). Quoted in Arapostathis and Gooday, "Electrical Technoscience," 205.

17. Ives, Drummond, and Schwarz, *First Civic University*, 82–83; Executive Committee Minutes, p. 26, UC 4/iii/9, Cadbury Research Library; further correspondence JC12/1/17, 8, 11.

18. Executive Committee Minutes, p. 39, UC 4/iii/9, Cadbury Research Library.

19. UA Building Committee, 1/1 and 1/19, Cadbury Research Library.

20. Simon Gunn, *The Public Culture of the Victorian Middle Class: Ritual and Authority in the English Industrial City, 1840–1914* (Manchester: Manchester University Press, 2008).

21. Peter T. Marsh, "Chamberlain, Joseph [Joe] (1836–1914)," *Oxford Dictionary of National Biography*, last revised 2013, https://doi.org/10.1093/ref:odnb/32350.

22. Ives, Drummond, and Schwarz, *First Civic University*, 82; and Eric W. Vincent and P. Hinton, *The University of Birmingham, Its History and Significance* (Birmingham: Cornish Brothers, 1947), 82.

23. Ives, Drummond, and Schwarz, *First Civic University*, 143; Principal's and Vice Chancellor's Letter Books, p. 35, UC 14/ii, Cadbury Research Library.

24. Ives, Drummond, and Schwarz, *First Civic University*, 147; Marsh, *Joseph Chamberlain*, 13. Details of the deputation to the Birmingham Chamber of Commerce that led to the establishment of the Board of Studies are in Report on the Curriculum in the University of Birmingham (n.d.), UC 7/iv/4/27, Cadbury Research Library.

25. See D. K. Drummond, "The University of Birmingham and the Industrial Spirit: Reasons for Local Support of Joseph Chamberlain's Campaign to Found the University of Birmingham, 1897–1900," *History of Universities* 19 (1999): 247–64. See also Drummond, "'Power of Our Provincial Cities' and 'Opportunities for the Highest Culture': The Call for a University of Birmingham and University Scholarship in Late Victorian Britain," in *Scholarship in Victorian Britain*, ed. Martin Hewitt, Leeds Working Papers in Victorian Studies 1 (Leeds: Leeds Centre for Victorian Studies, 1998), 53–65.

26. Advisory Subcommitees, 4/iii/10, Cadbury Research Library, and Departments of Mining and Metallurgy, UC 8/iv/2/1, Cadbury Research Library.

27. Ives, Drummond, and Schwarz, *First Civic University*, 144; Chamberlain to Beale, January 13, 1909, UC 14/ii, Cadbury Research Library.

28. Ives, Drummond, and Schwarz, *First Civic University*, 147; S. P. Keeble, "University Education and Business Management, 1880–1950s" (PhD diss., London School of Economics, 1984), 196.

29. "The University of Birmingham," *Birmingham Daily Post*, May 9, 1900, 4.

30. UA Building Committee, Cadbury Research Library; and E.W. Ives, *Image of a University, the Great Hall at Edgbaston: 1900-1908: an inaugural lecture delivered in the University of Birmingham on 9 May 1988* (Birmingham: University of Birmingham, 1988).

31. "Proposed curriculum of the Birmingham Jewellers' and Silversmiths' Association," Birmingham Archives and Collections; Ives, Drummond, and Schwarz, *First Civic University*, 49.

32. Marsh, *Joseph Chamberlain*, 443–46.

33. "The Birmingham University," *Engineering* 79 (August 25, 1905): 240.

34. Building Committee Minutes, January 9, 1900, p. 3, UC 4/iii/20, Cadbury Research Library.

35. Adrian Desmond, "Huxley, Thomas Henry (1825–1895)," *Oxford Dictionary of National Biography*, last revised May 28, 2015, https://doi.org/10.1093/ref:odnb/14320.

36. *Mason College Calendar 1880–1881* (Birmingham: Mason College, 1881), 94–110; Huxley, "Science and Culture," 3–4.

37. Oliver Lodge, *Past Years* (London: Hodder and Stoughton, 1931), 314–15.

38. Papers of Joseph Chamberlain, April 1889, JC L Add, JC 12/1/1/20–22 2, Cadbury Research Library.

39. Ives, Drummond, and Schwarz, *First Civic University*, 119; and UC 7/iv/8/39, Cadbury Research Library.

40. Ives, Drummond, and Schwarz, *First Civic University*, 120.

41. Academic Board Minutes, UC 4/iii/9, 38–39, Cadbury Research Library; Marsh, *Joseph Chamberlain*, 490–92; Lodge, *Past Years*, 243–46.

FOUR. THE ALTERNATIVE PATH

1. Oliver Lodge, *Lightning Conductors and Lightning Guards* (London: Whittaker, 1892), 111, and Oliver Lodge, *Past Years: An Autobiography* (London: Hodder and Stoughton, 1931), 185.

2. Hugh G. J. Aitken, *Syntony and Spark: The Origins of Radio* (New York: John Wiley and Sons, 1976), 106–9.

3. On efforts to distinguish between "pure" and "applied" science, and on Lodge's views on the question, see Graeme Gooday, "'Vague and Artificial': The Historically Elusive Distinction between Pure and Applied Science," *Isis* 103 (2012): 546–54; and Stathis Arapostathis and Graeme Gooday, "Electrical Technoscience and Physics in Transition, 1880–1920," *Studies in History and Philosophy of Science A* 44, no. 2 (2013): 202–11.

4. In their accounts of Lodge's work on lightning protection and oscillating currents, Hugh Aitken, Peter Rowlands, and Ido Yavetz all begin by mentioning the Mann Lectures but quickly move on to focus on Lodge's experiments on electromagnetic waves, and in Yavetz's case on Lodge's later dispute with Preece. Note that Aitken mistakenly places the Mann Lectures in 1885. See Aitken, *Syntony and Spark*, 85; Peter Rowlands, *Oliver Lodge and the Liverpool Physical Society* (Liverpool: Liverpool University Press, 1990), 17–25; and Peter Rowlands, "Radiowaves," in *Oliver Lodge and the Invention of Radio*, ed. Peter Rowlands and J. Patrick Wilson (Liverpool: PD Publications, 1994), 49; and Ido Yavetz, "A Victorian Thunderstorm: Lightning Protection and Technological Pessimism in the Nineteenth Century," in *Technology, Pessimism, and Postmodernism*, ed. Yaron Ezrahi, Everett Mendelsohn, and Howard P. Segal (Amherst: University of Massachusetts Press, 1995), 56.

5. [R. J. Mann], *The Atlantic Telegraph: A History of Preliminary Experimental Proceedings* (London: Jarrold and Sons, 1857); on Mann's authorship, see R. J. Mann to Latimer Clark, November 12, 1880, Wheeler Collection, New York Public Library.

6. J. Malcolm Walker, "Mann, Robert James (1817–1886)," *Oxford Dictionary of National Biography*, last revised September 23, 2004, https://doi.org/10.1093/ref:odnb/17947.

7. On Mann's widow endowing lectures in his honor on lightning protection, see a note in the *Journal of the Society of Arts* 36 (June 29, 1888): 898.

8. Oliver J. Lodge, *Modern Views of Electricity* (London: Macmillan, 1889), composed mainly of a series of articles that appeared in *Nature* between October 6, 1887, and January 31, 1889.

9. W. P. Jolly, *Sir Oliver Lodge* (London: Constable, 1974), 79.

10. Oliver J. Lodge, "Dust," *Nature* 31, no. 795 (January 22, 1885): 265–69; Oliver Lodge, *Advancing Science: Being Personal Reminiscences of the British Association in the Nineteenth Century* (London: Ernest Benn, 1931), 64. Lodge's son Lionel later developed the process into a successful electrodeposition business; see J. Patrick Wilson, "The Technological Heritage of Oliver Lodge," in Rowlands and Wilson, *Lodge and the Invention of Radio*, 188–91; Rowlands, this volume.

11. G. J. Symons, ed., *Report of the Lightning Rod Conference* (London: E and F. N. Spon, 1882).

12. Lodge recounted many of the experiments he performed in February and March 1888 in *Lightning Conductors and Lightning Guards*, 274–365. See also Rowlands, "Radiowaves," esp. 49–52 and figures 21 and 22, which reproduce pages from Lodge's laboratory notebook.

13. Lodge was following a long tradition of model experiments on lightning; see Willem D. Hackmann, "The Lightning Rod: A Case Study of Eighteenth-Century Model Experiments," in *Playing with Fire: Histories of the Lightning Rod*, ed. Peter Heering, Oliver Hochadel, and David J. Rhees (Philadelphia: American Philosophical Society, 2009), 209–29.

14. Lodge, *Past Years*, 154, 181–84.

15. Christa Jungnickel and Russell McCormmach, *Cavendish: The Experimental Life*

(Lewisburg, PA: Bucknell University Press, 1999), 245–48; William J. Turkel, *Spark from the Deep: How Shocking Experiments with Strongly Electric Fish Powered Scientific Discovery* (Baltimore: Johns Hopkins University Press, 2013), 59–64.

16. On Froude's ship models, as well as Cavendish's model torpedo and Benjamin Wilson's models of lightning, see Simon Schaffer, "Fish and Ships: Models in the Age of Reason," in *Models: The Third Dimension of Science*, ed. Soraya de Chadarevian and Nick Hopwood (Stanford, CA: Stanford University Press, 2004), 71–105.

17. Stephen Pumfrey, *Latitude and the Magnetic Earth* (Cambridge: Icon Books, 2001), 111–12.

18. Oliver Lodge, "Dr. Mann Lectures. Protection of Buildings from Lightning. Lecture I—Delivered March 10th 1888," *Journal of Society of Arts* 36 (June 15, 1888): 867–74, and "Dr. Mann Lectures. Protection of Buildings from Lightning. Lecture II—Delivered March 17th 1888," *Journal of the Society of Arts* 36 (June 22, 1888): 880–93; these were reprinted in *Electrician* 21 (June 22, 1888): 204–7; (June 29, 1888): 234–36; (July 6, 1888): 273–76; and (July 13, 1888): 302–3; and, with some additions and revisions, in Lodge, *Lightning Conductors*, 1–73.

19. Lodge, *Lightning Conductors*, 54.

20. Yavetz notes that Lodge estimated the resistance of the air column to be no more than a few hundred ohms ("Victorian Thunderstorm," 73n10).

21. Lodge's sons Alec and Brodie later exploited this kind of sudden discharge, or "B spark," for the automobile spark plugs produced by their successful company, Lodge Plugs. See Wilson, "Technological Heritage of Oliver Lodge," 184–88; and Lodge, *Past Years*, 187.

22. Lodge, *Lightning Conductors*, 15–18, 38.

23. Lodge, *Modern Views*, 186. On Maxwellians' views on the primacy of the field, see Bruce J. Hunt, *The Maxwellians* (Ithaca, NY: Cornell University Press, 1991).

24. Lodge, *Lightning Conductors*, 367.

25. E. T. Whittaker, *A History of the Theories of Aether and Electricity, from the Age of Descartes to the Close of the Nineteenth Century* (London: Longmans, Green, 1910), 253–54.

26. Lodge, *Lightning Conductors*, 40, 170–71.

27. Lodge, *Advancing Science*, 96; on modern understanding of lightning, see C. B. Moore, G. D. Aulich, and William Rison, "A Modern Assessment of Benjamin Franklin's Lightning Rods," in Heering, Hochadel, and Rhees, *Playing with Fire*, 256–68, and works cited there.

28. See in particular the statement in Lodge, *Lightning Conductors*, 32, that "high conductivity appears to be an actual objection" in lightning conductors. Lodge criticized "rash statements" in the *Report of the Lightning Rod Conference* more directly in Oliver Lodge, "Lightning, Lightning Conductors, and Lightning Protectors," *Journal of the Institution of Electrical Engineers* 18 (April 25, 1889): 407–11; this section was reprinted in Lodge, *Lightning Conductors*, 184–88.

29. Lodge, *Lightning Conductors*, 71–73, 207–11.

30. Yavetz suggests that Lodge contributed to "technological pessimism" by arguing that no rods or conductors could completely protect one from lightning ("Victorian Thunderstorm," 57).

31. Lodge, *Advancing Science*, 88. See also Rowlands, this volume.

32. On Lodge's friendship with Heaviside, see Hunt, *Maxwellians*, 149–51 and 171–72.

33. D. W. Jordan, "The Adoption of Self-Induction by Telephony," *Annals of Science* 39, no. 5 (1982): 433–61; Bruce J. Hunt, "'Practice vs. Theory': The British Electrical Debate, 1888–1891," *Isis* 74 (1983): 341–55.

34. Hunt, *Maxwellians*, 138–43.

35. Oliver Heaviside to Oliver Lodge, June 5, 1888, MS ADD 89/78–9, University College London Archives.

36. Oliver Heaviside to Oliver Lodge, June 27, 1888, MS ADD 89/78–9, University College London Archives.

37. Lodge, "Lecture II," 885. He dropped the words from "eccentric" to "repellent" when the lectures were reprinted in *Lightning Conductors* (47).

38. Oliver Heaviside to Oliver Lodge, September 25, 1888, MS ADD 89/78–9, University College London Archives.

39. Yavetz suggests that the criticisms the Maxwellians G. F. FitzGerald and H. A. Rowland offered of Lodge's claims about lightning protection may have been motivated by concerns that Lodge's "technological pessimism" might undermine confidence in the practical value of scientific progress ("Victorian Thunderstorm," 63). It seems more likely that they, like Heaviside, simply (and quite rightly) thought that Lodge had gone too far when he claimed that his Leyden jar discharges were a good proxy for real lightning.

40. W. H. Preece, "Presidential Address to Section G," in *Report of the Fifty-Eighth Meeting of the British Association for the Advancement of Science* (London: John Murray, 1889), 791.

41. On plans for the "joint discussion," see a note in *Electrician* 21 (August 3, 1888): 397.

42. "Discussion on Lightning Conductors," *Electrician* 22 (September 21, 1888): 646 and "Discussion on Lightning Conductors," *Electrician* 22 (September 28, 1888): 673–80.

43. "The British Association," *Times* (London), September 14, 1888, 6.

44. On Lodge's patented lightning guards, see Lodge, *Lightning Conductors*, 419–26, and Stathis Arapostathis and Graeme Gooday, *Patently Contestable: Electrical Technologies and Inventor Identities on Trial in Britain* (Cambridge, MA: MIT Press, 2013), 148–49; on the Lodge-Muirhead Syndicate, see Aitken, *Syntony and Spark*, 143–44 and 158–63.

45. Lodge, *Lightning Conductors*, 60.

46. On Chattock's contribution, see Lodge, *Past Years*, 183.

47. Hunt, *Maxwellians*, 30–33.

48. Hunt, *Maxwellians*, 33–44; G. F. FitzGerald, "On a Method of Producing Electromagnetic Disturbances of Comparatively Short Wave-Lengths," in *Report of the Fifty-Third Meeting of the British Association for the Advancement of Science* (London: John Murray, 1884), 405.

49. G. F. FitzGerald to J. J. Thomson, December 23, 1884, quoted in Hunt, *Maxwellians*, 45.

50. Lodge, *Lightning Conductors*, 62n.

51. Oliver Lodge, "On the Theory of Lightning Conductors," *Philosophical Magazine* 26 (July 1888): 230; see also Lodge, "Measurement of Electro-Magnetic Wave Length," *Electrician* 21 (September 14, 1888): 607–9, reprinted in Lodge, *Lightning Conductors*, 108–13.

52. G. F. FitzGerald, "Presidential Address to Section A," in *Report of the Fifty-Eighth Meeting of the British Association for the Advancement of Science* (London: John Murray, 1889), 558.

53. Lodge, *Lightning Conductors*, 111.

54. Oliver J. Lodge, "The Discharge of a Leyden Jar," *Electrician* 22 (March 15, 1889): 532. After Ludwig Boltzmann took offense at this crack, Lodge omitted it from subsequent reprintings of the lecture; see Hunt, *Maxwellians*, 154n6. On Lodge's first meeting with Hertz, see Lodge, *Past Years*, 154.

55. A. W. Ewing, *The Man of Room 40: The Life of Sir Alfred Ewing* (London: Hutchinson, 1939), 91.

56. Heinrich Hertz, *Memoirs, Letters, Diaries*, ed. Johanna Hertz, trans. Lisa Brinner, Mathilde Hertz, and Charles Susskind, 2nd ed. (San Francisco, CA: San Francisco Press,

1977), 307–11; J. G. O'Hara and W. Pricha, *Hertz and the Maxwellians* (London: Peter Peregrinus, 1987), 85–101.

57. Oliver J. Lodge, *Modern Views of Electricity*, 2nd ed. (London: Macmillan, 1892), 339; Lodge said that the word was suggested to him by A. T. Myers. See also Aitken, *Syntony and Spark*, 106.

58. Oliver J. Lodge, *The Work of Hertz and Some of His Successors* (London: Electrician Printing and Publishing, 1894); Lodge's lecture, which was also published in *Nature* and *Electrician*, strongly influenced Augusto Righi, Guglielmo Marconi, Alexander Popov, Jagadish Chandra Bose, and other pioneers of wireless telegraphy. See Arapostathis and Gooday, *Patently Incontestable*, 152–53; Rowlands, this volume.

59. On such a relationship between research driven by curiosity and that driven by practical aims, see Donald E. Stokes, *Pasteur's Quadrant: Basic Science and Technological Innovation* (Washington, DC: Brookings Institution Press, 1997).

FIVE. LODGE AND MATHEMATICS

1. Oliver Lodge, "The Geometrisation of Physics, and its Supposed Basis on the Michelson-Morley Experiment," *Nature* 106, no. 2677 (February 17, 1921): 799.

2. Lodge, "Geometrisation of Physics," 800.

3. For example, Bruce J. Hunt, *The Maxwellians* (Ithaca, NY: Cornell University Press, 1991), 26–28 and 205–6.

4. Oliver Lodge, *Advancing Science: Being Personal Reminiscences of the British Association in the Nineteenth Century* (London: Ernest Benn, 1931), 20–21.

5. Andrew Warwick, *Masters of Theory: Cambridge and the Rise of Mathematical Physics* (Chicago: University of Chicago Press, 2003), 296–99, and chapter 6.

6. Oliver Lodge, *Past Years* (London: Hodder and Stoughton, 1931), 88.

7. Lodge, *Advancing Science*, 26.

8. Lodge, *Past Years*, 71, 81–82.

9. Adrian Rice, "Henrici, Olaus Magnus Friedrich Erdmann (1840–1918)," *Oxford Dictionary of National Biography*, last revised September 23, 2004, https://doi.org/10.1093/ref:odnb/39487.

10. Lodge, *Past Years*, 104, 85.

11. Hunt, *Maxwellians*.

12. Oliver Lodge, "Modern Views of Electricity," *Nature* 36, no. 936 (October 6, 1887): 533.

13. Oliver Lodge, "Modern Views of Electricity," *Nature* 36, no. 937 (October 13, 1887): 561.

14. Oliver Lodge, "Clerk Maxwell and Wireless Telegraphy," in *James Clerk Maxwell: A Commemoration Volume, 1831–1931* (New York: MacMillan, 1931), 126.

15. Oliver Lodge, "Steps toward a New Principia," *Nature* 70, no. 1804 (May 26, 1904): 73; Lodge, *Advancing Science*, 90.

16. Oliver Lodge, *Elementary Mechanics: Including Hydrostatics and Pneumatics* (London: W. and R. Chambers, 1879), 29–30.

17. "The Teaching of Physics in Schools," *Proceedings of the Physical Society of London*, 30 (1918), 3S.

18. Oliver Lodge, *Easy Mathematics, Chiefly Arithmetic* (London: Macmillan, 1906), viii.

19. Oliver Lodge, *School Teaching and School Reform* (London: Williams and Norgate, 1905), 111.

20. Janet Delve, "The College of Preceptors and the *Educational Times*: Changes for British Mathematics Education in the Mid-Nineteenth Century," *Historia Mathematica* 30, no. 2 (2003): 140–72.

21. Oliver Lodge, "Modern Views of Electricity," *Nature* 41, no. 1048 (November 28, 1889): 80.

22. Oliver Lodge, *Energy* (New York: Robert M. McBride, 1929), 26–27.

23. Oliver Lodge, "The Modern Theory of Light," in *Annual Report of the Board of Regents of the Smithsonian Institution, 1889* (Washington, DC: Government Printing Office, 1890), 442.

24. Lodge, *Energy*, 26–27, 25.

25. Lodge, *School Teaching and School Reform*, 89.

26. Lodge, *Easy Mathematics*, 425.

27. Lodge, "Modern Views of Electricity," (October 6, 1887), 532.

28. Lodge, *Easy Mathematics*, 169.

29. Oliver Lodge, *Electrons* (London: G. Bell and Sons, 1919).

30. Oliver Lodge, "The Relation between Electricity and Light," *Nature* 23, no. 587 (January 29, 1881): 302–3.

31. Lodge, *Easy Mathematics*, 192.

32. Lodge, *School Teaching and School Reform*, 96.

33. Michael Price, "Mathematics in English Education, 1860–1914: Some Questions and Explanations in Curriculum History," *History of Education* 12, no. 4 (1983): 282–84. On Victorian primary and secondary mathematics education, see A. G. Howson, *A History of Mathematics Education in England* (Cambridge: Cambridge University Press, 1982).

34. Lodge, *School Teaching and School Reform*, 34.

35. Price, "Mathematics in English Education," 274.

36. Lodge, *Advancing Science*, 14–16.

37. Lodge, "Teaching of Physics in Schools," 42S.

38. Amirouche Moktefi, "Geometry: The Euclid Debate," in *Mathematics in Victorian Britain*, ed. Raymond Flood, Adrian Rice, and Robin Wilson (Oxford: Oxford University Press, 2011), 326.

39. Lodge, *School Teaching and School Reform*, 37.

40. Moktefi, "Geometry," 326.

41. Price, "Mathematics in English Education," 274–78.

42. Lodge, *School Teaching and School Reform*, 38.

43. Lodge, *Advancing Science*, 33.

44. Lodge, *School Teaching and School Reform*, 112.

45. Lodge, *Easy Mathematics*, vii.

46. Lodge, *School Teaching and School Reform*, 75–77.

47. Lodge, *Easy Mathematics*, 198.

48. A. G. Greenhill, "The Flying to Pieces of a Whirling Ring," *Nature* 43, no. 1116 (March 19, 1891): 462.

49. Oliver Lodge, "The Meaning of Algebraic Symbols in Applied Mathematics," *Nature* 43, no. 1118 (April 2, 1891): 513.

50. Oliver Lodge, "Thoughts of the Bifurcation of the Sciences," *Nature* 48, no. 1250 (October 12, 1893): 566.

51. Oliver Lodge, "The Meaning of Symbols in Applied Algebra," *Nature* 55, no. 1420 (January 14, 1897): 247.

52. Oliver Lodge, "The Progress of Physics," *Nature* 87, no. 2186 (September 21, 1911): 375.

53. Lodge, *Energy*, 61–62.

54. Lodge, *Easy Mathematics*, 184.

55. Oliver Lodge, "Use and Abuse of Empirical Formulae, and of Differentiation, by Chemists," *Nature* 40, no. 1029 (July 18, 1889): 273.

56. Oliver Lodge, *Relativity: A Very Elementary Exposition*, 3rd ed. (New York: George Doran, 1926), 41.

57. Oliver Lodge, "The New Theory of Gravity," *Nineteenth Century and After* 86 (1919): 1195.

58. Lodge, *Past Years*, 350.

59. Lodge, *Energy*, 66–67.

60. Lodge, *Past Years*, 350.

61. Oliver Lodge, "Mathematics and Physics," *Times* (London), April 15, 1922, 13.

62. Oliver Lodge, "Popular Relativity and the Velocity of Light," *Nature* 106, no. 2662 (November 4, 1920): 325.

63. Lodge, "New Theory of Gravity," 1201.

64. Oliver Lodge, "Einstein's Real Achievement," *Fortnightly Review* 110 (1921): 370–72.

65. Lodge, "Geometrisation of Physics," 795–96.

66. Lodge, *Relativity*, 42.

67. Lodge, "Geometrisation of Physics," 796.

68. Oliver Lodge, "Some Elementary Considerations Connected with Modern Physics," *Philosophical Magazine* 15 (1933): 719–20.

69. Lodge, "Einstein's Real Achievement," 370–72.

70. Oliver Lodge to A. S. Eddington, January 25, 1929, MS ADD 89 (Eddington), Library Services, Special Collections, University College London.

71. On this distinction see Stathis Arapostathis and Graeme Gooday, "Electrical Technoscience and Physics in Transition, 1880–1920," *Studies in History and Philosophy of Science A* 44, no. 2 (2013): 202–11; and Graeme Gooday, "'Vague and Artificial': The Historically Elusive Distinction between Pure and Applied Science," *Isis* 103 (2012): 546–54.

72. Lodge, "Clerk Maxwell and Wireless Telegraphy," 129; Lodge, *Advancing Science*, 167–71.

73. Lodge, *Advancing Science*, 29–30.

74. Oliver Lodge, "On Some Problems Connected with the Flow of Electricity in a Plane," *Philosophical Magazine*, 5th series, 1 (1876): 378.

75. Oliver Lodge, "Hertz's Equations," *Nature* 39, no. 1016 (April 18, 1889): 583.

76. Lodge, *Easy Mathematics*, 63.

77. Lodge, "Meaning of Symbols in Applied Algebra," 247.

SIX. THE RETIRING POPULARIZER

I am indebted to Richard Noakes his helpful suggestions for revision. The work for this article was undertaken while the author was on a Templeton Religion Trust grant.

1. Oliver Lodge, *Past Years: An Autobiography* (London: Hodder and Stoughton, 1931), 342–43.

2. Bernard Lightman, *Victorian Popularizers of Science: Designing Nature for New Audiences* (Chicago: University of Chicago Press, 2007), 64–71, 490.

3. David B. Wilson, "The Thought of Late Victorian Physicists: Oliver Lodge's Ethereal Body," *Victorian Studies* 15, no. 1 (September 1971): 30.

4. Lodge, *Past Years*, 65–66.

5. See Stanley, this volume.

6. Lodge, *Past Years*, 168.

7. Wilson, "Thought of Late Victorian Physicists," 31.

8. W. P. Jolly, *Sir Oliver Lodge* (London: Constable, 1974), 58.

9. Peter J. Bowler, *Reconciling Science and Religion: The Debate in Early Twentieth-Century Britain* (Chicago: University of Chicago Press, 2001), 96.

10. Jolly, *Sir Oliver Lodge*, 222.

11. Bruce J. Hunt, *The Maxwellians* (Ithaca, NY: Cornell University Press, 1991), 24–25.

12. Bowler, *Reconciling Science and Religion*, 49–50, 95–97; Peter J. Bowler, *Science for All: The Popularization of Science in Early Twentieth-Century Britain* (Chicago: University of Chicago Press, 2009), 36, 95, 218–20.

13. Roger Cooter and Stephen Pumfrey, "Separate Spheres and Public Places: Reflections of the History of Science Popularization and Science in Popular Culture," *History of Science* 32, no. 3 (1994): 237–67; Lightman, *Victorian Popularizers of Science*, 9–17.

14. Oliver Lodge, *Ether and Reality* (London: Hodder and Stoughton, 1930), viii.

15. Oliver Lodge, *Atoms and Rays: An Introduction to Modern Views on Atomic Structure and Radiation* (London: Ernest Benn, 1924), v–vii.

16. I am indebted to Richard Noakes for this insight.

17. Lodge, *Atoms and Rays*, 12, 14, 20–21.

18. Oliver Lodge, *Phantom Walls* (London: Hodder and Stoughton, 1929), 94.

19. Crosbie Smith, *The Science of Energy: A Cultural History of Energy Physics in Victorian Britain* (London: Athlone, 1998), 172.

20. Oliver Lodge, *The Immortality of the Soul* (Boston: Ball, 1908), 46–48.

21. Oliver Lodge, *Making of Man: A Study in Evolution* (London: Hodder and Stoughton, 1924), 73.

22. Oliver Lodge, *Evolution and Creation* (London: Hodder and Stoughton, 1926), 15, 24, 26, 104, 106, 145.

23. Although Raia acknowledges that evolution is a part of Lodge's thinking, she does not assign it a prominent role. See Courtenay Raia, "From Ether Theory to Ether Theology: Oliver Lodge and the Physics of Immortality," *Journal of the History of the Behavioral Sciences* 43 (2007): 19–43.

24. Lodge, *Evolution and Creation*, 21, 67–69, 71, 74, 77.

25. Peter J. Bowler, *The Non-Darwinian Revolution: Reinterpreting a Historical Myth* (Baltimore: Johns Hopkins University Press, 1988); Peter J. Bowler, *The Eclipse of Darwinism: Anti-Darwinian Evolution Theories in the Decades around 1900* (Baltimore: Johns Hopkins University Press, 1983).

26. Lodge, *Evolution and Creation*, 103–4.

27. Lodge, *Making of Man*, 141.

28. Oliver Lodge, *Science and Human Progress* (London: George Allen and Unwin, 1927), 47.

29. Oliver Lodge, *Modern Scientific Ideas* (London: Ernest Benn, 1927), 76–77.

30. In his brief discussion of Lodge, Bowler insightfully refers to it as "a coherent presentation of the new natural theology." I have tried to flesh out Bowler's point in far more detail. Bowler, *Reconciling Science and Religion*, 50.

31. Jolly, *Sir Oliver Lodge*, 231–32.

32. See Hendy, this volume; and Mussell, this volume.

33. For example, Bowler is somewhat dismissive of the content of Lodge's books. See Bowler, *Science for All*, 36, 145.

34. Oliver Lodge, "The New Theory of Gravity," *Nineteenth Century and After* 86 (1919): 1196, 1200, 1201.

35. Oliver Lodge, "Einstein's Real Achievement," *Fortnightly Review* 110 (1921): 353, 364, 366, 371.

36. It was published four years later as a short book of forty-one pages. See Oliver Lodge, *Relativity: A Very Elementary Exposition* (London: Methuen, 1925). A note on the first page states that the lecture is based on a report by a stenographer from shorthand notes taken on the evening of the address.

37. Oliver Lodge, "The Geometrisation of Physics, and Its Supposed Basis on the Michelson-Morley Experiment," *Nature* 106, no. 2677 (February 17, 1921): 795–800; Oliver Lodge, "Remarks on Simple Relativity and the Relative Velocity of Light," *Nature* 107, no. 2701 (August 4, 1921): 716–19; Oliver Lodge, "Remarks on Simple Relativity and the Relative Velocity of Light," *Nature* 107, no. 2702 (August 11, 1921): 748–51; Oliver Lodge, "Further Remarks on Relativity," *Nature* 107, no. 2703 (August 18, 1921): 784–85; Oliver Lodge, "Remarks on Gravitational Relativity," *Nature* 107, no. 2704 (August 25, 1921): 814–18.

38. Oliver Lodge, *Beyond Physics: Or, the Idealisation of Mechanism* (London: George Allen and Unwin, 1930): 101–2.

39. Lodge, *Ether and Reality*, 123.

40. Albert Einstein, *Sidelights on Relativity*, trans. G. B. Jeffery and Wilfrid Perrett (London: Methuen, 1922), 15–16, 18, 23.

41. Lodge emphasized where he and Einstein agreed, ignoring that they disagreed on how to conceive of the ether. Einstein did not agree with Lodge and others that the ether had some kind of internal rotational or vortical motion. I am indebted to Richard Noakes on this point.

42. Lodge, *Ether and Reality*, 82.

43. Lodge, *Science and Human Progress*, 117.

44. Oliver Lodge, *My Philosophy: Representing My Views on the Many Functions of the Ether of Space* (London: Ernest Benn, 1933), 158.

45. Lodge, *Beyond Physics*, 57.

46. Lodge, *Phantom Walls*, vii.

47. Lodge, *Atoms and Rays*, 134.

48. Lodge, *Beyond Physics*, 102.

49. Lodge, *Atoms and Rays*, 63–64.

50. Lodge, *Phantom Walls*, 130.

51. Lodge, *Beyond Physics*, 103.

52. Lodge, *My Philosophy*, 59.

53. Lodge, *Science and Human Progress*, 145.

54. Lodge, *Ether and Reality*, 16.

55. Lodge, *My Philosophy*, 138.

56. See Bowler, *Science for All*, 36, 219.

57. Wilson, "Thought of Late Victorian Physicists," 48.

58. John D. Root, "Science, Religion, and Psychical Research: The Monistic Thought of Sir Oliver Lodge," *Harvard Theological Review* 71, no. 3/4 (1978): 258.

59. "The Competition," *Spectator* 144 (May 31, 1930): 905.

60. "The Five Best Brains," *Spectator* 144 (June 14, 1930): 979.

61. Richard Noakes, "Ethers, Religion and Politics in Late-Victorian Physics: Beyond the Wynne Thesis," *History of Science* 43, no. 4 (2005): 31.

SEVEN. THE FORGOTTEN CELEBRITY OF MODERN PHYSICS

1. This dichotomy is evident in Helge Kragh, *Quantum Generations* (Princeton, NJ: Princeton University Press, 1999); David Knight, *Public Understanding of Science: A History of Com-*

municating Scientific Ideas (London: Routledge, 2006), chapter 12; Jochen Büttner, Jürgen Renn and Matthias Schemmel, "Exploring the Limits of Classical Physics: Planck, Einstein, and the Structure of a Scientific Revolution," *Studies in the History and Philosophy of Modern Physics* 34, no. 1 (2003): 37–59.

2. While Richard Staley traces these terms back to 1911, Graeme Gooday and Daniel Mitchell have argued that in the British case "classical physics" first appeared in Eddington's 1928 Gifford Lecture. Graeme Gooday and Daniel Jon Mitchell, "Rethinking 'Classical Physics,'" in *The Oxford Handbook of the History of Physics*, ed. Jed Z. Buchwald and Robert Fox (Oxford: Oxford University Press, 2013), 721–64; Richard Staley, "On the Co-creation of Classical and Modern Physics," *Isis* 96, no. 4 (2005): 530–58; Arthur Stanley Eddington, *The Nature of the Physical World* (Cambridge: Cambridge University Press, 1928). I have elsewhere argued for a multiplicity of definitions of modern physics throughout the 1920s and 1930s. Imogen Clarke, "Negotiating Progress: Promoting 'Modern' Physics in Britain, 1900–1940" (PhD diss., University of Manchester, 2012).

3. Many studies explore the transition from classical to modern physics by taking a category of "modern" physics as their starting point and then looking back to the origins of these ideas. For example, Jed Z. Buchwald, *From Maxwell to Microphysics: Aspects of Electromagnetic Theory in the Last Quarter of the Nineteenth Century* (Chicago: University of Chicago Press, 1985); Olivier Darrigol, *Electrodynamics from Ampère to Einstein* (Oxford: Oxford University Press, 2000).

4. Staley, in what he admits is itself a reception study, argues that such accounts dominated historical studies, obscuring many of the subtleties by which relativity was developed. Richard Staley, "On the Histories of Relativity: The Propagation and Elaboration of Relativity Theory in Participant Histories in Germany, 1905–1911," *Isis* 89, no. 2 (1998): 264.

5. Andrew Warwick, "Cambridge Mathematics and Cavendish Physics: Cunningham, Campbell and Einstein's Relativity 1905–1911. Part I: The Uses of Theory," *Studies in History and Philosophy of Science* 23, no. 4 (1992): 625–56; Andrew Warwick, "Cambridge Mathematics and Cavendish Physics: Cunningham, Campbell and Einstein's Relativity 1905–1911. Part II: Comparing Traditions in Cambridge Physics," *Studies in History and Philosophy of Science*, 24, no. 1 (1993): 1–25; Andrew Warwick, *Masters of Theory: Cambridge and the Rise of Mathematical Physics* (Chicago: University of Chicago Press, 2003).

6. Jeff Hughes has argued that our conception of "modern" physics in early twentieth-century Britain has been influenced by an emphasis on the work undertaken in the Cavendish, while Benoit Lelong has described an "international diaspora of Cavendish physicists." Jeff Hughes, "Radioactivity and Nuclear Physics," in *The Cambridge History of Science*, vol. 5, *The Modern Physical and Mathematical Sciences*, ed. Mary Jo Nye (Cambridge: Cambridge University Press, 2002), 350–51. Benoit Lelong, "Translating Ion Physics from Cambridge to Oxford: John Townsend and the Electrical Laboratory, 1900–24," in *Physics in Oxford 1839–1939: Laboratories, Learning, and College Life*, ed. Robert Fox and Graeme Gooday (Oxford: Oxford University Press, 2005), 212.

7. See Robert Fox and Graeme Gooday's reassessment of Oxford University, which, while an elite institution, has historically been viewed as something of a comparative failure with regards to late nineteenth- and early twentieth-century physics. Fox and Gooday, eds., *Physics in Oxford*.

8. Graeme Gooday, "Precision Measurement and the Genesis of Physics Teaching Laboratories in Victorian Britain," *British Journal for the History of Science* 23 (1990): 25–51; Graeme Gooday, "The Questionable Matter of Electricity: The Reception of J. J. Thomson's 'Corpuscle' among Electrical Theorists and Technologists," in *Histories of the Electron: The Birth of*

Microphysics, ed. Jed Z. Buchwald and Andrew Warwick (Cambridge, MA: MIT Press, 2001), 101–34.

9. On the eclipse expedition, see Alistair Sponsel, "Constructing a 'Revolution in Science': The Campaign to Promote a Favourable Reception for the 1919 Solar Eclipse Experiments," *British Journal of the History of Science* 35 (2002): 439–67; Matthew Stanley, "'An expedition to heal the wounds of war': The 1919 Eclipse and Eddington as Quaker Adventurer," *Isis* 94, no. 1 (2003): 57–89; John Earman and Clark Glymour, "Relativity and Eclipses: the British Eclipse Expeditions of 1919 and their Predecessors," *Historical Studies in the Physical Sciences* 11 (1980): 49–85.

10. Michael Whitworth, "The Clothbound Universe: Popular Physics Books, 1919–39," *Publishing History* 40 (1996): 55–82.

11. Peter J. Bowler, *Science for All: The Popularization of Science in Early Twentieth-Century Britain* (Chicago: University of Chicago Press, 2009), 36, 137.

12. For example, studies exploring the effect of Einstein on literature: Alan J. Friedman and Carol C. Donley, *Einstein as Myth and Muse* (Cambridge: Cambridge University Press, 1985); Michael Whitworth, *Einstein's Wake: Relativity, Metaphor, and Modernist Literature* (Oxford: Oxford University Press, 2001). Price's more inclusive study moves beyond an elite focus, but again does so within the context of relativity theory: Katy Price, *Loving Faster than Light: Romance and Readers in Einstein's Universe* (London: University of Chicago Press, 2012). Clarke and Henderson have warned against an approach that focuses exclusively on atomic, quantum, and relativistic theories, arguing that ideas of the ether still held considerable cultural influence. Bruce Clarke and Linda Dalrymple Henderson, "Ether and Electromagnetism: Capturing the Invisible," in *From Energy to Information: Representation in Science and Technology, Art, and Literature*, ed. Bruce Clarke and Linda Dalrymple Henderson (Stanford, CA: Stanford University Press, 2002), 95–97.

13. Warwick, *Masters of Theory*, 483–86.

14. Einstein spoke about the ether in a 1920 conference in Leiden, subsequently published in English as Albert Einstein, *Sidelights on Relativity*, trans. G. B. Jeffery and Wilfrid Perrett (London: Methuen, 1922). Eddington discussed the ether at length in *Nature of the Physical World*. Conversely, as early as 1914, O. W. Richardson declared the ether to be "a superfluous hypothesis": Owen Willans Richardson, *The Electron Theory of Matter* (Cambridge: Cambridge University Press, 1914), 325.

15. Oliver Lodge, "The Ether Versus Relativity," *Fortnightly Review* 107 (1920): 58.

16. "DD," "Introduction," *Nature* 106, no. 2677 (February 17, 1921): 781.

17. Oliver Lodge, "The Geometrisation of Physics, and Its Supposed Basis on the Michelson-Morley Experiment," *Nature* 106, no. 2677 (February 17, 1921): 795–800.

18. The editor in 1921 was the astronomer Richard A. Gregory, but the introduction of this collection was initialed "DD."

19. For an overview of the British Empire Exhibition see Donald R. Knight and Alan D. Sabey, *The Lion Roars at Wembley: British Empire Exhibition, 60th Anniversary 1924–1925* (New Barnet: D.R. Knight, 1984).

20. The exhibits and committee are listed in the exhibition's official handbook, *British Empire Exhibition 1924: Handbook to the Exhibition of Pure Science: Galleries 3 and 4 British Government Pavilion* (London: Royal Society, 1924). There is an overview of the Pure Science exhibit in Alan Q. Morton, "The Electron Made Public: The Exhibition of Pure Science in the British Empire Exhibition, 1924–5," in *Exposing Electronics*, ed. Bernard Finn, Robert Bud, and Helmuth Trischler (Amsterdam: Harwood Academic, 2000), 25–44. Morton does not distinguish between the 1924 and 1925 runs of the exhibition, which can cause confusion.

21. T. Martin (secretary of the British Empire Exhibition) to O. Lodge, October 21, 1925, British Empire Exhibition 1924, box 2, Royal Society Archives, London.

22. Lodge's appointment (and the article he was to write) is referred to in a letter from Martin to F. E. Smith, February 25, 1925, F.E. Smith, British Empire Exhibition 1924 Correspondence, Royal Society Archives. The decision to appoint him came from the Publications Subcommittee; see Publications Subcommittee, British Empire Exhibition 1924 Correspondence.

23. *Phases of Modern Science: Published in Connexion with the Science Exhibit Arranged by a Committee of the Royal Society in the Pavilion of His Majesty's Government at the British Empire Exhibition, 1925* (London: Royal Society, 1925).

24. For Mitchell, see D. P. Crook, "Peter Chalmers Mitchell and Antiwar Evolutionism in Britain during the Great War," *Journal of the History of Biology* 22, no. 2 (1989): 325–56; P. D. Duncan, "Newspaper Science: The Presentation of Science in Four British Newspapers During the Interwar Years, 1919–1939" (MPhil thesis, University of Sussex, 1980).

25. Peter Chalmers Mitchell, *My Fill of Days* (London: Faber and Faber, 1937), 274–75.

26. Peter Chalmers Mitchell to Oliver Lodge, August 9, 1927, Lodge Papers, MS ADD 89 (Lodge), University College London Library Services, Special Collections.

27. [Peter Chalmers Mitchell], "The Progress of Science. Transmutation of Metals. 'Synthetic Gold,'" *Times* (London), December 20, 1921, 8.

28. [Peter Chalmers Mitchell], "The Progress of Science. Low-Temperature Research. Cryogenic Laboratories," *Times* (London), August 7, 1923, 8.

29. [Peter Chalmers Mitchell], "The Progress of Science. Atoms and their Nuclei. Hydrogen as Primitive Matter," *Times* (London), March 30, 1925, 9.

30. [Peter Chalmers Mitchell], "The Progress of Science. Atomic Systems. Disintegration of Matter," *Times* (London), April 27, 1925, 7.

31. [Peter Chalmers Mitchell], "The Progress of Science. Ultimate Facts of the Universe. The Quantum Theory," *Times* (London), October 27, 1924, 19.

32. Edward Neville da Costa Andrade, *The Structure of the Atom* (London: Bell, 1923).

33. Peter Chalmers Mitchell to Oliver Lodge, November 9, 1923, Lodge Papers, MS ADD 89 (Lodge), University College London Library Services, Special Collections.

34. [Peter Chalmers Mitchell], "Progress of Physical Science. Sir Oliver Lodge on Television," *Times* (London), March 15, 1927, 14.

35. Oliver Lodge, *Atoms and Rays: An Introduction to Modern Views on Atomic Structure and Radiation* (London: Ernest Benn, 1924).

36. Unsigned review of *Atoms and Rays*, by Oliver Lodge, *Yorkshire Post*, July 23, 1924, 4.

37. Andrade, *Structure of the Atom.*

38. Alan Cottrell, "Edward Neville da Costa Andrade. 1887–1971," *Biographical Memoirs of Fellows of the Royal Society* 18 (1972): 1–20.

39. Edward Neville da Costa Andrade, "Books of the Day: The New Physics," *Observer*, August 10, 1924, 5.

40. Lodge, *Atoms and Rays*, 202.

41. Andrade, "Books of the Day," 5.

42. Oliver Lodge to Edward Andrade, August 13, 1924, and August 23, 1924, MS ADD 89 (Lodge), University College London Library Services, Special Collections.

43. Andrade, "Books of the Day," 5.

44. "News and Views," *Nature* 119, no. 3005 (June 4, 1927): 827.

45. "News and Views," 827; V. A. Pullin, "Benn's Sixpenny Library: First Scientific Titles," *Discovery* 9 (1928): 163–65.

46. Pullin, "Benn's Sixpenny Library," 165.

47. Jaume Navarro, "Ether and Wireless: An Old Medium into New Media," *Historical Studies in the Natural Sciences* 46, no. 4 (2016): 460–89.

48. Oliver Lodge, *Ether and Reality: A Series of Discourses on the Many Functions of the Ether of Space* (London: Hodder and Stoughton, 1925).

49. "Ether and Reality," *Wireless World*, July 1, 1925, 16.

50. "Modern Physics and the Engineer," *Engineer* 154 (November 11, 1932): 485–86.

51. Sir James Jeans, "The New World-Picture of Modern Physics," *Engineer* 158 (September 7, 1934): 238–39.

52. "The Old and the New Physics," *Engineer* 158 (September 7, 1934): 237.

53. "Sir Oliver Lodge," *Times* (London), August 23, 1940, 4.

54. "Sir Oliver Lodge: A Great Scientist," *Times* (London), August 23, 1940, 7.

55. "Sir Oliver Joseph Lodge D.Sc. Sc.D. LL.D. F.R.S. (1940) XXXII. Obituary," *Philosophical Magazine*, 7th series, 30 (1940): 341–43; "Sir Oliver Lodge. A Great Scientific Teacher," *Electrical Review* 127 (1940): 169; R. A. Gregory and A. Ferguson, "Oliver Joseph Lodge (1851–1940)," *Obituary Notices of Fellows of the Royal Society* 3 (1941): 551–74.

56. W. Bragg, "Science and National Welfare," *Nature* 146, no. 3710 (December 7, 1940): 731.

57. Jaume Navarro, *A History of the Electron* (Cambridge: Cambridge University Press, 2012). Thomson refers to the ether in J. J. Thomson, "Electronic Waves," *Philosophical Magazine*, 7th series, 27 (1939): 1–32.

58. "Sir J. J. Thomson, O.M. The Discover of the Electron," *Times* (London), August 31, 1940, 7.

59. "Sir J. J. Thomson: The King's Message of Sympathy," *Times* (London), September 3, 1940, 7.

60. See G. E. Smith, "J. J. Thomson and the Electron, 1897–1899," in *Histories of the Electron*, ed. Jed Z. Buchwald and Andrew Warwick (Cambridge, MA: MIT Press, 2001), 21–76. Isobel Falconer, "Corpuscles, Electrons and Cathode Rays: J. J. Thomson and the 'Discovery of the Electron,'" *British Journal for the History of Science* 20 (1987): 241–76.

61. Oliver Lodge, *Electrons; or, the Nature and Properties of Negative Electricity* (London: George Bell and Sons, 1906).

62. Isobel Falconer, "Corpuscles to Electrons," in Buchwald and Warwick, *Histories of the Electron*, 77–100. See also Gooday, "Questionable Matter of Electricity."

63. Geoffrey Cantor, "The Scientist as Hero: Public Images of Michael Faraday," in *Telling Lives in Science: Essays on Scientific Biography*, ed. Michael Shortland and Richard Yeo (Cambridge: Cambridge University Press, 2008), 190.

64. Navarro, *History of the Electron*, 169.

65. H. Kant, "Lodge, Oliver Joseph," in *Biographical Encyclopedia of Astronomers*, ed. T. Hockey et al., 2nd ed. (New York: Springer, 2014), 1341.

66. Gooday, "Questionable Matter of Electricity."

EIGHT. GLORIFYING MECHANISM

This chapter grew out of a paper presented at a workshop on Lodge and wireless, held at the Royal Society in 2014. It develops material published in Richard Noakes, *Physics and Psychics: The Occult and the Sciences in Modern Britain* (Cambridge: Cambridge University Press, 2019). For their comments on earlier versions of this chapter and their extraordinary patience I would like to express my profound gratitude to the editors. For permission to quote from unpublished material in their collections I thank the Syndics of Cambridge University Library.

1. Oliver Lodge, *Past Years: An Autobiography* (London: Hodder and Stoughton, 1931), 315–16.

2. Theodore Besterman, *A Bibliography of Sir Oliver Lodge F.R.S.* (Oxford: Oxford University Press, 1935).

3. Peter Bowler, *Reconciling Science and Religion: The Debate in Early Twentieth-Century Britain* (Chicago: Chicago University Press, 2001), 95–96; Peter J. Bowler, *Science for All: The Popularization of Science in Early Twentieth-Century Britain* (Chicago: University of Chicago Press, 2009), 218.

4. On the historical development of the ether see G. N. Cantor and M. J. S. Hodge, eds., *Conceptions of Ether: Studies in the History of Ether Theories, 1740–1900* (Cambridge: Cambridge University Press, 1981).

5. Oliver Lodge, *My Philosophy: Representing My Views on the Many Functions of the Ether of Space* (London: Ernest Benn, 1933), 5.

6. This contrast was made by Lord Rayleigh in "Obituary: Sir Oliver Lodge F.R.S. Sir J.J. Thomson, O.M., F.R.S.," *Proceedings of the Society for Psychical Research* 46 (1940–41): 216.

7. Matthew Stanley, *Huxley's Church and Maxwell's Demon: From Theistic Science to Naturalistic Science* (Chicago: University of Chicago Press, 2015).

8. Oliver Lodge, "Scope and Tendencies of Physics," *The 19th Century: A Review of Progress*, by in A. G. Sedgwick et al. (London: G. P. Putnam's, 1901), 354.

9. John Tyndall, "The Belfast Address," in John Tyndall, *Fragments of Science: A Series of Detached Essays, Addresses, and Reviews* (London: Longmans, Green, 1889), 2:180–81. For an excellent study of this address see Ruth Barton, "John Tyndall, Pantheist: A Rereading of the Belfast Address," *Osiris* 3 (1987): 111–34; and Ursula DeYoung, *A Vision of Modern Science: John Tyndall and the Role of the Scientist in Victorian Culture* (Basingstoke: Palgrave Macmillan, 2011), 89–130.

10. Bernard Lightman, "Victorian Sciences and Religions: Discordant Harmonies," *Osiris* 16 (2001): 346. For further discussion of scientific naturalism see Frank M. Turner, *Contesting Cultural Authority: Essays in Victorian Intellectual Life* (Cambridge: Cambridge University Press, 1993), 131–228; Gowan Dawson and Bernard Lightman, eds., *Victorian Scientific Naturalism: Community, Identity, Continuity* (Chicago: University of Chicago Press, 2014).

11. Lorraine J. Daston, "British Responses to Psycho-physiology, 1860–1900," *Isis* 69, no. 2 (1978): 198; Roger J. Smith, *Free Will and the Human Sciences in Britain, 1870–1910* (London: Pickering and Chatto, 2013), 17–33.

12. Oliver Lodge, *Advancing Science: Being Personal Reminiscences of the British Association in the Nineteenth Century* (London: Ernest Benn, 1931), 35–36. See also Amigoni, this volume.

13. Lodge, *Past Years*, 53, 65, 70, 76–78, 83, 168, 345. Lodge once described Tyndall as one of the "heroes" of his youth; see Oliver Lodge, "The Books of My Youth," *Living Age*, May 8, 1926, 332.

14. On the early SPR see Alan Gauld, *The Founders of Psychical Research* (London: Routledge and Kegan Paul, 1968) and Janet Oppenheim, *The Other World: Spiritualism and Psychical Research in England, 1850–1914* (Cambridge: Cambridge University Press, 1985), 111–58.

15. For the Piper séances see Gauld, *Founders of Psychical Research*, 251–68.

16. Oliver Lodge, "A Record of Observations of Certain Phenomena of Trance," *Proceedings of the Society for Psychical Research* 6 (1889–90): 443.

17. Oliver Lodge, *Conviction of Survival: Two Discourses in Memory of F.W.H. Myers* (London: Methuen, 1930), 9–10; Lodge, *Past Years*, 277. One of Lodge's scientific heroes, William Thomson (Lord Kelvin), remarked in 1893 that he had nothing to do with spiritualism, mes-

merism, psychical research, and other subjects classed as "borderland" because "nearly everything in hypnotism and clairvoyance is imposture and the rest bad observation." Kelvin quoted in "The Response to the Appeal," *Borderland* 1 (1893): 17.

18. By the time he wrote *Past Years*, however, Lodge was open about his conviction in the genuineness of some of Palladino's performances: Lodge, *Past Years*, 290–313. For analysis of Lodge's investigations of Palladino see Richard Noakes, "Haunted Thoughts of the Careful Experimentalist: Psychical Research and the Troubles of Experimental Physics," *Studies in the History and Philosophy of the Biological and Biomedical Sciences* 48 (2014): 46–56.

19. Lodge, *Past Years*, 345.

20. Lodge, *Past Years*, 220; Oliver Lodge, "A Scheme of Vital Faculty," *Nature* 68, no. 1755 (June 18, 1903): 145; Oliver Lodge, "The Life-Work of My Friend F.W.H. Myers," *Nature* 144, no. 3660 (December 23, 1939): 1027–28.

21. The bulk of the Lodge-Myers correspondence comprises nearly three hundred letters from Myers covering the period 1889–1900: SPR.MS.35/1298–1572, Oliver Lodge Papers, Society for Psychical Research Archive, Cambridge University Library. Henceforth this collection is abbreviated to OJL-SPR.

22. Gauld, *Founders of Psychical Research*, 275–312; Trevor Hamilton, *Immortal Longings: FWH Myers and the Victorian Search for Life after Death* (Exeter: Imprint Academic, 2009).

23. Lodge's understanding of these implications of Myers's writing is evident in Oliver Lodge, "The Survival of Personality," *Quarterly Review* 198 (1903): 220.

24. Oliver Lodge, introduction, in *Essays on Man's Place in Nature*, by T. H. Huxley (London: J. M. Dent, 1906), ix; Oliver Lodge, introduction, in *Lectures and Lay Sermons*, by T. H. Huxley (London: J. M. Dent, 1910), xi.

25. Lodge, introduction, in Huxley, *Essays*, xi.

26. O.J.L. [Oliver Lodge], "John Tyndall," in *Encyclopaedia Britannica*, 9th/10th ed. (Edinburgh: A. and C. Black, 1902–3), 33:521. For discussion of the hostile reaction of Tyndall's widow and close colleagues to this see W. P. Jolly, *Sir Oliver Lodge* (London: Constable, 1974), 180–82.

27. P. M. Harman, *The Natural Philosophy of James Clerk Maxwell* (Cambridge: Cambridge University Press, 1998), 197–208; Crosbie Smith, *The Science of Energy: A Cultural History of Energy Physics in Victorian Britain* (London: Athlone, 1998), 239–67; Crosbie Smith and M. Norton Wise, *Energy and Empire: A Biographical Study of Lord Kelvin* (Cambridge: Cambridge University Press, 1989), 612–45; Stanley, *Huxley's Church and Maxwell's Demon*, 194–241; David B. Wilson, *Kelvin and Stokes: A Comparative Study in Victorian Physics* (Bristol: Adam Hilger, 1987), 90–94.

28. For excellent discussion of this argument see Harman, *Natural Philosophy*, 197–208; Stanley, *Huxley's Church and Maxwell's Demon*, 194–241; Wilson, *Kelvin and Stokes*, 90–94.

29. Stanley, *Huxley's Church and Maxwell's Demon*, 237.

30. On the Synthetic Society see William C. Lubenow, "Intimacy, Imagination and the Inner Dialectics of Knowledge Communities: The Synthetic Society, 1896–1908," in *The Organisation of Knowledge in Victorian Britain*, ed. Martin J. Daunton (Oxford: Oxford University Press, 2005), 357–70.

31. Oliver Lodge, untitled essay, in *Papers Read before the Synthetic Society, 1896–1908* ([London]: Spottiswoode, 1909), 386.

32. Lodge, untitled essay, 386.

33. On Balfour see L. S. Jacyna, "Science and Social Order in the Thought of Arthur Balfour," *Isis* 71 (1980): 11–34; J. D. Root, "The Philosophical and Religious Thought of Arthur James Balfour (1848–1930)," *Journal of British Studies* 19, no. 2 (1980): 120–41. On Ward see

Daston, "British Responses," 204–8; Smith, *Free Will*, 49–55, 65–75; Frank Miller Turner, *Between Science and Religion: The Reaction to Scientific Naturalism in Late Victorian England* (New Haven, CT: Yale University Press, 1974), 228–35.

34. J. L. Heilbron, "Fin-de-siècle Physics," in *Science, Technology and Society in the Time of Alfred Nobel*, ed. Carl Gustaf Bernhard, Elisabeth Crawford, and Per Sörbom (Oxford: Oxford University Press, 1982), 51–73; Richard Staley, *Einstein's Generation: The Origins of the Relativity Revolution* (Chicago: University of Chicago Press, 2008), 347–96.

35. Oliver Lodge, "Supplement to the Discussion on Mr. Balfour's Paper," in *Papers Read before the Synthetic Society, 1896–1908*, 336.

36. Lodge, "Supplement," 336–37.

37. Lodge, "Supplement," 338.

38. Oliver Lodge, *Life and Matter: A Criticism of Professor Haeckel's "Riddle of the Universe"* (London: Williams and Norgate, 1905), 24.

39. Lodge, *Life and Matter*, 172. On the historical context of this argument see Smith, *Free Will*, 81–101.

40. Lodge, "Supplement," 334.

41. Lodge, diary entry for October 3, 1872, OJL2/3/1, Oliver Lodge Papers, Cadbury Research Library, University of Birmingham. The collection of essays was the first (1871) edition of Tyndall's *Fragments of Science*.

42. John Tyndall, "Prayer as a Form of Natural Law" (1861), in Tyndall, *Fragments of Science*, 2:5.

43. Lodge, "Supplement," 334. See also Oliver Lodge, "The Outstanding Controversy between Science and Faith," *Hibbert Journal* 1 (1902–3): 51.

44. Lodge, "Supplement," 336.

45. Lodge, "Outstanding Controversy," 50.

46. Lodge, untitled essay, in *Synthetic Society*, 391.

47. See, for example, E. W. Hobson, "Sir O. Lodge and the Conservation of Energy," *Nature* 67, no. 1748 (April 30, 1903): 611–12; E. W. Hobson, "Psychophysical Interaction," *Nature* 68, no. 1752 (May 28, 1903): 77; George M. Minchin, "The Glorification of Energy," *Nature* 68, no. 1750 (May 14, 1903): 31–32; Evan McLennan, "Force and Determinism," *Nature* 44, no. 1131 (July 2, 1891): 198; William McDougall, "Psychophysical Interaction," *Nature* 68, no. 1758 (1903): 32–33; Conwy Lloyd Morgan, "Force and Determinism," *Nature* 43, no. 1120 (April 16, 1891): 558; Conwy Lloyd Morgan, "Force and Determinism," *Nature* 44, no. 1136 (August 6, 1891): 319.

48. McDougall, "Psychophysical Interaction."

49. Oliver Lodge, "Psychophysical Interaction," *Nature* 68, no. 1750 (May 14, 1903): 33.

50. Oliver Lodge, "The Interaction of Life and Matter," *Hibbert Journal* 29 (1931): 399.

51. For a fuller discussion of Lodge's proposed psychic functions for the ether see Richard Noakes, "Making Space for the Soul: Oliver Lodge, Maxwellian Psychics and the Etherial Body," in *Ether and Modernity: The Recalcitrance of an Epistemic Object in the Early Twentieth Century*, ed. Jaume Navarro (Oxford: Oxford University Press, 2018), 88–106.

52. [Balfour Stewart and Peter Guthrie Tait], *The Unseen Universe; or, Physical Speculations on a Future State* (London: Macmillan, 1875). For insightful analysis of *Unseen Universe* see Graeme Gooday, "Sunspots, Weather, and the Unseen Universe: Balfour Stewart's Anti-Materialist Representations of 'Energy' in British Periodicals," in *Science Serialized: Representations of the Sciences in Nineteenth-Century Periodicals*, ed. Geoffrey Cantor and Sally Shuttleworth (Cambridge, MA: MIT Press, 2004), 111–47; P. M. Heimann, "The *Unseen Universe*: Physics and the Philosophy of Nature in Victorian Britain," *British Journal for the History of*

Science 6 (1972): 73–79; Elisabeth Lewis, "P. G. Tait, Balfour Stewart, and *The Unseen Universe*," in *Mathematicians and Their Gods: Interactions between Mathematics and Religious Beliefs*, ed. Snezana Lawrence and Mark McCartney (Oxford: Oxford University Press, 2015), 213–48.

53. [Stewart and Tait], *Unseen Universe*, xv.

54. James Clerk Maxwell, "Ether," in *The Scientific Papers of James Clerk Maxwell*, ed. W. D. Niven (Cambridge: Cambridge University Press, 1890), 2:775; Maxwell, "Paradoxical Philosophy," in *Scientific Papers of James Clerk Maxwell*, 2:756–62. For discussion see Smith and Wise, *Energy and Empire*, 630–31; Smith, *Science of Energy*, 255. On the tradition of drawing out religious significance from the ether see Geoffrey Cantor, "The Theological Significance of Ethers," in Cantor and Hodge, *Conceptions of Ether*, 135–55.

55. Oliver Lodge, "The Ether of Space," *Contemporary Review* 93 (1908): 540. Lodge certainly knew about the book from James Clerk Maxwell's reference to it in his entry on ether in the ninth edition of the *Encyclopaedia Britannica*, published between 1875 and 1889. Lodge probably read the version of Maxwell's entry as reprinted in *The Scientific Papers of James Clerk Maxwell* (Maxwell, "Ether"). In 1890 fellow physicist and SPR member William Fletcher Barrett told Lodge that the idea, then being discussed by Lodge and Myers, of a higher cosmic consciousness that could perceive the past, present, and future was similar to Stewart and Tait's notion of the "memory of the universe." Barrett to Lodge, October 18, 1890, SPR. MS.35/60, Lodge Papers, Society for Psychical Research Archive, Cambridge University Library.

56. Oliver Lodge, "The Ether and its Functions," *Nature* 27, no. 691 (January 25, 1883): 304–6 and "The Ether and its Functions," *Nature* 27, no. 692 (February 1, 1883): 328–30.

57. Oliver Lodge, "Address," in *Report of the Sixty-First Meeting of the British Association for the Advancement of Science, Held at Cardiff in August 1891* (London: John Murray, 1892), 552.

58. Oliver Lodge, "The Interstellar Ether," *Fortnightly Review* 53 (1893): 862.

59. On the late nineteenth-century interpretation of Maxwell's work see Jed Z. Buchwald, *From Maxwell to Microphysics: Aspects of Electromagnetic Theory in the Last Quarter of the Nineteenth Century* (Chicago: University of Chicago Press, 1985); Olivier Darrigol, *Electrodynamics from Ampère to Einstein* (Oxford: Oxford University Press, 2000), 177–264; Bruce J. Hunt, *The Maxwellians* (Ithaca, NY: Cornell University Press, 1991).

60. For a succinct overview of this tradition see Peter Dear, *The Intelligibility of Nature: How Science Makes Sense of the World* (Chicago: University of Chicago Press, 2006), 115–40.

61. Lodge, "Interstellar Ether," 862. On Lodge's ether drag experiments see Bruce J. Hunt, "Experimenting on the Ether: Oliver Lodge and the Great Whirling Machine," *Historical Studies in the Physical Sciences* 16 (1986): 111–34.

62. Lodge, "Interstellar Ether," 862.

63. F. W. H. Myers to Oliver Lodge, December 20, 1893, SPR.MS.35/1298, Lodge Papers, Society for Psychical Research Archive, Cambridge University Library.

64. Some of these discourses are touched on in Cantor, "Theological Significance."

65. F. W. H. Myers, *Human Personality and Its Survival of Bodily Death* (London: Longmans, Green, 1903), 1:215–16.

66. Lodge, "Survival of Personality," 225–26.

67. Oliver Lodge, *Ether and Reality: A Series of Discourses on the Many Functions of the Ether of Space* (London: Hodder and Stoughton, 1925), 179.

68. See, for example, Lodge, "Ether of Space," 540 and 543.

69. See for example, H.L. [Horace Lamb], "The Ether of Space," *Nature* 82, no. 2097 (January 6, 1910): 271; Owen Willans Richardson, "The Structure of the Ether," *Nature* 76, no. 1960

(May 23, 1907): 78. Lodge's awareness of these problems is evident in Lodge, "Ether of Space," 543.

70. Oliver Lodge, "Continuity," in *Report of the Eighty-Third Meeting of the British Association for the Advancement of Science, Birmingham: 1913* (London: John Murray, 1914), 6.

71. Oliver Lodge, "Einstein's Real Achievement," *Fortnightly Review* 110 (1921): 369.

72. Einstein quoted in Lodge, *Ether and Reality*, 123. The quotation is from Albert Einstein, *Sidelights on Relativity* (New York: E. P. Dutton, 1922), 23–24.

73. Lodge, "Einstein's Real Achievement," 369.

74. Lodge, "Scope and Tendencies," 354.

75. In 1930 Lodge described the electrical theory of matter as the "greatest scientific event" of the twentieth century: Oliver Lodge, *Beyond Physics: Or, the Idealisation of Mechanism* (London: George Allen and Unwin, 1930), 36. On electrical and electrical theories of matter see Darrigol, *Electrodynamics*, 314–50; and Hunt, *Maxwellians*, 209–39.

76. Oliver Lodge, *Electrons; or, the Nature and Properties of Negative Electricity* (London: George Bell, 1906), vii–viii.

77. Lodge, "Scope and Tendencies," 352.

78. Lodge, "Scope and Tendencies," 355.

79. Oliver Lodge, *Phantom Walls* (London: Hodder and Stoughton, 1929), 139.

80. Lodge, *Beyond Physics*, 113; Lodge, *Phantom Walls*, 57.

81. Lodge first publicly declared his belief in survival in 1902: Oliver Lodge, "Address by the President," *Proceedings of the Society for Psychical Research* 17 (1901–3): 49. The psychical investigations that played a significant part in moving Lodge toward conviction of survival were sittings with the medium Mrs. Thompson in 1901. These revealed communications from Lodge's close friend Myers, who had died a month earlier.

82. Lodge, *Phantom Walls*, 136, 141, 201.

83. Oliver Lodge, *Modern Problems* (London: Methuen, 1912), 3. This originally appeared as Oliver Lodge, "Free Will and Determinism," *Cosmopolitan Journal* 1 (1903): 23–24. In many ways, Lodge was drawing on Myers's idea that the "metetherial" world was as much law-bound as the world of matter and ether. The "order of nature" governed the "will of spirits" and disembodied personalities that resided in this world as it did the "will of men" in the material world; see F. W. H. Myers, "The Drift of Psychical Research," *National Review* 24 (1894): 206; F. W. H. Myers, "The Subliminal Consciousness," *Proceedings of the Society for Psychical Research* 8 (1892): 534.

84. Lodge, *Beyond Physics*, 8, 20.

85. Lodge, *Phantom Walls*, 94.

86. On Lodge's tendency to "brood" see Hunt, *Maxwellians*.

87. Lodge, *My Philosophy*, 234. Earlier discussions of the concept can be found in Oliver Lodge, "Address by the President," *Proceedings of the Society for Psychical Research* 17 (1901–2): 47 and Oliver Lodge, "Ether, Matter and the Soul," *Hibbert Journal* 17 (1918–19): 252–60. For Lodge's discussion with Raymond Lodge about etheric bodies see "Sir Oliver Lodge on the Possibilities of the Human Spirit," *Light*, April 16, 1927, 182–85. For analysis of Lodge's concept of the etherial body see David B. Wilson, "The Thought of Late Victorian Physicists: Oliver Lodge's Ethereal Body," *Victorian Studies* 15, no. 1 (1971): 29–48.

88. Lodge, *My Philosophy*, 221, 238.

89. G. B. Brown, review of *Beyond Physics*, by Oliver Lodge, *Philosophy* 5 (1930): 624–26; W. H. Mallock, "Sir Oliver Lodge on Life and Matter," *Fortnightly Review* 80 (1906): 33–47; Eleanor Mildred Sidgwick to Oliver Lodge, October 4, 1918, SPR.MS.35/2255, Lodge Papers, Society for Psychical Research archive.

90. Lodge, *My Philosophy*, 222.

91. Oliver Lodge, "The Nineteenth Kelvin Lecture. 'The Revolution in Physics,'" *Journal of the Institution of Electrical Engineers* 66 (1928): 1012.

92. Lodge, *My Philosophy*, 82.

93. Lodge, *Beyond Physics*, 145–46.

94. On Eddington's idealism see Bowler, *Reconciling Science and Religion*, 101–10.

95. Oliver Lodge, "Eddington's Philosophy," *Nineteenth Century* 105 (1929): 360–69; Lodge, *My Philosophy*, 173; Lodge, *Beyond Physics*, 77.

96. Lodge, *Beyond Physics*, 73.

97. J. Arthur Hill, *Letters from Sir Oliver Lodge Psychical, Religious, Scientific and Personal* (London: Cassell and Company, 1932), 49.

98. See, for example, "Necromancy," *Church Times* 53 (February 18, 1910): 225; unsigned review of *Raymond, or Life and Death*, by Oliver Lodge, *Saturday Review*, February 3, 1917, 110–11; Edward Clodd, "The Revival of the Dangerous Cult of Spiritualism," *Graphic* 101 (February 14, 1920): 222; Edwin H. Hall, "Sir Oliver Lodge's British Association Address," *Harvard Theological Review* 8 (1915): 238–51; Paul Hookham, *"Raymond": A Rejoinder Questioning the Validity of Certain Evidence and of Sir Oliver Lodge's Conclusions Regarding It* (Oxford: Blackwell, 1917); Charles Mercier, *Spiritualism and Sir Oliver Lodge* (London: Watts, 1917); Ivor Tuckett, "Psychical Researchers and 'The Will to Believe,'" *Bedrock* 1 (1912–13): 180–204.

99. See, for example, "Science," *Athenaeum* no. 4321 (November 28, 1908): 686–87; "Professor Lodge's Theology," *Church Times* 50 (December 4, 1908): 767; Joseph McCabe, *The Religion of Sir Oliver Lodge* (London: Watts, 1914); E. S. Talbot, "Sir Oliver Lodge on 'The Reinterpretation of Christian Doctrine,'" *Hibbert Journal* 2 (1903–4): 649–61. For discussion see Root, "Science, Religion and Psychical Research."

100. "E.N.Da.C.A." [Edward N. Da Costa Andrade], "A Veteran's View of Modern Physics," *Nature* 114, no. 2869 (October 25, 1924): 600. Lodge defended his use of such terms as *inertia* and *density* by insisting that they were merely analogies to unknown properties of the ether: Lodge, *My Philosophy*, 190.

101. "The Ether, Life and Mind," *Times* (London), May 15, 1925, 10; "Space, Matter, Mind and God," *Church Times* 103 (June 13, 1930): 759–60; G. Calver, "Sir Oliver Lodge and the Ether," *English Mechanic and World of Science* 109 (January 31, 1919): 21; [J. W. N. Sullivan], "Ether and Reality," *Times Literary Supplement*, May 14, 1925, 325; [J. W. N. Sullivan], "Beyond Physics," *Times Literary Supplement*, June 19, 1930, 504; [Ivor Thomas], "Sir O. Lodge's Philosophy," *Times Literary Supplement*, June 22, 1933, 421.

102. This is evident in "Behind the Veil," *Church Times* 102 (November 22, 1929): 631; Shaw Desmond, "Phantom Walls," *Bookman* 77 (1929): 210–11.

103. Horace J. Bridges, "Sir Oliver Lodge and the Public Mind," *Forum* 51 (May 1914): 695–705; McCabe, *Religion of Sir Oliver Lodge*. For discussion of this point see Bowler, *Reconciling Science and Religion*, 255.

104. "Ether and Human Survival," *Light*, April 25, 1925, 198; "My Philosophy," *Psychic Science: Quarterly Transactions of the British College of Psychic Science* 12 (1933–34): 187–201; H. A. Dallas, *Leaves from the Psychic Note Book* (London: Rider, 1927), 84–85; J. A. Findlay, *On the Edge of the Etheric Being an Investigation of Psychic Phenomena* (London: Rider, 1931), 33–47; H. F. Prevorst Battersby, "A Scientist's Philosophy," review of *My Philosophy*, by Oliver Lodge, *Light*, July 7, 1933, 425–26. In 1916 Lodge implied that his ideas about the etheric body were partly inspired by Plotinus, the ancient Neoplatonic philosopher whose writings informed so much of nineteenth- and twentieth-century spiritualist discourses: Oliver Lodge, *Raymond: Or Life and Death* (London: Methuen, 1916), 336.

105. E. E. Free, "Radio and Relativity," *Popular Radio*, April 1923, 243–53; P. J. Risdon, *Wireless*, 2nd ed. (London: Ward, Lock, 1924[?]), 34; Ralph Stranger, *The Outline of Wireless for the Man on the Street* (London: George Newnes, 1932), 201–5, 220.

106. See, for example, "Ether and Reality," *Wireless World*, July 1, 1925, 16; "Philosophy of Sir Oliver Lodge," *Derby Evening Telegraph*, June 22, 1933, 4; J. H. T. Roberts, "Sir Oliver Lodge's New Book," *Popular Wireless*, April 1927, 440–41.

107. "Sir Oliver and Ether," *Spectator*, June 29, 1933, 951.

108. On *Raymond* see Georgina Byrne, *Modern Spiritualism and the Church of England, 1850–1939* (Woodbridge: Boydell and Brewer, 2010), 75–79; George M. Johnson, *Mourning and Mysticism in First World War Literature and Beyond* (Basingstoke: Palgrave Macmillan, 2015), 60–85. See also chapters by Byrne, Ferguson, and Mussell, this volume.

109. Lodge, *Raymond*, 298.

110. Basil De Selincourt, "Sir Oliver Lodge's Philosophy," *Observer*, June 25, 1933, 4.

NINE. THE CASE OF FLETCHER

1. Cited in Oliver Lodge, *Raymond: Or Life and Death* (London: Methuen, 1916), 39.

2. Fiona Reid, *Broken Men: Shell Shock, Treatment and Recovery in Britain, 1914–30* (London: Continuum, 2010), 26. Lieutenant Raymond Lodge died on September 14, 1915.

3. Lieutenant Eric Graham Fletcher died on July 3, 1916.

4. The diagnostic history of shell shock and its development into the category of posttraumatic stress disorder is treated in Edgar Jones and Simon Wessely, *Shell Shock to PTSD: Military Psychiatry from 1900 to the Gulf War* (Hove: Psychology, 2005). For more on the bestseller status of *Raymond* during the war, see Jay Winter, *Sites of Memory, Sites of Mourning: The Great War in European Cultural History* (Cambridge: Cambridge University Press, 1995), 62.

5. These otherworldly facilities and opportunities are described in *Raymond*, 196–98, 209.

6. Winter, *Sites of Memory*, 54.

7. More information on the origins and development of Lodge's involvement with the Society for Psychical Research can be found in Oliver Lodge, *Past Years: An Autobiography* (London: Hodder and Stoughton, 1931) and W. P. Jolly, *Sir Oliver Lodge* (London: Constable, 1974).

8. For examples of these different critiques of Lodge's Edwardian spiritualist writing, see Henry Sulley's curiously anachronistic *What is the Substance of Faith? A Reply to Sir Oliver Lodge* (London: Simpkin, Marshall, Hamilton, Kent, 1909), which attacks Lodge's beliefs from the standpoint of biblical literalism, and Joseph McCabe's *The Religion of Sir Oliver Lodge* (London: Watts, 1914), which presents Lodge as an isolated holdover from a more gullible age who lacks the power to assess the so-called evidence of spirit existence.

9. Oliver Lodge, *Man and the Universe: A Study of the Influence of the Advance in Scientific Knowledge upon Our Understanding of Christianity* (London: Methuen, 1908), 179.

10. Lodge, *Raymond*, 83.

11. In the famous "Faunus" incident, Boston medium Leonora Piper claimed to have received a spirit message from F. W. H. Myers on August 8, 1915, that predicted Raymond's death and implied that Lodge would receive comfort in the form of postlife communication from his son. Her spirit channel "R. Hodgson" claimed that the dead Myers would take the role of "Faunus" in relation to Lodge; Lodge later interpreted this as an allusion to lines xvii, 27–30 of Horatio's Ode II, which describe the deity as saving the poet from death beneath a falling tree. "I perceived," he writes, "that the meaning was that some blow was going to fall, or was likely to fall, though I didn't know of what kind, and that Myers would intervene, appar-

ently to protect me from it." See Lodge *Raymond,* 93. For the group photograph proof, see chapter 4 of the "supernormal" section.

12. Lodge, *Raymond,* 194–95, 197–98.

13. Charles Mercier, *Spiritualism and Sir Oliver Lodge* (London: Watts, 1917), 85.

14. Lodge, *Raymond,* 172.

15. The connections between *Raymond, Gone West,* and other spirit soldier memoirs such as W. Tudor-Pole's *Private Dowding: A Plain Record of After Death Experiences of a Soldier Killed in Battle* (London: John M. Watkins, 1917), are discussed in Hereward Carrington's *Psychical Phenomena and the War* (London: T. Werner, 1918) and in James Ingall Wedgwood's *Spiritualism and the Great War* (London: Theosophical Publishing House, 1919).

16. Unsigned review of *Raymond, or Life and Death,* by Oliver Lodge, *Saturday Review,* February 3, 1917, 111; Mercier, *Spiritualism and Sir Oliver Lodge,* vii.

17. The exact number of shell shock sufferers among the Allied forces is difficult to determine with any finality, but as one historian notes, "any estimate . . . must err on the low side." The only statistics indexed to the condition, she continues, were linked to hospital admissions and pensions for neurasthenia; from these we know that "by the end of the war, 80,000 cases of war neuroses had passed through the army hospitals." Joanna Bourke, *Dismembering the Male: Men's Bodies, Britain, and the Great War* (London: Reaktion, 1996), 109.

18. Although it had been used in a broader and more colloquial sense previously, the term was first introduced as an official diagnostic category in Sir Charles Myers's 1915 article "A Contribution to the Study of Shell Shock: Being an Account of Three Cases of Memory, Vision, Smell, and Taste, Admitted into the Duchess of Westminster's War Hospital Le Touquet," *Lancet* 185 (February 13, 1915): 316–20. For more on the prediagnostic uses of the expression, see Tracey Loughran, "Shell Shock, Trauma, and the First World War: The Making of a Diagnosis and Its Histories," *Journal of the History of Medicine and Allied Sciences* 67, no. 1 (2010): 105. For the figures, see Anthony Babington, *Shell-Shock: A History of the Changing Attitudes to War Neurosis* (Barnsley: Pen and Sword, 2003), 87; and Reid, *Broken Men,* 13.

19. Frederick Mott, *The Effects of High Explosives upon the Central Nervous System,* (London: Harrison and Sons, 1916), 70.

20. Reid, *Broken Men,* 72.

21. "Those Who Know," in *The Hidden Side of the War: Some Revelations and Prophecies* (London: Elliot Stock, 1918), 43.

22. Tudor-Pole, *Private Dowding,* 7–8.

23. Carrington, *Psychical Phenomena and the War,* 90.

24. Oliver Lodge, *Christopher: A Study in Human Personality,* (London: Cassell, 1918), 3.

25. Winter, *Sites of Memory,* 76.

26. Arthur Machen, "The Bowmen" (1914), in *The White People and Other Stories,* ed. S. T. Joshi (London: Penguin, 2011), 224.

27. For more on story's reception within and impact on the British wartime imagination, see David Clarke, "Rumours of Angels: A Legend of the First World War," *Folklore* 113, no. 2 (October 2002): 151–73; and Richard J. Bleiler, *The Strange Case of "The Angels of Mons": Arthur Machen's World War I Story, the Insistent Believers, and His Refutations* (Jefferson, NC: McFarland, 2015).

28. John Garnier, *The Visions of Mons and Ypres: Their Meaning and Purpose* (London: Robert Banks and Son, 1916), 18. Machen's vexed relationship with the modern spiritualist movement is treated in Christine Ferguson, "Reading with the Occultists: Arthur Machen,

A. E. Waite, and the Ecstasies of Popular Fiction," *Journal of Victorian Culture* 21, no. 1 (2016): 40–55.

29. Machen, "Bowmen," 224.

30. Lodge, *Raymond*, vii.

31. For more on the form of *Raymond*, see James Mussell's chapter in this volume.

32. Ruth Leys, *Trauma: A Genealogy* (Chicago: University of Chicago Press, 2000), 2. Examples of this popular spiritualist trope can be found in Elsa Barker, *War Letters from the Living Dead Man* (London; William Rider and Son, 1915), 119; J. S. M. Ward, *A Subaltern in Spirit Land: A Sequel to "Gone West"* (London: William Rider and Son, 1920), 69; Harriet McCrory Grove and Mattie Mitchell Hunt, *A Soldier Gone West* (London: Kegan, Paul, Trench, Trubner, 1920), 44.

33. Lodge, *Raymond*, 47.

34. For more on Lodge's later relationship with Chalmers Mitchell, see Imogen Clarke's chapter in this volume.

35. See, for example, the *Saturday Review*'s discussion of *Raymond*, which contrasts the clarity of the séance messages about unverifiable facts with their hesitancy about the actual facts of Raymond Lodge's life. Unsigned review of *Raymond, or Life and Death*, 111.

36. Bernard Sickert, "Spiritualism and Its New Revelations II," *English Review* 27 (1918): 340.

37. Lodge, *Raymond*, 182.

38. The channeled messages repeatedly attribute their mistakes or confusion to the interference of competing, if not necessarily evil, spirits. See Lodge, *Raymond*, 182–83.

39. Barker, *War Letters*, 182. Italics mine.

40. One spirit soldier memoir, for example, claims that Britain's dead servicemen were in fact elevated "old souls" who returned to earth specifically for the purpose of elevating the world through a final "glorious sacrifice." L. Kelway Bamber, ed., *Claude's Book* (London: Methuen, 1918), 32. Another declared that "the soldiers are the pick of humanity. The young, brave, blameless manhood . . . [has] been brought to its majority on earth so that it may form an ideal democracy in this existence." Grace Duffie Boylan, *Thy Son Liveth: Messages from a Soldier to His Mother* (1918; repr., London: Frederick Muller, 1944), 12.

41. Lodge, *Raymond*, 322.

42. Lodge, *Christopher*, 65.

43. Alfred Martin, *Psychic Tendencies of To-Day: An Exposition and Critique of New Thought, Christian Science, Psychical Research (Oliver Lodge) and Modern Materialism in Relation to Immortality* (London: D. Appleton, 1918), 107.

44. Boylan, *Thy Son Liveth*, 47.

45. Lodge, *Raymond*, 185, 6.

46. A spirit doctor in Prentiss Tucker's *In the Land of the Living Dead*, for example, cheerfully explains to a new arrival: "If you had had your arm blown off and had come over here with only one arm you could have replaced [it] with . . . ease. . . . Matter on this side of the veil is remarkably amenable to the power of will." Tucker, *In the Land of the Living Dead: An Occult Story* (Oceanside, CA: Rosicrucian Fellowship, 1921), 44.

47. Lodge, *Raymond*, 196.

48. Leonard's attempts to feel out the location of Raymond Lodge's injuries here were unsuccessful; the shell that killed him hit him on the lower left side of his back. Lodge, *Raymond*, 76.

49. Victoria Stewart, "War Memoirs of the Dead: Writing and Remembrance in the First World War," *Literature and History* 14, no. 2 (2005): 50.

50. See Boylan, *Thy Son Liveth* (1918); Barker, *War Letters from the Living Dead Man* (1915); Tudor-Pole, *Private Dowding* (1917); Martin, *Psychic Tendencies of To-Day* (1918); and Grove and Hunt, *A Soldier Gone West* (1920) for examples of these propositions.

51. Lodge, *Raymond*, 171.

52. Ford Madox Ford, "Fun?—It's Heaven" (1915), in *War Prose*, ed. Max Saunders (Manchester: Carcanet Press, 1999), 152.

TEN. BEYOND *RAYMOND*

1. For a full survey of the beginnings of modern spiritualism in late nineteenth-century England, see Georgina Byrne, *Modern Spiritualism and the Church of England, 1850–1939* (Woodbridge: Boydell and Brewer, 2010), esp. 18–22, 39–79.

2. Oliver Lodge, *Raymond: Or Life and Death* (London: Methuen, 1916), viii.

3. *Official Report of the Church Congress Held in Leicester* (London: Nisbet, 1919), 114.

4. Mass Observation, *Puzzled People: A Study in Popular Attitudes to Religion, Ethics, Progress and Politics in a London Borough* (London: Victor Golancz, 1947), 32. Mass Observation began as a social research project in 1937, recording aspects of everyday life in Britain through diaries, questionnaires and conversations with volunteers. The archive is housed at the University of Sussex.

5. Gladys Osborne Leonard, *My Life in Two Worlds* (London: Cassell, 1931), 52.

6. Arthur Conan Doyle, *A Full Report of a Lecture on Spiritualism Delivered by Sir Arthur Conan Doyle at the Connaught Hall, Worthing on Friday, July 11th, 1919*, facsimile ed. (1919; repr., London: Rupert Books, 1997), 5.

7. See Archbishop's Committee on Spiritualism, "Report of the Committee to the Archbishop of Canterbury" (unpublished, 1939), 6, Lambeth Palace Library Manuscript Collection, London.

8. A fuller account of the theology of spiritualism is given in Byrne, *Modern Spiritualism*, 80–108.

9. "The Philosophy of Death," *Spiritualist*, November 19, 1869, 1; J. S. M. Ward, *Gone West: Three Narratives of After-Death Experiences* (London: William Rider and Son, 1917), 29. See also Byrne, *Modern Spiritualism*, 83–85.

10. Oliver Lodge, *Raymond Revised* (London: Methuen, 1922), 191. Christine Ferguson, in this volume, connects this narrative of disorientation in *Raymond* with the affliction of shell shock, although it is a theme in spirit communications that pre-dates the war.

11. Lodge, *Raymond Revised*, 192.

12. George Vale Owen, *Life beyond the Veil* (1922; repr., London: Thornton Butterworth, 1926), 1:17.

13. Lodge, *Raymond*, 264.

14. R. H. Benson, *Spiritualism* (London: Catholic Truth Society, 1911), 9. One departed spirit was reported to communicate at a séance that "the fearful place the Church tells you about has no existence beyond their fevered imaginations." "The Testimony of a Spirit," *Spiritualist*, December 31, 1869, 25.

15. Lodge, *Raymond*, 234.

16. For a good account of the judgment see Geoffrey Rowell, *Hell and the Victorians: A Study of the Nineteenth-Century Theological Controversies Concerning Eternal Punishment and the Future Life* (1974; repr., Oxford: Oxford University Press, 2000), 118–23. Further comments and petitions favoring the Athanasian Creed were made until 1873, until debates ceased in Convocation. The issue was revived again in 1927 as the Prayer Book was revised, but

caused far less controversy. Even so, some favored maintaining the traditional clauses. See Church of England (Church Assembly), *Report of Proceedings*, July 6, 1927.

17. Eleven pages, for example, are given to a complicated description of a photograph and where exactly Raymond was positioned in it. *Raymond*, 105–16.

18. Oliver Lodge, *Past Years: An Autobiography* (London: Hodder and Stoughton, 1931), 348.

19. An unnamed bishop made this comment to Lodge. Vernon Storr, *Do Dead Men Live Again?* (London: Hodder and Stoughton, 1932), 189.

20. Viscount Halifax [Charles Lindley Wood], *"Raymond": Some Criticisms [On the Work of that Name by Sir Oliver J. Lodge]* (London: Mowbray, 1917), 12.

21. *Official Report of the Church Congress Held in Leicester* (London: Nisbet, 1919), 90–91.

22. Lodge, *Raymond Revised*, 190.

23. It was condemned outright by the Roman Catholic Church in 1898.

24. John Henry Elliott, *A Refutation of Modern Spiritualism* (London: William Freeman, 1866), 3.

25. George Longridge, *Spiritualism and Christianity* (London: Mowbray, 1919), 37, 39.

26. See, for example, C. F. Hogg, *Spiritism in the Light of Scripture* (London: Pickering and Inglis, 1923); W. J. L. Sheppard, *Messages from the Dead* (Stirling: Drummond's Tract Depot, 1926).

27. "Editor's Table," *Church Times*, December 8, 1916, 515.

28. Rene Kollar, *Searching for Raymond: Anglicanism, Spiritualism and Bereavement between the Two World Wars* (Lanham, MD: Lexington, 2000), 156. A 1939 report into spiritualism, following investigations by a committee chaired by the dean of Rochester, Francis Underhill, remained unpublished until the 1980s. The conclusions of the majority were that, with some reservations, it was possible that departed spirits did communicate with the living. See Byrne, *Modern Spiritualism*, 177–81.

29. In May 1855, the editors of the *Yorkshire Spiritual Telegraph*, for example, noted that clergymen from all parties and sects of the Church of England, as well as ministers belonging to "almost all the various dissenting denominations" shared in the manifestations of spiritualism. In other words, they either tried the spirits themselves, or were witnesses at séances. "Our Object," *Yorkshire Spiritual Telegraph*, May 1855, 20. A letter to another spiritualist journal described a meeting in Notting Hill nearly thirty years later, at which, of the twenty-four people present, twenty-one were clergy from the Church of England. G. Damiani, "Spiritualism amongst the Clergy," *Light*, December 17, 1881, 398.

30. Percy Dearmer and Nancy Dearmer, *The Fellowship of the Picture: An Automatic Script Taken Down by Nancy Dearmer* (London: Nisbet, 1920), 8.

31. W. R. Matthews, *Memories and Meanings* (London: Hodder and Stoughton, 1969), 344.

32. This is the conclusion of Ruth Clayton Windscheffel in "Politics, Religion and Text: W. E. Gladstone and Spiritualism," *Journal of Victorian Culture* 11, no. 1 (2006): 1–29.

33. See Arthur Conan Doyle, *The History of Spiritualism* (1926; repr., London: Psychic Press, 1989), 2:222; George Vale Owen, *On Tour in USA* (London: Hutchinson, 1924), 144.

34. Correspondence between Bishop of Guildford and Archbishop's Chaplain, Lang Papers, 133 / 297 and 299, Lambeth Palace Library Manuscript Collection, London.

35. For details of other clergy spiritualists, see Byrne, *Modern Spiritualism*, 156–77.

36. See Philip Almond, *Heaven and Hell in Enlightenment England* (Cambridge: Cambridge University Press, 1994), 38; D. P. Walker, *The Decline of Hell: Seventeenth Century Discussions of Eternal Torment* (London: Routledge and Kegan Paul, 1964), 113.

37. Frederick Maurice, *Theological Essays* (1853; repr., London: James Clarke, 1957), 137.

38. Frederick Maurice, *The Life of Frederick Denison Maurice Chiefly Told in His Own Letters*, 2nd ed. (London: Macmillan, 1884), 2:210.

39. See especially R. L. Ottley, "Christian Ethics," *Lux Mundi: A Series of Studies in the Religion of the Incarnation*, ed. Charles Gore (1889; repr., London: John Murray, 1890), 468–520.

40. Hastings Rashdall, *The Idea of Atonement in Christian Theology: Being the Bampton Lectures for 1915* (London: Macmillan, 1920), 458.

41. Burnett H. Streeter et al., *Immortality: An Essay in Discovery, Co-ordinating Scientific, Psychical and Biblical Research* (London: Macmillan, 1917), 135.

42. Leslie Weatherhead, *After Death: A Popular Statement of the Modern Christian View of Life Beyond the Grave* (London: James Clarke, 1923), 28.

43. Storr, *Do Dead Men Live Again*, 211.

44. Weatherhead, *After Death*, 49.

45. Storr, *Do Dead Men Live Again*, 73.

46. W. R. Inge, "Survival and Immortality," *Hibbert Journal* 15 (1917): 595.

47. John Baillie, *And the Life Everlasting* (London: Oxford University Press, 1934), 236–37.

48. Oliver Lodge, *Man and the Universe: A Study of the Influence of the Advance in Scientific Knowledge upon Our Understanding of Christianity* (London: Methuen, 1908), 234.

49. Oliver Lodge, *The Reality of the Spiritual World* (London: E. Benn, 1930), 25. This volume was part of a series titled Affirmations of God in the Modern World, edited by Percy Dearmer.

50. W. R. Matthews, *God in Christian Thought and Experience* (London: Nisbet, 1930), vii.

51. James Welldon, "The Nature of Immortality," in *Life after Death According to Christianity and Spiritualism*, ed. James Marchant (London: Cassell, 1925), 4.

52. Storr, *Do Dead Men Live Again*, 47.

53. Weatherhead, *After Death*, 17.

54. A. F. Winnington-Ingram, introduction, in Marchant, *Life after Death*, viii.

55. Storr, *Do Dead Men Live Again*, 102.

56. Lodge, *Raymond*, 312.

57. Storr, *Do Dead Men Live Again*, 211.

58. A. F. Winnington-Ingram, *The Spirit of Peace* (London: Gardner, Darton, 1921), 159.

59. Cyril William Emmet, "The Bible and Hell," in Streeter et al., *Immortality*, 216.

60. Lodge, *Raymond*, 322, 324.

61. *Doctrine in the Church of England: The Report of the Commission on Christian Doctrine Appointed by the Archbishops of Canterbury and York in 1922* (London: SPCK, 1938), 4.

ELEVEN. OLIVER LODGE'S ETHER AND THE BIRTH OF BRITISH BROADCASTING

Epigraph: Susan Douglas, *Listening In: Radio and the American Imagination* (New York: Times Books, 1999), 40.

1. Gillian Beer, "'Wireless': Popular Physics, Radio and Modernism," in *Cultural Babbage: Technology, Time and Invention*, ed. Francis Spufford and Jenny Uglow (London: Faber and Faber, 1996), 149.

2. "Wireless at Sea," *Broadcaster*, August 1922, 32.

3. Harry Ricketts, *The Unforgiving Minute: A Life of Rudyard Kipling* (London: Chatto and Windus, 1999), 247–58.

4. Rudyard Kipling, "Wireless," *Traffics and Discoveries* (London: Macmillan, 1904), 211–39.

5. Jeffrey Sconce, *Haunted Media: Electronic Presence from Telegraphy to Television* (Durham, NC: Duke University Press, 2000), 4–8. See also: Jeffrey Sconce, "The Voice from the Void: Wireless, Modernity, and the Distant Dead," *International Journal of Cultural Studies* 1, no. 2 (1998): 211–32.

6. John Durham Peters, *Speaking into the Air: A History of the Idea of Communication* (Chicago: University of Chicago Press, 1999), 100.

7. David Trotter, *Literature in the First Media Age: Britain between the Wars* (Cambridge, MA: Harvard University Press, 2014). See also Richard Menke, *Telegraphic Realism: Victorian Fiction and Other Information Systems* (Stanford, CA: Stanford University Press, 2008).

8. Sconce, *Haunted Media*, 63.

9. Sconce, "Voice from the Void," 213.

10. Oliver Lodge, *Past Years: An Autobiography* (London: Hodder and Stoughton, 1931), 34.

11. Lodge, *Past Years*, 66, 78.

12. Oliver Lodge, *My Philosophy: Representing My Views on the Many Functions of the Ether of Space* (London: Ernest Benn, 1933), 5. Emphasis added.

13. Oliver Lodge, *Continuity: The Presidential Address to the British Association for 1913* (London: J. M. Dent and Sons, 1913), 19.

14. Lodge, *My Philosophy*, 261.

15. Oliver Lodge, *Raymond: Or Life and Death* (London: Methuen, 1916), viii.

16. Lodge, *My Philosophy*, 35–38.

17. Lodge, *Past Years*, 172.

18. Lodge, *Raymond*, 341–42.

19. David B. Wilson, "The Thought of Late Victorian Physicists: Oliver Lodge's Ethereal Body," *Victorian Studies* 15, no. 1 (September 1971): 33–34.

20. "Signalling throughout the Ages: From Aeschylus to Marconi," *Wireless World*, April 1913, 29, 34.

21. "Amateur Notes," *Wireless World*, December 1913, 587.

22. "Amateur Notes," *Wireless World*, February 1914, 712.

23. Dwayne Winseck and Robert Pike, "The Global Media and the Empire of Liberal Internationalism, circa 1910–1930," *Media History* 15, no. 1 (2009): 35.

24. Winseck and Pike, "Global Media," 37. See also: Daniel R. Headrick, *The Invisible Weapon: Telecommunications and International Politics, 1851–1945* (Oxford: Oxford University Press, 1992).

25. Stathis Arapostathis and Graeme Gooday, *Patently Contestable: Electrical Technologies and Inventor Identities on Trial in Britain* (Cambridge, MA: MIT Press, 2013), 160.

26. "Wireless Telegraphy in the War," *Wireless World*, April 1917, 24–25.

27. "Wireless Telegraphy in the War," *Wireless World*, July 1917, 244–45.

28. "Notes of the Month," *Wireless World*, April 1917, 40.

29. "Amateur Notes," *Wireless World*, October 1913, 464.

30. Sconce, "Voice from the Void," 219.

31. Sconce, *Haunted Media*, 68.

32. James Curran and Jean Seaton, *Power without Responsibility: The Press and Broadcasting in Britain*, 5th ed. (London: Routledge, 1997), 113–15.

33. Richard Overy, *The Morbid Age: Britain between the Wars* (London: Allen Lane, 2009), 4.

34. Matthew Arnold, *Culture and Anarchy: An Essay in Political and Social Criticism*, 2nd ed. (London: Smith, Elder, 1875), 15–16, 20–21.

35. See Paul Fussell, *The Great War and Modern Memory* (Oxford: Oxford University Press, 2000), 316; Adrian Gregory, *The Last Great War: British Society and the First World War* (Cambridge: Cambridge University Press, 2008), 40–69.

36. "Aerial Voices," *London Standard*, December 28, 1912, 8, in Special Collections S236/18, BBC Written Archives, Caversham.

37. "Arthur Richard Burrows," n.d., Special Collections S236/21, BBC Written Archives.

38. "Arthur Richard Burrows," n.d.

39. "Broadcasting as a Means of Education," n.d., unpaginated, Special Collections S236/3, BBC Written Archives, Caversham.

40. "Broadcasting as a Means of Education," unpaginated; Speech on Educational Broadcasting, December 1923, Special Collections S236/4, BBC Written Archives.

41. P. Russell Mallinson, "When Wireless Dreams Come True," *Broadcaster*, August 1922, 15–17.

42. "Impatience," *Broadcaster*, September 1922, 86.

43. "Other People's Troubles," *Broadcaster*, August 1922, 64; "Wanted," *Broadcaster*, September 1922, 1922, 78.

44. Cecil A. Lewis, *Sagittarius Rising* (London: Peter Davies, 1936), 93–95; Cecil A. Lewis, *Never Look Back: An Attempt at Autobiography* (London: Hutchinson, 1974), 116.

45. Cecil Lewis, oral history interview, 1982, 10–11, BBC Written Archives.

46. Lance Sieveking, *The Stuff of Radio* (London: Cassell, 1934), 111–12.

47. John C. W. Reith, *Broadcast over Britain* (London: Hodder and Stoughton, 1924), 34.

48. Lodge, *Past Years*, 333.

49. Reith, *Broadcast over Britain*, 217–19.

50. Arnold, *Culture and Anarchy*, xiii.

51. Reith, *Broadcast over Britain*, 219.

52. Reith, *Broadcast over Britain*, 219.

53. "Vox Dei: Religion and Radio," *Vox: The Radio Critic and Broadcast Review*, November 9, 1929, 9.

54. Reith, *Broadcast over Britain*, 18.

55. Jürgen Habermas, *The Structural Transformation of the Public Sphere: An Inquiry into a Category of Bourgeois Society* (Cambridge: Polity, 1989).

56. See Peters, *Speaking into the Air*.

57. Beer, "'Wireless,'" 150.

58. Stefan Collini, *Absent Minds: Intellectuals in Britain* (Oxford: Oxford University Press, 2006), 435.

TWELVE. "BODY SEPARATES: SPIRIT UNITES"

1. Oliver Lodge, *Christopher: A Study in Human Personality* (London: Cassell, 1918), 5.

2. Oliver Lodge, *Past Years: An Autobiography* (London: Hodder and Staughton, 1931), 111.

3. That matter individuates is a founding principle of forensic science. See, for instance, Matthew G. Kirschenbaum, *Mechanisms: New Media and the Forensic Imagination* (Cambridge, MA: MIT Press, 2008), esp. chapter 1, 25–71.

4. Oliver Lodge, *Continuity: The Presidential Address to the British Association Birmingham 1913* (London: J. M. Dent and Sons, 1913), 15.

5. See also Hendy, this volume.

6. Lodge, *Continuity*, 5.

7. Lodge, *Past Years*, 194–204. See also Peter Rowlands, *Oliver Lodge and the Liverpool Physical Society* (Liverpool: Liverpool University Press, 1990), 66–79; Bruce J. Hunt, "Experimenting on the Ether: Oliver J. Lodge and the Great Whirling Machine," *Historical Studies in the Physical Sciences* 16 (1986): 111–34.

8. Lodge, *Continuity*, 54.

9. See, for instance, Oliver Lodge, *The Substance of Faith* (London: Methuen, 1907); Oliver Lodge, *Man and the Universe: A Study of the Influence of the Advance in Scientific Knowledge upon Our Understanding of Christianity* (London: Methuen, 1908); Oliver Lodge, *The Survival of Man: A Study in Unrecognised Human Faculty* (London: Methuen, 1909).

10. "The British Association (From Our Correspondent)," *Nature* 44 (1891): 385. For the 1891 address and discursivity, see Erhard Schuettpelz and Ehler Voss, "Fragile Balance: Human Mediums and Technical Media in Oliver Lodge's Presidential Address of 1891," *Communication+1* 4 (2015): 1–15. See also Noakes, this volume.

11. "British Association (From Our Correspondent)," 385.

12. Lodge, *Continuity*, 90

13. Lodge, *Continuity*, 90–91.

14. Lodge's hostility to materialism is set most fully in *Life and Matter* (London: Williams and Norgate, 1905). It can also be traced in his relationship to John Tyndall; see especially his remarks on Tyndall's Belfast Lecture, *Past Years*, 138–39. For Lodge and materialism more broadly, see Noakes, this volume.

15. Oliver Lodge, "Ether, Matter and the Soul," *Hibbert Journal* 17 (1918–9): 256.

16. Oliver Lodge, *Modern Views of Electricity* (London: Macmillan, 1889), x, xi.

17. For these models, see Bruce J. Hunt, *The Maxwellians* (Ithaca, NY: Cornell University Press, 1991), 87–95; Shawn Michael Bullock, "The Pedagogical Implications of Maxwellian Electromagnetic Models: A Case Study from Victorian-Era Physics," *Endeavour* 38, no. 3–4 (2014): 280–88.

18. Lodge, *Modern Views of Electricity*, 177–92, 193–216.

19. James Clerk Maxwell, "On the Physical Lines of Force," *Philosophical Magazine* 21 (1861): 161–75; Maxwell, "On the Physical Lines of Force," *Philosophical Magazine* 21 (1861): 281–91; Maxwell, "On the Physical Lines of Force," *Philosophical Magazine* 21 (1861): 338–48; Maxwell, "On the Physical Lines of Force," *Philosophical Magazine* 23 (1862): 12–24; and Maxwell, "On the Physical Lines of Force," *Philosophical Magazine* 23 (1862); 85–95; William Thomson, "On Vortex Atoms," *Proceedings of the Royal Society of Edinburgh* 6 (1867): 94–105. For Lodge on the history of these models, see Sir Oliver Lodge, *An Introduction to Modern Views on Atomic Structure and Radiation* (London: Ernest Benn, 1924), 15.

20. Lodge, *Modern Views of Electricity*, 356. The lecture was from 1882. By the time he came to write *Modern Views of Electricity*, Lodge was more convinced by Fitzgerald's sponge model, in which the ether was made of interlaced vortex filaments like a sponge (see xi). See also Hunt, *Maxwellians*, 96.

21. Oliver Lodge, *Modern Views of Electricity*, 3rd ed. (London: Macmillan, 1907), vii, viii.

22. Lodge, *Modern Views of Electricity*, 3rd ed., 323.

23. Oliver Lodge, *Modern Views on Matter: The Romanes Lecture 1903* (Oxford: Clarendon, 1903), 10.

24. Lodge, *Modern Views on Matter*, 8.

25. Lodge, *Modern Views of Electricity*, 3rd ed., 340–41. See also introduction, 3.

26. Lodge, *Modern Views of Electricity* (1889), 181.

27. Lodge, *Modern Views of Electricity*, 3rd ed., 412.

28. See "British Association," *Times* (London), August 5, 1907, 10. For a fuller account of

Lodge's position, see "Lord Kelvin's Philosophy," *Nature* 78, no. 2018 (July 2, 1908): 198–99.

29. Lodge, "Ether, Matter and the Soul," 256.

30. Lodge, *Modern Views of Electricity* (1889), viii, and *Modern Views of Electricity*, 3rd ed., vii.

31. Lodge, *Continuity*, 60.

32. Jacques Derrida, "Structure, Sign and Play in the Discourse of the Human Sciences," *Writing and Difference*, trans. Alan Bass (London: Routledge, 2002), 351–70, esp. 365, 367.

33. Oliver Lodge, *Atoms and Rays: An Introduction to Modern Views on Atomic Structure and Radiation* (London: Ernest Benn, 1924), 14–15.

34. Lodge, *Continuity*, 22.

35. Lodge, *Atoms and Rays*, 15.

36. Lodge, *Continuity*, 60.

37. Derrida, "Structure, Sign and Play," 353.

38. For Lodge, Balfour, and the Souls, see Lodge, *Past Years*, 220–24; W. P. Jolly, *Sir Oliver Lodge* (London: Constable, 1974), 115–19.

39. For Lodge's relationship with Chamberlain see *Past Years*, 314–19; Jolly, *Sir Oliver Lodge*, 129–41, 158; Eric Ives, Diane Drummond, and Leonard Schwarz, *The First Civic University: Birmingham, 1880–1980* (Birmingham: University of Birmingham Press, 2000), chapters 6–10.

40. Lodge, *Past Years*, 267–69. Lodge had gotten to know Ruskin in the 1880s and organized the testimonial for his eightieth birthday in 1899. Lodge wrote the introduction to the 1907 edition of *Sesame and Lilies*. See John Ruskin, *Sesame and Lilies, the Two Paths, and the King of the Golden River* (London: J. M. Dent, 1907).

41. Oliver Lodge, *Public Service versus Private Expenditure* (London: Fabian Society, 1905), reprinted in Sidney Webb, Sidney Ball, G. Bernard Shaw, and Oliver Lodge, *Socialism and Individualism* (London: A.C. Fifield, 1908), 92–102.

42. Oliver Lodge, *The Link between Matter and Matter* (London: British Science Guild, 1925), 15.

43. F. W. H. Myers, "The Subliminal Consciousness," *Proceedings of the Society for Psychical Research* 7 (1892): 301, 305, 306. See also Roger Luckhurst, *The Invention of Telepathy, 1870–1901* (Oxford: Oxford University Press, 2002), 107–12.

44. See Lodge, *Past Years*, 276–79; Jolly, *Sir Oliver Lodge*, 92–95.

45. Lodge, "Ether, Matter and the Soul," 257.

46. Lodge, *Raymond: Or, Life and Death* (London: Methuen, 1916), 89.

47. Lodge, *Past Years*, 285.

48. See *Raymond Revised: A New and Abbreviated Edition of "Raymond Or Life and Death"* (London: Methuen, 1922), n.p.

49. Lodge, *Raymond*, vii.

50. Oliver Lodge, "The Interstellar Ether," *Fortnightly Review* 53 (1893): 862. The essay was reprinted in *Modern Views of Electricity*, 3rd ed., 450–62.

51. Lodge often sent his wife, Mary, or one of his children, as he was too well known a sitter.

52. Lodge, *Raymond*, 231.

53. See, for instance, Eleanor Garnier, "Leicester Scientists Print Human Genome in 130 Books," *BBC News*, 2012, http://www.bbc.co.uk/news/uk-england-leicestershire-20520843.

54. Jolly, *Sir Oliver Lodge*, 237.

BIBLIOGRAPHY

ARCHIVES

BBC Written Archives, Caversham
Birmingham Archives and Collections, Library of Birmingham
Cadbury Research Library, University of Birmingham
Lambeth Palace Library Manuscript Collection, London
New York Public Library
Royal Society Archives, London
Society for Psychical Research Archive, Cambridge University Library
Special Collections and Archives, University of Liverpool
University College London Library Services, Special Collections

REFERENCES

"Address of the President Sir William Bragg, O.M., at the Anniversary Meeting, 30 November 1940." *Proceedings of the Royal Society A* 177 (December 31, 1940): 1–25.

"Address of the President Sir William Bragg, O.M., at the Anniversary Meeting, 30 November 1940." *Proceedings of the Royal Society B* 129 (December 31, 1940): 413–38.

Aitken, Hugh G. J. *Syntony and Spark: The Origins of Radio*. New York: John Wiley and Sons, 1976.

Almond, Philip. *Heaven and Hell in Enlightenment England*. Cambridge: Cambridge University Press, 1994.

"Amateur Notes." *Wireless World*, December 1913, 587–88.

"Amateur Notes." *Wireless World*, February 1914, 712–14.

"Amateur Notes." *Wireless World*, October 1913, 464–65.

Amigoni, David. "Between Medicine and Evolutionary Theory: Sympathy and Other Emotional Investments in Life Writings by and about Charles Darwin." In *After*

Darwin: Animals, Emotions and the Mind, edited by Angelique Richardson, 172–92. Amsterdam: Rodopi, 2013.

Anderson, Robert. *British Universities: Past and Present*. London: Bloomsbury, 2006.

Andrade, Edward Neville da Costa. *The Structure of the Atom*. London: Bell, 1923.

Edward Neville da Costa Andrade, "Books of the Day: The New Physics." *Observer*, August 10, 1924, 5.

Arapostathis, Stathis, and Graeme Gooday. "Electrical Technoscience and Physics in Transition, 1880–1920." *Studies in History and Philosophy of Science A* 44, no. 2 (2013): 202–11.

Arapostathis, Stathis, and Graeme Gooday. *Patently Contestable: Electrical Technologies and Inventor Identities on Trial in Britain*. Cambridge, MA: MIT Press, 2013.

Archbishop's Committee on Spiritualism. "Report of the Committee to the Archbishop of Canterbury." Unpublished manuscript, 1939.

Arnold, Matthew. *Culture and Anarchy: An Essay in Political and Social Criticism*. 2nd ed. London: Smith, Elder, 1875.

Babington, Anthony. *Shell-Shock: A History of the Changing Attitudes to War Neurosis*. Barnsley: Pen and Sword, 2003.

Baillie, John. *And the Life Everlasting*. London: Oxford University Press, 1934.

Baines, Thomas. *History of the Commerce and Town of Liverpool, and of the Rise of Manufacturing Industry in the Adjoining Counties*. London: Longman, 1852.

Bamber, L. Kelway, ed. *Claude's Book*. London: Methuen, 1918.

Barker, Elsa. *War Letters from the Living Dead Man*. London: William Rider and Son, 1915.

Barton, Ruth. "John Tyndall, Pantheist: A Rereading of the Belfast Address." *Osiris* 3 (1987): 111–34.

Battersby, H. F. Prevorst. "A Scientist's Philosophy." Review of *My Philosophy*, by Oliver Lodge. *Light*, July 7, 1933, 425–26.

Beer, Gillian. "'Wireless': Popular Physics, Radio and Modernism." In *Cultural Babbage: Technology, Time and Invention*, edited by Francis Spufford and Jenny Uglow, 149–66. London: Faber and Faber, 1996.

"Behind the Veil." *Church Times* 102 (November 22, 1929): 631.

Belchem, John. *Merseypride: Essays in Liverpool Exceptionalism*. Liverpool: Liverpool University Press, 2006.

Benson, R. H. *Spiritualism*. London: Catholic Truth Society, 1911.

Besterman, Theodore. *A Bibliography of Sir Oliver Lodge F.R.S.* Oxford: Oxford University Press, 1935.

"The Birmingham University." *Engineering* 79 (August 25, 1905): 239–42.

Bleiler, Richard J. *The Strange Case of "The Angels of Mons": Arthur Machen's World War I Story, the Insistent Believers, and His Refutations*. Jefferson, NC: McFarland, 2015.

Bose, Jagadish Chandra. "On the Determination of the Indices of Refraction of Various Substances for the Electric Ray. I. Index of Refraction of Sulphur." *Proceedings of the Royal Society* 59 (December 12, 1895): 60–67.

Bourke, Joanna. *Dismembering the Male: Men's Bodies, Britain, and the Great War*. London: Reaktion, 1996.

Bowler, Peter J. *The Eclipse of Darwinism: Anti-Darwinian Evolution Theories in the Decades around 1900*. Baltimore: Johns Hopkins University Press, 1983.

Bowler, Peter J. *The Non-Darwinian Revolution: Reinterpreting a Historical Myth*. Baltimore: Johns Hopkins University Press, 1988.

Bowler, Peter J. *Reconciling Science and Religion: The Debate in Early Twentieth-Century Britain*. Chicago: University of Chicago Press, 2001.

Bowler, Peter J. *Science for All: The Popularization of Science in Early Twentieth-Century Britain*. Chicago: University of Chicago Press, 2009.

Boylan, Grace Duffie. *Thy Son Liveth: Messages from a Soldier to His Mother*. London: Frederick Muller, 1944.

Bragg, William. "Science and National Welfare." *Nature* 146, no. 3710 (December 7, 1940): 731–32.

Bridges, Horace J. "Sir Oliver Lodge and the Public Mind." *Forum* 51 (May 1914): 695–705.

"The British Association (From Our Correspondent)." *Nature* 44 (1891): 371–87.

British Empire Exhibition 1924: Handbook to the Exhibition of Pure Science: Galleries 3 and 4 British Government Pavilion. London: Royal Society, 1924.

Brooks, Walter R. "New Books." *Outlook*, March 1932, 190–94.

Brown, G. B. Review of *Beyond Physics*, by Oliver Lodge. *Philosophy* 5 (1930): 624–26.

Buchwald, Jed Z. *From Maxwell to Microphysics: Aspects of Electromagnetic Theory in the Last Quarter of the Nineteenth Century*. Chicago: University of Chicago Press, 1985.

Bullock, Shawn Michael. "The Pedagogical Implications of Maxwellian Electromagnetic Models: A Case Study from Victorian-Era Physics." *Endeavour* 38, no. 3–4 (2014): 280–88.

Büttner, Jochen, Jürgen Renn, and Matthias Schemmel. "Exploring the Limits of Classical Physics: Planck, Einstein, and the Structure of a Scientific Revolution." *Studies in the History and Philosophy of Modern Physics* 34, no. 1 (2003): 37–59.

Byrne, Georgina. *Modern Spiritualism and the Church of England, 1850–1939*. Woodbridge: Boydell and Brewer, 2010.

Calver, G. "Sir Oliver Lodge and the Ether." *English Mechanic and World of Science* 109 (January 31, 1919): 21.

Cantor, Geoffrey. "The Scientist as Hero: Public Images of Michael Faraday." In *Telling Lives in Science: Essays on Scientific Biography*, edited by Michael Shortland and Richard Yeo, 171–94. Cambridge: Cambridge University Press, 2008.

Cantor, Geoffrey. "The Theological Significance of Ethers." In *Conceptions of Ether: Studies in the History of Ether Theories, 1740–1900*, edited by Geoffrey Cantor and M. J. S. Hodge, 135–55. Cambridge: Cambridge University Press, 1981.

Cantor, Geoffrey, and M. J. S. Hodge, eds. *Conceptions of Ether: Studies in the History of Ether Theories, 1740–1900*. Cambridge: Cambridge University Press, 1981.

Carrington, Hereward. *Psychical Phenomena and the War*. London: T. Werner, 1918.

Chesterton, G. K. *Eugenics and Other Evils*. London: Cassell, 1922.

Church of England (Church Assembly). *Report of Proceedings*, July 6, 1927.

Clarke, Bruce, and Linda Dalrymple Henderson. "Ether and Electromagnetism: Capturing the Invisible." In *From Energy to Information: Representation in Science and Technology, Art, and Literature*, edited by Bruce Clarke and Linda Dalrymple Henderson, 95–97. Stanford, CA: Stanford University Press, 2002.

Clarke, David. "Rumours of Angels: A Legend of the First World War." *Folklore* 113, no. 2 (October 2002): 151–73.

Clarke, Imogen. "Negotiating Progress: Promoting 'Modern' Physics in Britain, 1900–1940." PhD diss., University of Manchester, 2012.

Clodd, Edward. "The Revival of the Dangerous Cult of Spiritualism." *Graphic* 101 (February 14, 1920): 222.

Collini, Stefan. *Absent Minds: Intellectuals in Britain*. Oxford: Oxford University Press, 2006.

"The Commercial Ports of England." *Bankers' Magazine and Journal of the Money Market*, December 1851, 783–84.

"The Competition." *Spectator* 144 (May 31, 1930): 905.

Cook, Walter. *Reflections on "Raymond": An Appreciation and Analysis*. London: Grant Richards, 1917.

Cooter, Roger, and Stephen Pumfrey. "Separate Spheres and Public Places: Reflections of the History of Science Popularization and Science in Popular Culture." *History of Science* 32, no. 3 (1994): 237–67.

Crook, D. P. "Peter Chalmers Mitchell and Antiwar Evolutionism in Britain during the Great War." *Journal of the History of Biology* 22, no. 2 (1989): 325–56.

Cottrell, Alan. "Edward Neville da Costa Andrade. 1887–1971." *Biographical Memoirs of Fellows of the Royal Society* 18 (1972): 1–20.

Curran, James, and Jean Seaton. *Power Without Responsibility: The Press and Broadcasting in Britain*. 5th ed. London: Routledge, 1997.

Dallas, H. A. *Leaves from the Psychic Note Book*. London: Rider, 1927.

Damiani, G. "Spiritualism amongst the Clergy." *Light*, December 17, 1881, 398.

Darrigol, Olivier. *Electrodynamics from Ampère to Einstein*. Oxford: Oxford University Press, 2000.

Darwin, Charles. *Autobiography*. Edited by Nora Barlow. London: Collins, 1958.

Daston, Lorraine J. "British Responses to Psycho-physiology, 1860–1900." *Isis* 69, no. 2 (1978): 192–208.

Dawson, Gowan, and Bernard Lightman, eds. *Victorian Scientific Naturalism: Community, Identity, Continuity*. Chicago: University of Chicago Press, 2014.

"DD." "Introduction." *Nature* 106, no. 2677 (February 17, 1921): 781.

De Selincourt, Basil. "Sir Oliver Lodge's Philosophy." *Observer*, June 25, 1933, 4.Dear, Peter. *The Intelligibility of Nature: How Science Makes Sense of the World*. Chicago: University of Chicago Press, 2006.

Dearmer, Percy, and Nancy Dearmer. *The Fellowship of the Picture: An Automatic Script Taken Down by Nancy Dearmer*. London: Nisbet, 1920.

Delve, Janet. "The College of Preceptors and the *Educational Times*: Changes for British Mathematics Education in the Mid-Nineteenth Century." *Historia Mathematica* 30, no. 2 (2003): 140–72.

Derrida, Jacques. "Structure, Sign and Play in the Discourse of the Human Sciences." *Writing and Difference*, translated by Alan Bass, 351–70. London: Routledge, 2002.

Derrida, Jacques. *Of Grammatology*. Translated by Gayatri Chakravorty Spivak. Baltimore: Johns Hopkins Press, 2016.

Desmond, Adrian. "Huxley, Thomas Henry (1825–1895)." *Oxford Dictionary of National Biography*. Last revised May 28, 2015. https://doi.org/10.1093/ref:odnb/14320.

Desmond, Shaw. "Phantom Walls." *Bookman* 77 (1929): 210–11.

DeYoung, Ursula. *A Vision of Modern Science: John Tyndall and the Role of the Scientist in Victorian Culture*. Basingstoke: Palgrave Macmillan, 2011.

"Discussion on Lightning Conductors." *Electrician*, September 21, 1888, 644–48.

"Discussion on Lightning Conductors." *Electrician*, September 28, 1888, 673–80.

Doctrine in the Church of England: The Report of the Commission on Christian Doctrine Appointed by the Archbishops of Canterbury and York in 1922. London: SPCK, 1938.

Douglas, Susan. *Listening In: Radio and the American Imagination*. New York: Times Books, 1999.

Doyle, Arthur Conan. *A Full Report of a Lecture on Spiritualism Delivered by Sir Arthur Conan Doyle at the Connaught Hall, Worthing on Friday, July 11th, 1919*. 1919. Facsimile of the 1st edition. London: Rupert Books, 1997.

Doyle, Arthur Conan. *The History of Spiritualism*. 2 vols. 1926. Reprint, London: Psychic Press, 1989.

Drummond, D. K. "'Power of our Provincial Cities' and 'Opportunities for the Highest Culture': The Call for a University of Birmingham and University Scholarship in Late Victorian Britain." In *Scholarship in Victorian Britain*, edited by Martin Hewitt, 53–65. Leeds Working Papers in Victorian Studies 1. Leeds: Leeds Centre for Victorian Studies, 1998.

Drummond, D. K. "The University of Birmingham and the Industrial Spirit: Reasons for Local Support of Joseph Chamberlain's Campaign to Found the University of Birmingham, 1897–1900." *History of Universities* 19 (1999): 247–64.

Duncan, P. D. "Newspaper Science: The Presentation of Science in Four British Newspapers During the Interwar Years, 1919–1939." MPhil thesis, University of Sussex, 1980.

Earman, John, and Clark Glymour. "Relativity and Eclipses: The British Eclipse Expeditions of 1919 and Their Predecessors." *Historical Studies in the Physical Sciences* 11 (1980): 49–85.

Eddington, Arthur Stanley. *The Nature of the Physical World.* Cambridge: Cambridge University Press, 1928.

"Editor's Table." *Church Times,* December 8, 1916, 515.

Edwards, David. "A Victorian Polymath." In *Oliver Lodge and the Invention of Radio,* edited by Peter Rowlands and J. Patrick Wilson, 19–38. Liverpool: PD Publications, 1994.

Einstein, Albert. *Sidelights on Relativity*. Translated by G. B. Jeffery and Wilfrid Perrett. London: Methuen, 1922.

Elliott, John Henry. *A Refutation of Modern Spiritualism*. London: William Freeman, 1866.

Emmet, Cyril William. "The Bible and Hell." In *Immortality: An Essay in Discovery, Coordinating Scientific, Psychical and Biblical Research,* edited by Burnett H. Streeter et al., 167–217. London: Macmillan, 1917.

"E.N.Da.C.A." [Edward N. Da Costa Andrade]. "A Veteran's View of Modern Physics." *Nature* 114, no. 2869 (October 25, 1924): 599–601.

"Ether and Human Survival." *Light,* April 25, 1925, 198.

"Ether and Reality." *Wireless World,* July 1, 1925, 16.

"The Ether, Life and Mind." *Times* (London), May 15, 1925, 10.Ewing, A. W. *The Man of Room 40: The Life of Sir Alfred Ewing*. London: Hutchinson, 1939.

Falconer, Isobel. "Corpuscles, Electrons and Cathode Rays: J. J. Thomson and the 'Discovery of the Electron.'" *British Journal for the History of Science* 20 (1987): 241–76.

Falconer, Isobel. "Corpuscles to Electrons." In *Histories of the Electron,* edited by Jed Z. Buchwald and Andrew Warwick, 77–100. Cambridge, MA: MIT Press, 2001.

Ferguson, Christine. "Reading with the Occultists: Arthur Machen, A. E. Waite, and the Ecstasies of Popular Fiction." *Journal of Victorian Culture* 21, no. 1 (2016): 40–55.

Fichman, Martin. *An Elusive Victorian: The Evolution of Alfred Russel Wallace*. Chicago: University of Chicago Press, 2004.

Findlay, J. A. *On the Edge of the Etheric: Being an Investigation of Psychic Phenomena.* London: Rider, 1931.

FitzGerald, G. F. "Address." In *Report of the Fifty-Eighth Meeting of the British Association for the Advancement of Science,* 557–62. London: John Murray, 1889.

FitzGerald, G. F. "Address to the Mathematical and Physical Section of the British Association." In *The Scientific Writings of the Late George Francis Fitzgerald,* edited by Joseph Larmor, 229–40. London: Longmans, Green, 1902.

FitzGerald, G. F. "On a Method of Producing Electro-magnetic Disturbances of Comparatively Short Wave-Lengths." In *Report of the Fifty-Third Meeting of the British Association for the Advancement of Science,* 405. London: John Murray, 1884.

"The Five Best Brains." *Spectator,* June 14, 1930, 979.

Fleming, J. Ambrose. "Lodge and the Physical Society." *Proceedings of the Physical Society* 53 (1941): 57–58.

Fleming, J. Ambrose. "Obituary Sir Oliver Lodge, F.R.S." *Nature* 146, no. 3697 (September 7, 1940): 327–28.

Fleming, J. Ambrose. *Principles of Electric Wave Telegraphy.* London: Longmans, Green, 1906.

Ford, Ford Madox. "Fun!—It's Heaven." In *War Prose,* edited by Max Saunders, 149–53. Manchester: Carcanet, 1999.

Free, E. E. "Radio and Relativity." *Popular Radio,* April 1923, 243–53.

Friedman, Alan J., and Carol C. Donley. *Einstein as Myth and Muse.* Cambridge: Cambridge University Press, 1985.

Fussell, Paul. *The Great War and Modern Memory.* Oxford: Oxford University Press, 2000.

Gagnier, Regenia. *Subjectivities: A History of Self-Representation in Britain, 1832–1920.* Oxford: Oxford University Press, 1991.

Galton, Francis. "The History of Twins, as a Criterion of the Relative Powers of Nature and Nurture." *Fraser's Magazine,* November 1875, 566–76.

Garnier, Eleanor. "Leicester Scientists Print Human Genome in 130 Books." *BBC News,* 2012. http://www.bbc.co.uk/news/uk-england-leicestershire-20520843.

Garnier, John. *The Visions of Mons and Ypres: Their Meaning and Purpose.* London: Robert Banks and Son, 1916.

Gauld, Alan. *The Founders of Psychical Research.* London: Routledge and Kegan Paul, 1968.

Gooday, Graeme. "Precision Measurement and the Genesis of Physics Teaching Laboratories in Victorian Britain." *British Journal for the History of Science* 23 (1990): 25–51.

Gooday, Graeme. "The Questionable Matter of Electricity: The Reception of J. J. Thomson's 'Corpuscle' among Electrical Theorists and Technologists." In *Histories of the Electron: The Birth of Microphysics,* edited by Jed Z. Buchwald and Andrew Warwick, 101–34. Cambridge, MA: MIT Press, 2001.

Gooday, Graeme. "Sunspots, Weather, and the Unseen Universe: Balfour Stewart's Anti-materialist Representations of 'Energy' in British Periodicals." In *Science*

Serialized: Representations of the Sciences in Nineteenth-Century Periodicals, edited by Geoffrey Cantor and Sally Shuttleworth, 111–47. Cambridge, MA: MIT Press, 2004.

Gooday, Graeme. "'Vague and Artificial': The Historically Elusive Distinction between Pure and Applied Science." *Isis* 103 (2012): 546–54.

Gooday, Graeme, and Daniel Jon Mitchell. "Rethinking 'Classical Physics.'" In *The Oxford Handbook of the History of Physics,* edited by Jed Z. Buchwald and Robert Fox, 721–64. Oxford: Oxford University Press, 2013.

Greenhill, A. G. "The Flying to Pieces of a Whirling Ring." *Nature* 43, no. 1116 (March 19, 1891): 461–62.

Gregory, Adrian. *The Last Great War: British Society and the First World War.* Cambridge: Cambridge University Press, 2008.

Gregory, R. A., and Allan Ferguson. "Oliver Joseph Lodge (1851–1940)." *Obituary Notices of Fellows of the Royal Society* 3 (1941): 551–74.

Grove, Harriet McCrory, and Mattie Mitchell Hunt. *A Soldier Gone West.* London: Kegan, Paul, Trench, Trubner, 1920.

Gunn, Simon. *The Public Culture of the Victorian Middle Class: Ritual and Authority in the English Industrial City, 1840–1914.* Manchester: Manchester University Press, 2008.

Gurney, Edmund. "Thought-Transference and the Laws of Probability." *Proceedings of the Society for Psychical Research* 2 (1884): 239–64.

Habermas, Jürgen. *The Structural Transformation of the Public Sphere: An Inquiry into a Category of Bourgeois Society.* Cambridge: Polity, 1989.

Hackmann, Willem D. "The Lightning Rod: A Case Study of Eighteenth-Century Model Experiments." In *Playing with Fire: Histories of the Lightning Rod,* edited by Peter Heering, Oliver Hochadel, and David J. Rhees, 209–29. Philadelphia: American Philosophical Society, 2009.

Halifax, Viscount [Charles Lindley Wood]. *"Raymond": Some Criticisms [On the Work of that Name by Sir Oliver J. Lodge].* London: Mowbray, 1917.

Hall, Edwin H. "Sir Oliver Lodge's British Association Address." *Harvard Theological Review* 8 (1915): 238–51.

Hamilton, Trevor. *Immortal Longings: FWH Myers and the Victorian Search for Life after Death.* Exeter: Imprint Academic, 2009.

Harman, P. M. *The Natural Philosophy of James Clerk Maxwell.* Cambridge: Cambridge University Press, 1998.

Hatton, Brian. "Shifted Tideways: Liverpool's Changing Fortunes." *Architectural Review,* January 2008, 39–50.

Headrick, Daniel R. *The Invisible Weapon: Telecommunications and International Politics, 1851–1945.* Oxford: Oxford University Press, 1992.

Heilbron, J. L. "Fin-de-siècle Physics." In *Science, Technology and Society in the Time of Alfred Nobel*, edited by Carl Gustaf Bernhard, Elisabeth Crawford, and Per Sörbom, 51–73. Oxford: Oxford University Press, 1982.

Heimann, P. M. "The *Unseen Universe:* Physics and the Philosophy of Nature in Victorian Britain." *British Journal for the History of Science* 6 (1972): 73–79.

Henderson, W. O. "The Liverpool Office in London." *Economica* 1 (1933): 473–79.

Hertz, Heinrich. *Memoirs, Letters, Diaries.* Edited by Johanna Hertz. Translated by Lisa Brinner, Mathilde Hertz, and Charles Susskind. 2nd ed. San Francisco, CA: San Francisco Press, 1977.

Heyck, T. W. *The Transformation of Intellectual Life in Victorian England.* London: Croom Helm, 1982.

Hill, J. Arthur. *Letters from Sir Oliver Lodge Psychical, Religious, Scientific and Personal.* London: Cassell, 1932.

H. L. [Horace Lamb]. "The Ether of Space." *Nature* 82, no. 2097 (January 6, 1910): 271.

Hobson, E. W. "Psychophysical Interaction." *Nature* 68, no. 1752 (May 28, 1903): 77.

Hobson, E. W. "Sir O. Lodge and the Conservation of Energy." *Nature* 67, no. 1748 (April 30, 1903): 611–12.

Hogg, C. F. *Spiritism in the Light of Scripture.* London: Pickering and Inglis, 1923.

Hollis, H. P., and M. T. Brück. "Clerke, Agnes Mary (1842–1907)." *Oxford Dictionary of National Biography.* Last revised September 23, 2004. https://doi.org/10.1093/ref:odnb/32444.

Hookham, Paul. *"Raymond": A Rejoinder Questioning the Validity of Certain Evidence and of Sir Oliver Lodge's Conclusions Regarding It.* Oxford: B. H. Blackwell, 1917.

Howson, A. G. *A History of Mathematics Education in England.* Cambridge: Cambridge University Press, 1982.

Hughes, Jeff. "Radioactivity and Nuclear Physics." In *The Cambridge History of Science,* vol. 5, *The Modern Physical and Mathematical Sciences,* edited by Mary Jo Nye, 350–74. Cambridge: Cambridge University Press, 2002.

Hunt, Bruce J. "Experimenting on the Ether: Oliver Lodge and the Great Whirling Machine." *Historical Studies in the Physical Sciences* 16 (1986): 111–34.

Hunt, Bruce J. "'Practice vs. Theory': The British Electrical Debate, 1888–1891." *Isis* 74 (1983): 341–55.

Hunt, Bruce J. *The Maxwellians.* Ithaca, NY: Cornell University Press, 1991.

Huxley, T. H. "Autobiography." In *Autobiography and Selected Essays,* edited by Ada Snell, 1–14. Boston: Houghton Mifflin, 1909.

Huxley, T. H. "Science and Culture." In *Science and Culture and Other Essays,* 1–23. London: Macmillan, 1882.

"Impatience." *Broadcaster,* September 1922, 86.

Inge, W. R. "Survival and Immortality." *Hibbert Journal* 15 (1917): 585–97.

Ives, E. W. *Image of a University, the Great Hall at Edgbaston: 1900–1908: An Inaugural Lecture Delivered in the University of Birmingham on 9 May 1988*. Birmingham: University of Birmingham, 1988.

Ives, Eric, Diane Drummond, and Leonard Schwarz. *The First Civic University: Birmingham, 1880–1980*. Birmingham: University of Birmingham Press, 2000.

Jacyna, L. S. "Science and Social Order in the Thought of Arthur Balfour." *Isis* 71 (1980): 11–34.

Jeans, James. "The New World-Picture of Modern Physics." *Engineer* 158 (September 7, 1934): 238–39.

Johnson, George M. *Mourning and Mysticism in First World War Literature and Beyond*. Basingstoke: Palgrave Macmillan, 2015.

Jolly, W. P. *Marconi*. London: Constable, 1972.

Jolly, W. P. *Sir Oliver Lodge*. London: Constable, 1974.

Jones, D. R. *The Origins of Civic Universities: Manchester, Leeds, and Liverpool*. London: Routledge, 1988.

Jones, Edgar, and Simon Wessely. *Shell Shock to PTSD: Military Psychiatry from 1900 to the Gulf War*. Hove, East Sussex, UK: Psychology Press, 2005.

Jordan, D. W. "The Adoption of Self-Induction by Telephony." *Annals of Science* 39, no. 5 (1982): 433–61.

Jungnickel, Christa, and Russell McCormmach. *Cavendish: The Experimental Life*. Lewisburg, PA: Bucknell University Press, 1999.

Kaalund, Nanna Katrine Lüders. "A Frosty Disagreement: John Tyndall, James David Forbes, and the Early Formation of the X-Club." *Annals of Science* 74, no. 4 (2017): 282–98.

Kant, H. "Lodge, Oliver Joseph." In *Biographical Encyclopedia of Astronomers*, edited by T. Hockey et al., 1340–41. 2nd ed. New York: Springer, 2014.

Keeble, S. P. "University Education and Business Management, 1880–1950s." PhD diss., London School of Economics, 1984.

Kipling, Rudyard. "Wireless." In *Traffics and Discoveries*, 211–39. London: Macmillan, 1904.

Kirschenbaum, Matthew G. *Mechanisms: New Media and the Forensic Imagination*. Cambridge, MA: MIT Press, 2008.

Knight, David. *Public Understanding of Science: A History of Communicating Scientific Ideas*. London: Routledge, 2006.

Knight, Donald R., and Alan D. Sabey. *The Lion Roars at Wembley: British Empire Exhibition, 60th Anniversary 1924–1925*. New Barnet: D. R. Knight, 1984.

Know, E. A., and F. B. Smith. "Moorhouse, James (1826–1915)." *Oxford Dictionary of National Biography*. Last revised September 23, 2004. https://doi.org/10.1093/ref:odnb/35093.

Kollar, Rene. *Searching for Raymond: Anglicanism, Spiritualism and Bereavement between the Two World Wars*. Lanham, MD: Lexington, 2000.

Kragh, Helge. *Quantum Generations*. Princeton, NJ: Princeton University Press, 1999.

Langton, J. "Liverpool and Its Hinterland in the Late Eighteenth Century." In *Commerce, Industry and Transport: Studies in Economic Change on Merseyside*, edited by B. L. Anderson and P. J. M. Stoney, 1–25. Liverpool: Liverpool University Press, 1983.

Lelong, Benoit. "Translating Ion Physics from Cambridge to Oxford: John Townsend and the Electrical Laboratory, 1900–24." In *Physics in Oxford, 1839–1939: Laboratories, Learning, and College Life*, edited by Robert Fox and Graeme Gooday, 209–32. Oxford: Oxford University Press, 2005.

Leonard, Gladys Osborne. *My Life in Two Worlds*. London: Cassell, 1931.

Lewis, Cecil A. *Never Look Back: An Attempt at Autobiography*. London: Hutchinson, 1974.

Lewis, Cecil A. *Sagittarius Rising*. London: Peter Davies, 1936.

Lewis, Elizabeth F. "P. G. Tait, Balfour Stewart, and *The Unseen Universe*." In *Mathematicians and their Gods: Interactions between Mathematics and Religious Beliefs*, edited by Snezana Lawrence and Mark McCartney, 213–48. Oxford: Oxford University Press, 2015.

Leys, Ruth. *Trauma: A Genealogy*. Chicago: University of Chicago Press, 2000.

Lightman, Bernard. *Victorian Popularizers of Science: Designing Nature for New Audiences*. Chicago: University of Chicago Press, 2007.

Lightman, Bernard. "Victorian Sciences and Religions: Discordant Harmonies." *Osiris* 16 (2001): 343–66.

Lodge, Oliver. "Aberration Problems." *Philosophical Transactions of the Royal Society A* 184 (1893): 727–804.

Lodge, Oliver. *Advancing Science: Being Personal Reminiscences of the British Association in the Nineteenth Century*. London: Ernest Benn, 1931.

Lodge, Oliver. "An Account of Experiments in Thought-Transference." *Proceedings of the Society for Psychical Research* 2 (1884): 189–200.

Lodge, Oliver. "Address by the President." *Proceedings of the Society for Psychical Research* 17 (1901–3): 37–57.

Lodge, Oliver. "Address." In *Report of the Sixty-First Meeting of the British Association for the Advancement of Science*, 547–57. London: John Murray, 1892.

Lodge, Oliver. *Atoms and Rays: An Introduction to Modern Views on Atomic Structure and Radiation*. London: Ernest Benn, 1924.

Lodge, Oliver. *Beyond Physics: Or, the Idealisation of Mechanism*. London: George Allen and Unwin, 1930.

Lodge, Oliver. "The Books of My Youth." *Living Age*, May 8, 1926, 330–33.

Lodge, Oliver. *Christopher: A Study in Human Personality*. London: Cassell, 1918.

Lodge, Oliver. "Clerk Maxwell and Wireless Telegraphy." In *James Clerk Maxwell: A Commemoration Volume, 1831–1931*, 125–29. New York: MacMillan, 1931.

Lodge, Oliver. "Continuity." In *Report of the Eighty-Third Meeting of the British Association for the Advancement of Science, Birmingham: 1913*, 3–42. London: John Murray, 1914.

Lodge, Oliver. *Continuity: The Presidential Address to the British Association for 1913*. London: J. M. Dent and Sons, 1913.

Lodge, Oliver. *Conviction of Survival: Two Discourses in Memory of F.W.H. Myers*. London: Methuen, 1930.

Lodge, Oliver. "The Discharge of a Leyden Jar." *Electrician* 22 (March 15, 1889): 531–34.

Lodge, Oliver. "Dr. Mann Lectures. Protection of Buildings from Lightning. Lecture I—Delivered March 10th 1888." *Journal of the Society of Arts* 36 (June 15, 1888): 867–74.

Lodge, Oliver. "Dr. Mann Lectures. Protection of Buildings from Lightning. Lecture II—Delivered March 17th 1888." *Journal of the Society of Arts* 36 (June 22, 1888): 880–93.

Lodge, Oliver. "Dr. Mann Lectures. Protection of Buildings from Lightning." *Electrician* 21 (July 6, 1888): 273–76.

Lodge, Oliver. "Dr. Mann Lectures. Protection of Buildings from Lightning." *Electrician* 21 (July 13, 1888): 302–3.

Lodge, Oliver. "Dr. Mann Lectures. Protection of Buildings from Lightning." *Electrician* 21 (June 22, 1888): 204–7.

Lodge, Oliver. "Dr. Mann Lectures. Protection of Buildings from Lightning." *Electrician* 21 (June 29, 1888): 234–36.

Lodge, Oliver. "Dust." *Nature* 31, no. 795 (January 22, 1885): 265–69.

Lodge, Oliver. *Easy Mathematics, Chiefly Arithmetic*. London: Macmillan, 1906.

Lodge, Oliver. "Eddington's Philosophy." *Nineteenth Century* 105 (1929): 360–69.

Lodge, Oliver. "Einstein's Real Achievement." *Fortnightly Review* 110 (1921): 353–72.

Lodge, Oliver. "Electrical Accumulators or Secondary Batteries." *Engineer* 53 (May 19, 1882): 365.

Lodge, Oliver. "Electrical Accumulators or Secondary Batteries." *Engineer* 53 (May 26, 1882): 373.

Lodge, Oliver. "Electrical Accumulators or Secondary Batteries." *Engineer* 53 (June 16, 1882): 439.

Lodge, Oliver. "Electrical Accumulators or Secondary Batteries." *Engineer* 53 (June 26, 1882): 457.

Lodge, Oliver. "Electrical Accumulators or Secondary Batteries." *Engineer* 54 (July 7, 1882): 11.

Lodge, Oliver. "Electrical Accumulators or Secondary Batteries." *Engineer* 54 (July 14, 1882): 30–31.

Lodge, Oliver. "Electrical Accumulators or Secondary Batteries." *Engineer* 54 (December 8, 1882): 436.

Lodge, Oliver. "Electrical Accumulators or Secondary Batteries." *Engineer* 54 (October 6, 1882): 249–50.

Lodge, Oliver. "Electrical Accumulators or Secondary Batteries." *Engineer* 54 (September 29, 1882): 230.

Lodge, Oliver. *Electrons*. London: G. Bell and Sons, 1919.

Lodge, Oliver. *Electrons; or, the Nature and Properties of Negative Electricity*. London: George Bell and Sons, 1906.

Lodge, Oliver. *Elementary Mechanics: Including Hydrostatics and Pneumatics*. London: W. and R. Chambers, 1879.

Lodge, Oliver. *Energy*. New York: Robert M. McBride, 1929.

Lodge, Oliver. "The Ether and its Functions." *Nature* 27, no. 691 (January 25, 1883): 304–6.

Lodge, Oliver. "The Ether and its Functions." *Nature* 27, no. 692 (February 1, 1883): 328–30.

Lodge, Oliver. *Ether and Reality*. London: Hodder and Stoughton, 1930.

Lodge, Oliver. *Ether and Reality: A Series of Discourses on the Many Functions of the Ether of Space*. London: Hodder and Stoughton, 1925.

Lodge, Oliver. "Ether, Matter and the Soul." *Hibbert Journal* 17 (1918–1919): 252–60.

Lodge, Oliver. "The Ether of Space." *Contemporary Review* 93 (1908): 536–46.

Lodge, Oliver. *The Ether of Space*. London: Harper and Brothers, 1909.

Lodge, Oliver. "The Ether Versus Relativity." *Fortnightly Review* 107 (1920): 54–59.

Lodge, Oliver. *Evolution and Creation*. London: Hodder and Stoughton, 1926.

Lodge, Oliver. "An Experiment in Thought-Transference." *Nature* 30, no. 763 (June 12, 1884): 145.

Lodge, Oliver. "Experiments on the Discharge of Leyden Jars." *Philosophical Transactions of the Royal Society of London* 50 (1892): 2–39.

Lodge, Oliver. "Free Will and Determinism." *Cosmopolitan Journal* 1 (1903): 23–24.

Lodge, Oliver. "Further Remarks on Relativity." *Nature* 107, no. 2703 (August 18, 1921): 784–85.

Lodge, Oliver. "The Geometrisation of Physics, and Its Supposed Basis on the Michelson-Morley Experiment." *Nature* 106, no. 2677 (February 17, 1921): 795–800.

Lodge, Oliver. "Hertz's Equations." *Nature* 39, no. 1016 (April 18, 1889): 583.

Lodge, Oliver. *The Immortality of the Soul*. Boston: Ball, 1908.

Lodge, Oliver. "The Interaction of Life and Matter." *Hibbert Journal* 29 (1931): 385–400.

Lodge, Oliver. "The Interstellar Ether." *Fortnightly Review* 53 (1893): 856–62.

Lodge, Oliver. Introduction. In *Essays on Man's Place in Nature*, by T. H. Huxley, ix–xvii. London: J. M. Dent, 1906.

Lodge, Oliver. Introduction. In *Lectures and Lay Sermons*, by T. H. Huxley, vii–xv. London: J. M. Dent, 1910.

Lodge, Oliver. *Introductory Address Delivered at the Opening of the Liverpool Royal Infirmary School of Medicine as the Medical Faculty of University College, Liverpool, on Monday, 3rd October, 1881*. London: Harrison and Sons, 1881.

Lodge, Oliver. "The Lessons of Radium." In *Modern Views of Electricity*, 3rd ed., 463–77. London: Macmillan, 1907.

Lodge, Oliver. *Life and Matter: A Criticism of Professor Haeckel's "Riddle of the Universe"*. London: Williams and Norgate, 1905.

Lodge, Oliver. "The Life-Work of My Friend F.W.H. Myers." *Nature* 144, no. 3660 (December 23, 1939): 1027–28.

Lodge, Oliver. *Lightning Conductors and Lightning Guards: A Treatise on the Protection of Buildings, of Telegraph Instruments and Submarine Cables and of Electrical Installations Generally from Damage by Atmospheric Discharges*. London: Whittaker, 1892.

Lodge, Oliver. "Lightning, Lightning Conductors, and Lightning Protectors." *Journal of the Institution of Electrical Engineers* 18 (April 25, 1889): 386–430.

Lodge, Oliver. *The Link between Matter and Matter*. London: British Science Guild, 1925.

Lodge, Oliver. *Making of Man: A Study in Evolution*. London: Hodder and Stoughton, 1924.

Lodge, Oliver. *Man and the Universe: A Study of the Influence of the Advance in Scientific Knowledge upon Our Understanding of Christianity*. London: Methuen, 1908.

Lodge, Oliver. "Mathematics and Physics." *Times* (London), April 15, 1922, 13.

Lodge, Oliver. "The Meaning of Algebraic Symbols in Applied Mathematics." *Nature* 43, no. 1118 (April 2, 1891): 513.

Lodge, Oliver. "The Meaning of Symbols in Applied Algebra." *Nature* 55, no. 1420 (January 14, 1897): 246–47.

Lodge, Oliver. *Modern Problems*. London: Methuen, 1912.

Lodge, Oliver. *Modern Scientific Ideas*. London: Ernst Benn, 1927.

Lodge, Oliver. "The Modern Theory of Light." In *Annual Report of the Board of Regents of the Smithsonian Institution, 1889*, 441–47. Washington, DC: Government Printing Office, 1890.

Lodge, Oliver. *Modern Views of Electricity*. London: Macmillan, 1889.

Lodge, Oliver. *Modern Views of Electricity*. 2nd ed. London: Macmillan, 1892.

Lodge, Oliver. *Modern Views of Electricity*. 3rd ed. London: Macmillan, 1907.

Lodge, Oliver. "Modern Views of Electricity." *Nature* 36, no. 936 (October 6, 1887): 532–36.

Lodge, Oliver. "Modern Views of Electricity." *Nature* 36, no. 937 (October 13, 1887): 559–61.

Lodge, Oliver. "Modern Views of Electricity." *Nature* 41, no. 1048 (November 28, 1889): 80.

Lodge, Oliver. *Modern Views on Matter: The Romanes Lecture 1903*. Oxford: Clarendon, 1903.

Lodge, Oliver. *My Philosophy: Representing My Views on the Many Functions of the Ether of Space*. London: Ernest Benn, 1933.

Lodge, Oliver. "The New Theory of Gravity." *Nineteenth Century and After* 86 (1919): 1189–1201.

Lodge, Oliver. "The Nineteenth Kelvin Lecture. 'The Revolution in Physics.'" *Journal of the Institution of Electrical Engineers* 66 (1928): 1005–20.

Lodge, Oliver. "On Some Problems Connected with the Flow of Electricity in a Plane." *Philosophical Magazine*, 5th series, 1 (1876): 373–89.

Lodge, Oliver. "On the Question of Absolute Velocity, and on the Mechanical Functions of an Æther, with Some Remarks on the Pressure of Radiation." *Philosophical Magazine*, 5th series, 46 (1898): 414–26.

Lodge, Oliver. "On the Theory of Lightning-Conductors." *Philosophical Magazine*, 5th series, 26 (1888): 217–30.

Lodge, Oliver. "The Outstanding Controversy between Science and Faith." *Hibbert Journal* 1 (1902–3): 32–61.

Lodge, Oliver. *Past Years: An Autobiography*. London: Hodder and Stoughton, 1931.

Lodge, Oliver. *Phantom Walls*. London: Hodder and Stoughton, 1929.

Lodge, Oliver. *Pioneers of Science*. London: Macmillan, 1893.

Lodge, Oliver. "Popular Relativity and the Velocity of Light." *Nature* 106, no. 2662 (November 4, 1920): 325–26.

Lodge, Oliver. "The Progress of Physics." *Nature* 87, no. 2186 (September 21, 1911): 375–77.

Lodge, Oliver. "Psychophysical Interaction." *Nature* 68, no. 1750 (May 14, 1903): 33.

Lodge, Oliver. *Public Service versus Private Expenditure*. London: Fabian Society, 1905. Reprinted in Sidney Webb, Bernard Shaw, Sidney Ball, and Oliver Lodge. *Socialism and Individualism*. London: A.C. Fitfield, 1908.

Lodge, Oliver. "Radium and Its Lessons." *Nineteenth Century and After* 54 (1903): 78–85.

Lodge, Oliver. *Raymond Revised: A New and Abbreviated Edition of "Raymond Or Life and Death"*. London: Methuen, 1922.

Lodge, Oliver. *Raymond: Or Life and Death*. London: Methuen, 1916.

Lodge, Oliver. *The Reality of the Spiritual World*. London: E. Benn, 1930.

Lodge, Oliver. "A Record of Observations of Certain Phenomena of Trance." *Proceedings of the Society for Psychical Research* 6 (1889–90): 443–649.

Lodge, Oliver. "The Relation between Electricity and Light." *Nature* 23, no. 587 (January 27, 1881): 302–4.

Lodge, Oliver. *Relativity: A Very Elementary Exposition*. London: Methuen, 1925.

Lodge, Oliver. *Relativity: A Very Elementary Exposition*. 3rd ed. New York: George Doran, 1926.

Lodge, Oliver. "Remarks on Gravitational Relativity." *Nature* 107, no. 2704 (August 25, 1921): 814–18.

Lodge, Oliver. "Remarks on Simple Relativity and the Relative Velocity of Light." *Nature* 107, no. 2701 (August 4, 1921): 716–19.

Lodge, Oliver. "Remarks on Simple Relativity and the Relative Velocity of Light." *Nature* 107, no. 2702 (August 11, 1921): 748–51.

Lodge, Oliver. "A Scheme of Vital Faculty." *Nature* 68, no. 1755 (June 18, 1903): 145–47.

Lodge, Oliver. *School Teaching and School Reform*. London: Williams and Norgate, 1905.

Lodge, Oliver. *Science and Human Progress*. London: George Allen and Unwin, 1927.

Lodge, Oliver. "Scope and Tendencies of Physics." In *The 19th Century: A Review of Progress*, by A. G. Sedgwick et al., 348–57. London: G. P. Putnam's, 1901.

Lodge, Oliver. *Signalling across Space without Wires: Being a Description of the Work of Hertz & His Successors*. London: The Electrician Printing and Publishing Company, 1898.

Lodge, Oliver. "Some Elementary Considerations Connected with Modern Physics." *Philosophical Magazine*, 7th series, 15 (1933): 706–26.

Lodge, Oliver. "Steps toward a New Principia." *Nature* 70, no. 1804 (May 26, 1904): 73–76.

Lodge, Oliver. *The Substance of Faith*. London: Methuen, 1907.

Lodge, Oliver. "Supplement to the Discussion on Mr. Balfour's Paper." In *Papers Read before the Synthetic Society, 1896–1908*, 334–40. [London]: Spottiswoode, 1909.

Lodge, Oliver. *The Survival of Man: A Study in Unrecognised Human Faculty*. London: Methuen, 1909.

Lodge, Oliver. "The Survival of Personality." *Quarterly Review* 198 (1903): 211–29.

Lodge, Oliver. "Thoughts of the Bifurcation of the Sciences." *Nature* 48, no. 1250 (October 12, 1893): 564–66.

Lodge, Oliver. *University Development, and Survey of the Sciences: Appendix to University Development: Consideration of the Relation of the University of Birmingham to its Central and Suburban Sites by the Principal*. Privately printed pamphlet, 1902.

Lodge, Oliver. Untitled essay. In *Papers Read before the Synthetic Society, 1896–1908*, 385–92. [London]: Spottiswoode, 1909.

Lodge, Oliver. "Use and Abuse of Empirical Formulae, and of Differentiation, by Chemists." *Nature* 40, no. 1029 (July 18, 1889): 273.

Lodge, Oliver. *The Work of Hertz and Some of His Successors*. London: Electrician Printing and Publishing, 1894.

Lodge, Oliver. "The Work of Hertz." *Nature* 50, no. 1284 (June 7, 1894): 133–39.

Lodge, Oliver. "The Work of Hertz." *Nature* 50, no. 1285 (June 14, 1894): 160–61.

Loftus, Donna. "The Self in Society: Middle-Class Men and Autobiography." In *Life Writing and Victorian Culture*, edited by David Amigoni, 67–85. Aldershot: Ashgate, 2006.

Longridge, George. *Spiritualism and Christianity*. London: Mowbray, 1919.

"Lord Kelvin's Philosophy." *Nature* 78, no. 2018 (July 2, 1908): 198–99.

Loughran, Tracey. "Shell Shock, Trauma, and the First World War: The Making of a Diagnosis and Its Histories." *Journal of the History of Medicine and Allied Sciences* 67, no. 1 (2010): 94–119.

Lubenow, William C. "Intimacy, Imagination and the Inner Dialectics of Knowledge Communities: The Synthetic Society, 1896–1908." In *The Organisation of Knowledge in Victorian Britain*, edited by Martin J. Daunton, 357–70. Oxford: Oxford University Press, 2005.

Luckhurst, Roger. *The Invention of Telepathy, 1870–1901*. Oxford: Oxford University Press, 2002.

Machen, Arthur. "The Bowmen." In *The White People and Other Stories*, edited by S. T. Joshi, 223–26. London: Penguin, 2011.

Mallinson, P. Russell. "When Wireless Dreams Come True." *Broadcaster*, August 1922, 15–17.

Mandell, Laura. *Breaking the Book: Print Humanities in the Digital Age*. Malden, MA: Wiley-Blackwell, 2015.

[Mann, R. J.]. *The Atlantic Telegraph: A History of Preliminary Experimental Proceedings*. London: Jarrold and Sons, 1857.

Mallock, W. H. "Sir Oliver Lodge on Life and Matter." *Fortnightly Review* 80 (1906): 33–47.

Marsh, Peter T. *Joseph Chamberlain: Entrepreneur in Politics*. New Haven, CT: Yale University Press, 1994.

Marsh, Peter T. "Chamberlain, Joseph [Joe] (1836–1914)." *Oxford Dictionary of National Biography*. https://doi.org/10.1093/ref:odnb/32350.

Martin, Alfred. *Psychic Tendencies of To-Day: An Exposition and Critique of New Thought, Christian Science, Psychical Research (Oliver Lodge) and Modern Materialism in Relation to Immortality*. London: D. Appleton, 1918.

Mason College Calendar 1880–1881. Birmingham: Mason College, 1881.

Mason, Madeline. "A Noble Life." *Saturday Review of Literature*, April 30, 1932, 696.

Mass Observation. *Puzzled People: A Study in Popular Attitudes to Religion, Ethics, Progress and Politics in a London Borough*. London: Victor Golancz, 1947.

Matthews, W. R. *God in Christian Thought and Experience*. London: Nisbet, 1930.

Matthews, W. R. *Memories and Meanings*. London: Hodder and Stoughton, 1969.

Maurice, Frederick. *The Life of Frederick Denison Maurice Chiefly Told in His Own Letters*. 2 vols. 2nd ed. London: Macmillan, 1884.

Maurice, Frederick. *Theological Essays*. 1853. London: James Clarke, 1957.

Maxwell, James Clerk. "A Dynamical Theory of the Electromagnetic Field." *Philosophical Transactions of the Royal Society* 155 (1865): 459–512.

Maxwell, James Clerk. "Ether." In *The Scientific Papers of James Clerk Maxwell*, edited by W. D. Niven, 2:762–75. Cambridge: Cambridge University Press, 1890.

Maxwell, James Clerk. "On the Physical Lines of Force." *Philosophical Magazine*, 4th series, 21 (1861): 161–75.

Maxwell, James Clerk. "On the Physical Lines of Force." *Philosophical Magazine*, 4th series, 21 (1861): 281–91.

Maxwell, James Clerk. "On the Physical Lines of Force." *Philosophical Magazine*, 4th series, 21 (1861): 338–48.

Maxwell, James Clerk. "On the Physical Lines of Force." *Philosophical Magazine*, 4th series, 23 (1862): 12–24.

Maxwell, James Clerk. "On the Physical Lines of Force." *Philosophical Magazine*, 4th series, 23 (1862): 85–95.

Maxwell, James Clerk. "Paradoxical Philosophy." In *Scientific Papers of James Clerk Maxwell*, edited by W. D. Niven, 2:756–62. Cambridge: Cambridge University Press, 1890.

McCabe, Joseph. *The Religion of Sir Oliver Lodge*. London: Watts, 1914.

McDougall, William. "Psychophysical Interaction." *Nature* 68, no. 1750 (1903): 32–33.

McLennan, Evan. "Force and Determinism." *Nature* 44, no. 1131 (July 2, 1891): 198.

Menke, Richard. *Telegraphic Realism: Victorian Fiction and Other Information Systems*. Stanford, CA: Stanford University Press, 2008.

Mercier, Charles. *Spiritualism and Sir Oliver Lodge*. London: Watts, 1917.

Minchin, George M., "The Glorification of Energy," *Nature* 68, no. 1750 (May 14, 1903), 31–32.

Mitchell, Peter Chalmers. *My Fill of Days*. London: Faber and Faber, 1937.

[Peter Chalmers Mitchell]. "The Progress of Science. Transmutation of Metals. 'Synthetic Gold.'" *Times* (London), December 20, 1921, 8.

[Peter Chalmers Mitchell]. "The Progess of Science. Low-Temperature Research. Cryogenic Laboratories." *Times* (London), August 7, 1923, 8.

[Peter Chalmers Mitchell]. "The Progress of Science. Ultimate Facts of the Universe. The Quantum Theory." *Times* (London), October 27, 1924, 19.

[Peter Chalmers Mitchell]. "The Progress of Science. Atoms and their Nuclei. Hydrogen as Primitive Matter." *Times* (London), March 30, 1925, 9.

[Peter Chalmers Mitchell]. "The Progress of Science. Atomic Systems. Disintegration of Matter." *Times* (London), April 27, 1925, 7.

[Peter Chalmers Mitchell]. "Progress of Physical Science. Sir Oliver Lodge on Television." *Times* (London), March 15, 1927, 14."Modern Physics and the Engineer." *Engineer* 154 (November 11, 1932): 485–86.

Moktefi, Amirouche. "Geometry: The Euclid Debate." In *Mathematics in Victorian Britain*, edited by Raymond Flood, Adrian Rice, and Robin Wilson, 321–58. Oxford: Oxford University Press, 2011.

Moore, C. B., G. D. Aulich, and William Rison. "A Modern Assessment of Benjamin Franklin's Lightning Rods." In *Playing with Fire: Histories of the Lightning Rod*, edited by Peter Heering, Oliver Hochadel, and David J. Rhees, 256–68. Philadelphia: American Philosophical Society, 2009.

Morgan, Conwy Lloyd. "Force and Determinism." *Nature* 43, no. 1120 (April 16, 1891): 558.

Morgan, Conwy Lloyd. "Force and Determinism." *Nature* 44, no. 1136 (August 6, 1891): 319.

Morton, Alan Q. "The Electron Made Public: The Exhibition of Pure Science in the British Empire Exhibition, 1924–5." In *Exposing Electronics*, edited by Bernard Finn, Robert Bud and Helmuth Trischler, 25–44. Amsterdam: Harwood Academic, 2000.

Mott, Frederick. *The Effects of High Explosives upon the Central Nervous System*. London: Harrison and Sons, 1916.

"Mr Chamberlain on a University of Birmingham." *Times* (London), November 19, 1898, 10.

"My Philosophy." *Psychic Science: Quarterly Transactions of the British College of Psychic Science* 12 (1933–34): 187–201.

Myers, Charles. "A Contribution to the Study of Shell Shock: Being an Account of Three Cases of Memory, Vision, Smell, and Taste, Admitted into the Duchess of Westminster's War Hospital Le Touquet." *Lancet* 185 (February 13, 1915): 316–20.

Myers, Frederic. "The Drift of Psychical Research." *National Review* 24 (1894): 190–209.

Myers, Frederic. *Human Personality and its Survival of Bodily Death*. 2 vols. London: Longmans, Green, 1903.

Myers, Frederic. "The Subliminal Consciousness." *Proceedings of the Society for Psychical Research* 7 (1892): 298–354.

Myers, Frederic. "The Subliminal Consciousness," *Proceedings of the Society for Psychical Research* 8 (1892): 436–535.

Navarro, Jaume. "Ether and Wireless: An Old Medium into New Media." *Historical Studies in the Natural Sciences* 46, no. 4 (2016): 460–89.

Navarro, Jaume. *A History of the Electron: J. J. and G. P. Thomson*. Cambridge: Cambridge University Press, 2012.

"Necromancy." *Church Times* 53 (February 18, 1910): 225.

"News and Views." *Nature* 119, no. 3005 (June 4, 1927): 827.

Noakes, Richard. "Ethers, Religion and Politics in Late-Victorian Physics: Beyond the Wynne Thesis." *History of Science* 43, no. 4 (2005): 1–41.

Noakes, Richard. "Haunted Thoughts of the Careful Experimentalist: Psychical Research and the Troubles of Experimental Physics." *Studies in the History and Philosophy of the Biological and Biomedical Sciences* 48 (2014): 46–56.

Noakes, Richard. "Making Space for the Soul: Oliver Lodge, Maxwellian Psychics and the Etherial Body." In *Ether and Modernity: The Recalcitrance of an Epistemic Object in the Early Twentieth Century,* edited by Jaume Navarro, 88–106. Oxford: Oxford University Press, 2018.

Noakes, Richard. "Thoughts and Spirits by Wireless: Imagining and Building Psychic Telegraphs in America and Britain, circa 1900–1930." *History and Technology* 32, no. 2 (2016): 137–58.

"Notes of the Month." *Wireless World,* April 1917, 40–41.

Official Report of the Church Congress Held in Leicester. London: Nisbet, 1919.

O'Hara, J. G., and W. Pricha. *Hertz and the Maxwellians*. London: Peter Peregrinus, 1987.

O.J.L. [Oliver Lodge]. "John Tyndall." In *Encyclopaedia Britannica,* 9th/10th ed., vol. 33, 517–21. Edinburgh: A. and C. Black, 1902–3.

"The Old and the New Physics." *Engineer* 158 (September 7, 1934): 237.

Oppenheim, Janet. *The Other World: Spiritualism and Psychical Research in England, 1850–1914*. Cambridge: Cambridge University Press, 1985.

"Other People's Troubles." *Broadcaster,* August 1922, 64.

Ottley, R. L. "Christian Ethics." *Lux Mundi: A Series of Studies in the Religion of the Incarnation,* edited by Charles Gore. 1889; London: John Murray, 1890.

"Our Object." *Yorkshire Spiritual Telegraph,* May 1855, 19–20.

Overy, Richard. *The Morbid Age: Britain between the Wars*. London: Allen Lane, 2009.

Owen, George Vale. *Life beyond the Veil*. 2 vols. 1922. London: Thornton Butterworth, 1926.

Owen, George Vale. *On Tour in USA*. London: Hutchinson, 1924.

Parker, K. R. "The History of Lodge Cottrell Limited: Development of Electrostatic Precipitation in the United Kingdom." Unpublished typescript, n.d. [1989?].

Paton, W. D. M., and C. G. Phillips. "E. H. J. Schuster (1879–1969)." *Notes and Records of the Royal Society of London* 28 (1973): 111–17.

Pearson, Karl. *The Life, Letters and Labours of Francis Galton*. Vol. 2, *Researches of Middle Life*. Cambridge: Cambridge University Press, 1924.

Peters, John Durham. *Speaking into the Air: A History of the Idea of Communication*. Chicago: University of Chicago Press, 1999.

Peterson, Linda H. *Victorian Autobiography: The Tradition of Self-Interpretation*. New Haven, CT: Yale University Press, 1986.

Phases of Modern Science: Published in Connexion with the Science Exhibit Arranged by a Committee of the Royal Society in the Pavilion of His Majesty's Government at the British Empire Exhibition, 1925. London: Royal Society, 1925.

"Philosophy of Death." *Spiritualist*, November 19, 1869, 1.

"Philosophy of Sir Oliver Lodge." *Derby Evening Telegraph*, June 22, 1933, 4.

Pocock, Rowland F. *The Early British Radio Industry*. Manchester: Manchester University Press, 1988.

Preece, W. H. "Presidential Address to Section G." In *Report of the Fifty-Eighth Meeting of the British Association for the Advancement of Science*, 781–92. London: John Murray, 1889.

Price, Katy. *Loving Faster than Light: Romance and Readers in Einstein's Universe*. London: University of Chicago Press, 2012.

Price, Leah. *How to Do Things with Books in Victorian Britain*. Princeton, NJ: Princeton University Press, 2012.

Price, Michael. "Mathematics in English Education, 1860–1914: Some Questions and Explanations in Curriculum History." *History of Education* 12, no. 4 (1983): 271–84.

"Professor Lodge's Theology." *Church Times* 50 (December 4, 1908): 767.

Pullin, V. A. "Benn's Sixpenny Library: First Scientific Titles." *Discovery* 9 (1928): 163–65.

Pumfrey, Stephen. *Latitude and the Magnetic Earth*. Cambridge: Icon Books, 2001.

Raia, Courtenay. "From Ether Theory to Ether Theology: Oliver Lodge and the Physics of Immortality." *Journal of the History of the Behavioral Sciences* 43 (2007): 19–43.

Rashdall, Hastings. *The Idea of Atonement in Christian Theology: Being the Bampton Lectures for 1915*. London: Macmillan, 1920.

Rayleigh, Lord [John William Strutt]. "Obituary: Sir Oliver Lodge F.R.S. Sir J. J. Thomson, O.M., F.R.S." *Proceedings of the Society for Psychical Research* 46 (1940–41): 209–18.

Reid, Fiona. *Broken Men: Shell Shock, Treatment and Recovery in Britain, 1914–30*. London: Continuum, 2010.

Reith, John C. W. *Broadcast over Britain*. London: Hodder and Stoughton, 1924.

"The Response to the Appeal." *Borderland* 1 (1893): 10–23.

Rice, Adrian. "Henrici, Olaus Magnus Friedrich Erdmann (1840–1918)." *Oxford Dic-*

tionary of National Biography. Last revised September 23, 2004. https://doi.org/10.1093/ref:odnb/39487.

Richardson, Owen Willans. *The Electron Theory of Matter*. Cambridge: Cambridge University Press, 1914.

Richardson, Owen Willans. "The Structure of the Ether." *Nature* 76, no. 1960 (May 23, 1907): 78.

Ricketts, Harry. *The Unforgiving Minute: A Life of Rudyard Kipling*. London: Chatto and Windus, 1999.

Risdon, P. J. *Wireless*. 2nd ed. London: Ward, Lock, 1924[?].

Roberts, J. H. T. "Sir Oliver Lodge's New Book." *Popular Wireless*, April 1927, 440–41.

Roberts, R. G. "The Training of an Industrial Physicist. Oliver Lodge and Benjamin Davies. 1882–1940." PhD diss., University of Manchester Institute of Science and Technology, 1984.

Root, J. D. "The Philosophical and Religious Thought of Arthur James Balfour (1848–1930)." *Journal of British Studies* 19, no. 2 (1980): 120–41.

Root, J. D. "Science, Religion, and Psychical Research: The Monistic Thought of Sir Oliver Lodge." *Harvard Theological Review* 71, no. 3/4 (1978): 245–63.

Rothblatt, Sheldon. *The Revolution of the Dons: Cambridge and Society in Victorian England*. London: Faber and Faber, 1968.

Rowell, Geoffrey. *Hell and the Victorians: A Study of the Nineteenth-Century Theological Controversies Concerning Eternal Punishment and the Future Life*. 1974. Oxford: Clarendon, 2000.

Rowlands, Peter. *Oliver Lodge and the Liverpool Physical Society*. Liverpool: Liverpool University Press, 1990.

Rowlands, Peter. "Radio Begins in 1894." In *Oliver Lodge and the Invention of Radio*, edited by Peter Rowlands and J. Patrick Wilson, 75–114. Liverpool: PD Publications, 1994.

Rowlands, Peter. "Radiowaves." In *Oliver Lodge and the Invention of Radio*, edited by Peter Rowlands and J. Patrick Wilson, 39–66. Liverpool: PD Publications, 1994.

Ruskin, John. Extract from *Fors Clavigera*. In *Theory of the Glaciers of Savoy*, by M. Le Chanoine Rendu, edited by George Forbes, translated by Alfred Wills, 199–210. London: Macmillan, 1874.

Ruskin, John. "La Douce Dame." In *Fors Clavigera: Letters to the Workmen and Labourers of Great Britain*, vol. 1, 624–43. London: George Allen, 1907. Reprinted in *Library Edition of John Ruskin*, edited by E. T. Cook and Alexander Wedderburn, vol. 27. London: George Allen, 1907.

Ruskin, John. *Sesame and Lilies, the Two Paths, and the King of the Golden River*. London: J. M. Dent, 1907.

Saturday Review. Unsigned review of *Raymond, or Life and Death,* by Oliver Lodge. February 3, 1917, 110–11.

Schaffer, Simon. "Fish and Ships: Models in the Age of Reason." In *Models: The Third Dimension of Science,* edited by Soraya de Chadarevian and Nick Hopwood, 71–105. Stanford, CA: Stanford University Press, 2004.

Schuettpelz, Erhard, and Ehler Voss. "Fragile Balance: Human Mediums and Technical Media in Oliver Lodge's Presidential Address of 1891." *Communication+ 1* 4 (2015): 1–15.

"Science." *Athenaeum,* no. 4321 (November 28, 1908): 686–87.

Sconce, Jeffrey. *Haunted Media: Electronic Presence from Telegraphy to Television.* Durham, NC: Duke University Press, 2000.

Sconce, Jeffrey. "The Voice from the Void: Wireless, Modernity, and the Distant Dead." *International Journal of Cultural Studies* 1, no. 2 (1998): 211–32.

Sheppard, W. J. L. *Messages from the Dead.* Stirling: Drummond's Tract Depot, 1926.

Sickert, Bernard. "Spiritualism and Its New Revelations II." *English Review* 27 (1918): 339–45.

Sieveking, Lance. *The Stuff of Radio.* London: Cassell, 1934.

"Signalling throughout the Ages: From Aeschylus to Marconi." *Wireless World,* April 1913, 28–32.

"Sir J.J. Thomson, O.M. The Discover of the Electron." *Times* (London), August 31, 1940, 7.

"Sir J.J. Thomson: the King's Message of Sympathy." *Times* (London), September 3, 1940, 7.

"Sir Oliver Joseph Lodge D.Sc. Sc.D. LL.D. F.R.S. (1940) XXXII. Obituary." *Philosophical Magazine,* 7th series, 30 (1940): 341–43.

"Sir Oliver Lodge." *Times* (London), August 23, 1940, 4.

"Sir Oliver Lodge: A Great Scientist." *Times* (London), August 23, 1940, 7."Sir Oliver Lodge. A Great Scientific Teacher." *Electrical Review* 127 (1940): 169.

"Sir Oliver and Ether." *Spectator* 150 (June 29, 1933): 951.

"Sir Oliver Lodge on the Possibilities of the Human Spirit." *Light,* April 16, 1927, 182–85.

Smith, B. M. D. *Business Education in the University of Birmingham, 1899–1965.* Birmingham: Birmingham University Business School, 1990.

Smith, Crosbie. *The Science of Energy: A Cultural History of Energy Physics in Victorian Britain.* London: Athlone, 1998.

Smith, Crosbie, and M. Norton Wise. *Energy and Empire: A Biographical Study of Lord Kelvin.* Cambridge: Cambridge University Press, 1989.

Smith, George E., "J. J. Thomson and the Electron, 1897–1899." In *Histories of the Electron,* edited by Jed Z. Buchwald and Andrew Warwick, 21–76. Cambridge, MA: MIT Press, 2001.

Smith, Roger J. *Free Will and the Human Sciences in Britain, 1870–1910*. London: Pickering and Chatto, 2013.

"Space, Matter, Mind and God." *Church Times* 103 (June 13, 1930): 759–60.

Sponsel, Alistair. "Constructing a 'Revolution in Science': The Campaign to Promote a Favourable Reception for the 1919 Solar Eclipse Experiments." *British Journal of the History of Science* 35 (2002): 439–67.

Staley, Richard. *Einstein's Generation: The Origins of the Relativity Revolution*. Chicago: University of Chicago Press, 2008.

Staley, Richard. "On the Co-creation of Classical and Modern Physics." *Isis* 96, no. 4 (2005): 530–58.

Staley, Richard. "On the Histories of Relativity: The Propagation and Elaboration of Relativity Theory in Participant Histories in Germany, 1905–1911." *Isis* 89, no. 2 (1998): 263–99.

Stanley, Matthew. "'An expedition to heal the wounds of war': The 1919 Eclipse and Eddington as Quaker Adventurer." *Isis* 94, no. 1 (2003): 57–89.

Stanley, Matthew. *Huxley's Church and Maxwell's Demon: From Theistic Science to Naturalistic Science*. Chicago: University of Chicago Press, 2015.

[Stewart, Balfour, and Peter Guthrie Tait]. *The Unseen Universe; or, Physical Speculations on a Future State*. London: Macmillan, 1875.

Stewart, Susan. *On Longing: Narratives of the Miniature, the Gigantic, the Souvenir, the Collection*. Durham, NC: Duke University Press, 1993.

Stewart, Victoria. "War Memoirs of the Dead: Writing and Remembrance in the First World War." *Literature and History* 14, no. 2 (2005): 37–52.

Stokes, Donald E. *Pasteur's Quadrant: Basic Science and Technological Innovation*. Washington, DC: Brookings Institution Press, 1997.

Storr, Vernon. *Do Dead Men Live Again?* London: Hodder and Stoughton, 1932.

Stranger, Ralph. *The Outline of Wireless for the Man on the Street*. London: George Newnes, 1932.

Streeter, Burnett H., et al. *Immortality: An Essay in Discovery, Co-ordinating Scientific, Psychical and Biblical Research*. London: Macmillan, 1917.

Sulley, Henry. *What is the Substance of Faith? A Reply to Sir Oliver Lodge*. London: Simpkin, Marshall, Hamilton, Kent, 1909.

[Sullivan, J. W. N.]. "Beyond Physics." *Times Literary Supplement*, June 19, 1930, 504.

[Sullivan, J. W. N.]. "Ether and Reality." *Times Literary Supplement*, May 14, 1925, 325.

Symons, G. J., ed. *Report of the Lightning Rod Conference*. London: E and F. N. Spon, 1882.

Talbot, E. S. "Sir Oliver Lodge on 'The Reinterpretation of Christian Doctrine.'" *Hibbert Journal* 2 (1903–4): 649–61.

Taylor, Charles. *Sources of the Self: The Making of the Modern Identity*. Cambridge: Cambridge University Press, 1992.

"The Testimony of a Spirit." *Spiritualist*, December 31, 1869, 25.

"The Teaching of Physics in Schools." *Proceedings of the Physical Society of London* 30 (1918): 1–43S.

[Thomas, Ivor]. "Sir O. Lodge's Philosophy." *Times Literary Supplement*, June 22, 1933, 421.

Thomson, J. J. "Electronic Waves." *Philosophical Magazine*, 7th series, 27 (1939): 1–32.

Thomson, William. "On Vortex Atoms." *Proceedings of the Royal Society of Edinburgh* 6 (1867): 94–105.

"Those Who Know." In *The Hidden Side of the War: Some Revelations and Prophecies*. London: Elliot Stock, 1918.

Trotter, David. *Literature in the First Media Age: Britain between the Wars*. Cambridge, MA: Harvard University Press, 2014.

Tucker, Prentiss. *In the Land of the Living Dead: An Occult Story*. Oceanside, CA: Rosicrucian Fellowship, 1921.

Tuckett, Ivor. "Psychical Researchers and 'The Will to Believe.'" *Bedrock* 1 (1912–13): 180–204.

Tudor-Pole, W. *Private Dowding: A Plain Record of After Death Experiences of a Soldier Killed in Battle*. London: John M. Watkins, 1917.

Turner, Frank Miller. *Between Science and Religion: The Reaction to Scientific Naturalism in Late Victorian England*. New Haven, CT: Yale University Press, 1974.

Turner, Frank M. *Contesting Cultural Authority: Essays in Victorian Intellectual Life*. Cambridge: Cambridge University Press, 1993.

Turkel, William J. *Spark from the Deep: How Shocking Experiments with Strongly Electric Fish Powered Scientific Discovery*. Baltimore: Johns Hopkins University Press, 2013.

Tyndall, John. *Address Delivered before the British Association Assembled at Belfast*. London: Longman, Green, 1874.

Tyndall, John. "The Belfast Address." In *Fragments of Science: A Series of Detached Essays, Addresses, and Reviews*, 2:135–201. London: Longmans, Green, 1889.

Tyndall, John. *Faraday as a Discoverer*. London: Longman, Green, 1868.

Tyndall, John. "Prayer as a Form of Natural Law." First published 1861. Reprinted in *Fragments of Science: A Series of Detached Essays, Addresses, and Reviews*, 2:1–7. London: Longmans, Green, 1889.

Tyndall, L. C. "Tyndall, John (1820–1893)." *Oxford Dictionary of National Biography* archive. Published in print 1898. https://doi.org/10.1093/odnb/9780192683120.013.27948.

Vincent, David. *Bread, Knowledge and Freedom: A Study of Nineteenth-Century Working-Class Autobiography*. London: Methuen, 1981.

Vincent, Eric W., and P. Hinton. *The University of Birmingham, Its History and Significance*. Birmingham: Cornish Brothers, 1947.

"Vox Dei: Religion and Radio." *Vox: The Radio Critic and Broadcast Review*, November 9, 1929, 9.

Walker, D. P. *The Decline of Hell: Seventeenth Century Discussions of Eternal Torment*. London: Routledge and Kegan Paul, 1964.

Walker, J. Malcolm. "Mann, Robert James (1817–1886)." *Oxford Dictionary of National Biography*. Last revised September 23, 2004. https://doi.org/10.1093/ref:odnb/17947.

Wallace, Alfred Russel. *My Life: A Record of Events and Opinions*. London: Chapman and Hall, 1908.

"Wanted." *Broadcaster*, September 1922, 78.

Ward, J. S. M. *Gone West: Three Narratives of After-Death Experiences*. London: William Rider and Son, 1917.

Ward, J.S.M. *A Subaltern in Spirit Land: A Sequel to "Gone West"*. London: William Rider and Son, 1920.

Warwick, Andrew. "Cambridge Mathematics and Cavendish Physics: Cunningham, Campbell and Einstein's Relativity 1905–1911. Part I: The Uses of Theory." *Studies in History and Philosophy of Science* 23, no. 4 (1992): 625–56.

Warwick, Andrew. "Cambridge Mathematics and Cavendish Physics: Cunningham, Campbell and Einstein's Relativity 1905–1911. Part II: Comparing Traditions in Cambridge Physics." *Studies in History and Philosophy of Science* 24, no. 1 (1993): 1–25.

Warwick, Andrew *Masters of Theory: Cambridge and the Rise of Mathematical Physics*. Chicago: University of Chicago Press, 2003.

Weatherhead, Leslie. *After Death: A Popular Statement of the Modern Christian View of Life Beyond the Grave*. London: James Clarke, 1923.

Wedgwood, James Ingall. *Spiritualism and the Great War*. London: Theosophical Publishing House, 1919.

Welldon, James. "The Nature of Immortality." In *Life after Death According to Christianity and Spiritualism*, edited by James Marchant, 3–73. London: Cassell, 1925.

Whittaker, E. T. *A History of the Theories of Aether and Electricity, from the Age of Descartes to the Close of the Nineteenth Century*. London: Longmans, Green, 1910.

Whitworth, Michael. "The Clothbound Universe: Popular Physics Books, 1919–39." *Publishing History* 40 (1996): 55–82.

Whitworth, Michael. *Einstein's Wake: Relativity, Metaphor, and Modernist Literature*. Oxford: Oxford University Press, 2001.

Williams, Raymond. *Culture and Society, 1780–1950*. 2nd ed. New York: Columbia University Press, 1987.

Wilson, David B. *Kelvin and Stokes: A Comparative Study in Victorian Physics*. Bristol: Adam Hilger, 1987.

Wilson, David B. "The Thought of Late Victorian Physicists: Oliver Lodge's Ethereal Body." *Victorian Studies* 15, no. 1 (September 1971): 29–48.

Wilson, J. Patrick. "The Technological Heritage of Oliver Lodge." In *Oliver Lodge and the Invention of Radio*, edited by Peter Rowlands and J. Patrick Wilson, 173–92. Liverpool: PD Publications, 1994.

Windscheffel, Ruth Clayton. "Politics, Religion and Text: W. E. Gladstone and Spiritualism." *Journal of Victorian Culture* 11, no. 1 (2006): 1–29.

Winnington-Ingram, A. F. *The Spirit of Peace*. London: Gardner, Darton, 1921.

Winnington-Ingram, A. F. Introduction. In *Life after Death According to Christianity and Spiritualism*, edited by James Marchant, vii–xiii. London: Cassell, 1925.

Winseck, Dwayne, and Robert Pike. "The Global Media and the Empire of Liberal Internationalism, circa 1910–1930." *Media History* 15, no. 1 (2009): 31–54.

Winter, Jay. *Sites of Memory, Sites of Mourning: The Great War in European Cultural History*. Cambridge: Cambridge University Press, 1995.

"Wireless at Sea." *Broadcaster*, August 1922, 32.

"Wireless Telegraphy in the War." *Wireless World*, April 1917, 24–27.

"Wireless Telegraphy in the War." *Wireless World*, July 1917, 243–47.

Wood, Charles Lindley. *"Raymond": Some Criticisms [On the Work of that Name by Sir Oliver J. Lodge]*. London: A. R. Mowbray, 1917.

Wordsworth, William. *Wordsworth: Poetical Works*. Edited by Tomas Hutchinson. Revised by Ernest de Selincourt. Oxford: Oxford University Press, 1969.

Yavetz, Ido. "A Victorian Thunderstorm: Lightning Protection and Technological Pessimism in the Nineteenth Century." In *Technology, Pessimism, and Postmodernism*, edited by Yaron Ezrahi, Everett Mendelsohn, and Howard P. Segal, 53–75. Amherst: University of Massachusetts Press, 1995.

CONTRIBUTORS

DAVID AMIGONI is professor of Victorian literature at Keele University. He is the author of *Colonies, Cults and Evolution* (2007), and more recently "Writing the Scientist: Biography and Autobiography," in *The Routledge Research Companion to Nineteenth-Century British Literature and Science* (2017). He is writing a book on theories of biological and cultural inheritance in writings across three generations of the Darwin, Huxley, and Bateson families.

GEORGINA BYRNE is an ordained priest in the Church of England and Residentiary Canon at Worcester Cathedral, having been previously the Diocesan Director of Ordinands. She is the author of *Modern Spiritualism and the Church of England, 1850–1939* (2010) and coeditor of *Life after Tragedy: Essays on Faith and the First World War Evoked by Geoffrey Studdert Kennedy* (2017).

IMOGEN CLARKE has a PhD in the history of science from the University of Manchester. She has published on early twentieth-century physics in Britain, with a particular focus on publishing practices and the relationships between science and broader culture.

DI DRUMMOND was formerly Reader in Modern History at Leeds Trinity University. Her research is in railway history, and she is currently investigating British-financed and British-built railways in Britain's empire, 1830–1950. Her books include *Crewe: Railway Town, Company and People, 1840–1914* (1995); *The First Civic University: Birmingham, 1880–1980* (coedited with E. W. Ives and Leonard Schwarz, 2000); and *Tracing Your Railway Ancestors: A Guide for Family Historians* (2010).

CHRISTINE FERGUSON is professor in English at the University of Stirling, where her research currently focuses on the literary production of the nineteenth-century occult revival. She is the author of two monographs—*Language, Science, and Popular Fiction in the Victorian Fin-de-Siècle: The Brutal Tongue* (2006) and *Determined Spirits: Eugenics, Heredity, and Racial Regeneration in Anglo-American Spiritualist Writing, 1848–1930* (2012)—and coeditor, along with Andrew Radford, of *The Occult Imagination in Britain, 1875–1947* (Routledge 2018).

GRAEME GOODAY is professor of the history of science and technology in the School of Philosophy, Religion and History of Science at the University of Leeds. He has participated in numerous Arts and Humanities Research Council (AHRC) projects, including "Owning and Disowning Invention: Intellectual Property, Authority and Identity in British Science and Technology, 1880–1920"; "Innovating in Combat: Telecommunications and Intellectual Property in the First World War"; and "Electrifying the Country House: Educational Resources on the History of Domestic Electricity." His books include *The Morals of Measurement: Accuracy, Irony and Trust in Late Victorian Electrical Practice* (2004); *Domesticating Electricity: Technology, Uncertainty and Gender in Late Nineteenth-Century Culture, 1880–1914* (2008); and, with Stathis Arapostathis, *Patently Contestable: Electrical Technologies and Inventor Identities on Trial in Britain* (2013). The last of these was awarded the Pickstone Prize in 2014 by the British Society for the History of Science.

DAVID HENDY is professor of media and cultural history at the University of Sussex. He has been a visiting fellow at Yale, Cambridge, and Indiana Bloomington. He is the author of four books about broadcasting and sound, including *Life on Air: A History of Radio Four* (2007), which won the Longman-History Today Book of the Year award. He has written and presented several programs for the BBC based on his research, including *Rewiring the Mind*, a five-part series about the modern media; *Noise: A Human History*, a thirty-part series about the evolving role of sound; and *Power of Three*, a seventy-part series about the history of the Third Programme and Radio 3. In 2011 he cowrote, with Adrian Bean, a BBC Radio 3 drama, *Between Two Worlds*, based on the life of Oliver Lodge. His next book, an authorized history of the BBC, will be published for the corporation's centenary in 2022.

BRUCE J. HUNT teaches the history of science and technology at the University of Texas. He is the author of *The Maxwellians* (1991) and *Pursuing Power and Light: Technology and Physics from James Watt to Albert Einstein* (2010), and of numerous articles on the history of nineteenth-century physics and electrical technology.

BERNARD LIGHTMAN is distinguished research professor of humanities at York University and current president of the History of Science Society. Lightman's research interests include nineteenth-century popular science and Victorian scientific naturalism. Among his most recent publications are *A Companion to the History of Science* (2016, editor) and *Science Museums in Transition* (2017, coedited with Carin Berkowitz). He is one of the general editors of the *John Tyndall Correspondence Project*, an international collaborative effort to obtain, digitalize, transcribe, and publish all surviving letters to and from Tyndall.

JAMES MUSSELL is associate professor in Victorian literature at the University of Leeds. He is the author of *Science, Time and Space in the Late Nineteenth-Century Periodical Press* (2007) and *The Nineteenth-Century Press in the Digital Age* (2012). He is one of the editors of the *Nineteenth-Century Serials Edition* (2008; 2018) and *W.T. Stead: Newspaper Revolutionary* (2012). He has written extensively on the nineteenth-century press, the place of science in nineteenth-century culture, and the materiality of archives.

RICHARD NOAKES is senior lecturer in history at the University of Exeter. His research focuses on the history of the physical sciences in the period roughly 1750–1930, the relationship between the sciences and occult, and the history of telecommunications since 1800. He is the coauthor, with K. C. Knox, of *From Newton to Hawking: A History of Cambridge University's Lucasian Professors of Mathematics* (2003) and, with Geoffrey Cantor, Gowan Dawson, and Graeme Gooday, of *Science in the Nineteenth-Century Periodical: Reading the Magazine of Nature* (2004). His first monograph is *Physics and Psychics: The Occult and the Sciences in Modern Britain* (2019).

PETER ROWLANDS is a theoretical physicist and science historian. A research fellow at the University of Liverpool since 1987, he is the author and editor of many books, including *Oliver Lodge and the Liverpool Physical Society* (1990); *Oliver Lodge and the Invention of Radio* (coeditor, 1994); *Zero to Infinity: The Foundations of Physics* (2007); *The Foundations of Physical Law* (2014); and *How Schrödinger's Cat Escaped the Box* (2015).

MATTHEW STANLEY teaches and researches the history and philosophy of science at New York University. He is the author of *Einstein's War: How Relativity Triumphed amid the Vicious Nationalism of World War I* (2019); *Practical Mystic: Religion, Science, and A. S. Eddington* (2007); and *Huxley's Church and Maxwell's Demon: From Theistic Science to Naturalistic Science* (2014). He currently runs the New York City History of Science Working Group. He cohosts the science podcast *What the If?*

INDEX